The Codex Alimentarius Commission and Its Standards

A sales edition of this dissertation is published by T.M.C. Asser Press, The Hague, The Netherlands, ISBN 978-90-6704-256-7.

Front cover: Floating food market, Damnoen Saduak, Thailand
Cover photographs by Mariëlle D. Masson-Matthee

All rights reserved.
© 2007, Mariëlle D. Masson-Matthee, Allogny, France

No part of the material protected by the copyright notice may be reproduced or utilized in any form or by any means, electronic or mechanical, including photocopying, recording, or by any information storage and retrieval system, without written permission from the copyright owner.

THE CODEX ALIMENTARIUS COMMISSION AND ITS STANDARDS

PROEFSCHRIFT

ter verkrijging van de graad van doctor aan de Universiteit Maastricht,
op gezag van de Rector Magnificus, Prof. mr. G.P.M.F. Mols
volgens het besluit van het College van Decanen,
in het openbaar te verdedigen
op vrijdag 26 oktober 2007 om 14.00 uur

door

Maria Elisabeth Dominica Matthee

Promotor:
Prof. dr. E.I.L. Vos

Beoordelingscommissie:
Prof. dr. P.L.H. Van den Bossche (voorzitter)
Prof. dr. N.M. Blokker (Universiteit Leiden)
Prof. dr. H.E.G.S. Schneider

Financieel ondersteund door:
T.M.C. Asser Instituut
Onderzoeksinstituut METRO

*To François and to my mother
to the memory of my father*

ACKNOWLEDGEMENTS

> *"Share your ideas in order to discover new ones"*
> (Words from a father to a daughter)

My gratitude goes to the many people, both to those who have enriched me with their ideas, views and insights, and to those who have helped me in any other way to advance in my research and to finish it. There is only place to acknowledge such thanks by naming some of them.

This research has been conducted at and financed by the T.M.C. Asser Institute, and my special thanks go to Rob Siekmann and Wybe Douma, who saw the potential of my research proposal. The T.M.C. Asser Institute was a great place to exchange ideas with other scholars, and I would like to thank some people in particular: Mark Jacobs for discussions on WTO-related issues, Jan Anne Vos, Bert Barnhoorn and my roommate Tanja Mehra for discussions on issues of international law, and Jan Koen Sluijs for discussions on any topic that came to mind.

Special thanks are also due to the people of the Dutch Delegation, who allowed me to participate as a member at several meetings of the Codex Alimentarius Commission and the subsidiary Codex Committees, in particular, Leo Hagedoorn, Suzanne Bont, Nathalie Scheidegger, Niek Schelling, Elfriede Adriaansz and Wim van Eck, for his clear explanations on some important scientific elements of Codex standards. Furthermore, I would like to thank Gabrielle Marceau, Gretchen Stanton, GianLuigi Burci and Nick Drager, who gave me some of their precious time in order to explain issues relating to the Codex Alimentarius Commission and to the WTO. I also would like to thank the participants to the 'Centre of Studies 2003' on the topic of food security at the Hague Academy of International Law, in particular, the Director of the English session, Professor Francis Snyder. I also owe a thanks to the people of the Peace Palace Library for their kind assistance and hospitality during my visits, and a thanks to Marieke van der Lugt and Kirstyn Inglis for reading an earlier draft of this book, and the members of the reading committee, Professor Peter Van den Bossche, Professor Niels Blokker and Professor Hildegard Schneider for their useful and constructive comments. The English text has been corrected by Chris Engert and I wish to express my thanks for his work. Likewise, I would like to thank the people of the Asser Press for their great assistance, in particular, Marjolijn Bastiaans and Mieke Eijdenberg for her accurateness.

There are several people, whose help has had a major impact on my research. I would like to mention Jan Staman (at that time, working for the Dutch Ministry of Agriculture, Food Safety and Nature) for his trust, assistance and creativity during the organisation of the 'Legal Platform for Consumers Concerns and International Trade in Food and Agriculture'. I would also like to thank Denise Prévost for the many discussions that we have had, her attentive reading of some earlier drafts, and her openness in sharing information and documentation. And of course, I wish to express my enormous gratitude to my promotor professor Ellen Vos for all her capacities that make her such a great promotor, for her feedback on so many aspects of the research, for the title of this book, and for simply listening.

I will end by thanking all my friends for their understanding and motivation. Last, but not least, I would like to thank my parents for believing in me, and, of course, my husband, François, for his love, patience and inspiration throughout these years.

Mariëlle MASSON-MATTHEE

TABLE OF CONTENTS

Acknowledgements		VII
Abbreviations		XV

Introduction
1.	The rise of the Codex Alimentarius Commission and its standards	1
2.	International governance and fragmentation of international law; questions as to the legal nature of Codex standards	3
3.	Legitimacy and the Codex Alimentarius Commission, its standard-setting procedure and its standards	6
4.	The structure of research	12

Chapter I
The Codex Alimentarius Commission: the Institutional Framework   13

1.	Introduction	13
2.	The origin of the Joint FAO/WHO Food Standards Programme	13
3.	A Joint FAO/WHO Institutional Framework	16
3.1	The Joint FAO/WHO Codex Alimentarius Commission	18
3.1.1	Powers and internal rules	18
3.1.1.1	The subsidiary character of the Codex Alimentarius Commission	19
3.1.1.2	Delegation of powers as basis for Codex Alimentarius Commission's authority	21
3.1.1.3	The Procedural Manual	24
3.1.2	Composition	25
3.1.2.1	Members of the Codex Alimentarius Commission	25
3.1.2.2	Observers of the Codex Alimentarius Commission	27
3.1.3	Meetings of the Codex Alimentarius Commission	28
3.1.4	Budgetary issues	30
3.2	The Executive Committee	31
3.2.1	Legal basis, powers and internal rules	31
3.2.2	Composition	32
3.2.3	Meetings and budgetary issues	33
3.3	The Codex Secretariat	33
3.4	The Codex Committees and Task Forces	34
3.4.1	Different types of Committees	35
3.4.1.1	General Codex Committees and Commodity Codex Committees	36
3.4.1.2	*Ad Hoc* Intergovernmental Task Forces	38
3.4.1.3	Regional Co-ordinating Committees	38

3.4.2	Powers and internal rules	39
3.4.3	Composition	39
3.4.4	Budgetary issues	40
3.5	Scientific expert bodies	42
3.5.1	Mandate and internal rules	42
3.5.2	Composition and meetings	44
3.5.3	Administration and finance	47
3.6	The Joint FAO/WHO Consultative Group for the Trust Fund	48
4.	Conclusions	49

Chapter II
The Codex Alimentarius: harmonisation through standard-setting — 51

1.	Introduction	51
2.	The structure of the Codex Alimentarius	52
2.1	From a vertical to a horizontal approach	53
2.2	World-wide and regional Codex standards	56
2.3	Other types of Codex measures	58
2.4	The relationship between the various measures	59
3.	The scope and specificity of the measures contained in the Codex Alimentarius	62
3.1	Ensuring the protection of health	63
3.2	Ensuring fair trade practices in foods	66
3.3	Addressing the diversity of national circumstances	67
3.3.1	Scientific 'uncertainty' due to variability	67
3.3.2	Diversity of national circumstances and 'non-scientific' factors	71
4.	The Codex standard-setting procedure and the acceptance procedure	73
4.1	A uniform 8-step standard-setting procedure	75
4.2	The emphasis on consensus-building during the standard-setting procedure	80
4.3	The publication and acceptance procedure	83
5.	The Codex Alimentarius: levels of harmonisation	86
5.1	Codex standards and other Codex measures	87
5.2	The acceptance procedure	89
6.	Legal status of the Codex Alimentarius under the acceptance procedure	91
7.	Conclusions	93

Chapter III
The Codex and the EC — 95

1.	Introduction	95
2.	Overview of the harmonisation process of the European Community and the Codex Alimentarius Commission compared	96
2.1	Prior to 1987	96
2.2	1987-1997	98

2.3	1997-2002	99
2.4	Post-2002	101
3.	The promotion of the acceptance of Codex standards and MRLs	102
4.	The use of Codex Alimentarius in secondary EC food legislation prior to 2002	103
4.1	The use of Codex Alimentarius in the preparation of EC food law	103
4.2	Reference to the Codex Alimentarius in secondary EC food legislation	106
5.	The Codex Alimentarius and the European Court of Justice	108
5.1	The European Court of Justice and its reference to the Codex Alimentarius in the context of Article 28 and 30 EC	108
5.1.1	The explicit role of the work of the Joint FAO/WHO Food Standard Programme in the jurisprudence on Article 30 EC	110
5.1.2	Developments reducing the role of the Codex Alimentarius under Article 30	115
5.2	The European Court of Justice and its use of the Codex Alimentarius in the context of secondary food legislation	116
5.3	The Codex Alimentarius, the European Court of Justice and the legality of EC measures adopted by the Community institutions	117
6.	2002 and beyond: Increasing resort to Codex standards?	120
6.1	General principles of food law and the reference to the Codex Alimentarius	121
6.2	EC food law and the consideration of Codex standards and other related texts	123
6.3	The European Food Safety Authority and the Joint FAO/WHO Food Standards Programme	126
7.	The changing inter-institutional relationship between the European Commission and the Codex Alimentarius Commission	129
8.	Conclusions	132

Chapter IV
The WTO Agreements and the Codex Alimentarius 135

1.	Introduction	135
2.	Harmonisation in the context of the SPS Agreement and the TBT Agreement	136
2.1	The objective and the function of harmonisation in the framework of the SPS Agreement and the TBT Agreement	137
2.2	The provisions on harmonisation under the SPS Agreement and the TBT Agreement	140
2.2.1	The scope of the obligation to harmonise	140
2.2.1.1	The terms 'based on' or 'use as a basis'	142
2.2.1.2	The right to deviate from international standards	144
2.2.2	Harmonisation as encouragement	152
3.	The status of Codex Measures under the SPS Agreement and the TBT Agreement	154

3.1	The Codex Alimentarius Commission as a recognised standard-setting body under the SPS Agreement and the TBT Agreement	155
3.2	The status of Codex standards as necessary measures to protect legitimate objectives	158
3.3	The status of international standards as binding norms	159
3.3.1	Explicit rejection of binding status of Codex standards by the Appellate Body	160
3.3.2	The application of the terms 'used as a basis' by panels	162
3.4	The status of international standards as appropriate and effective standards under the TBT Agreement	164
3.5	The status of international standards under Article 5 of the SPS Agreement	165
3.5.1	Article 5.1: Risk assessment techniques developed by the relevant international organisations	166
3.5.2	The role of the scientific basis of the international standards to interpret the scientific justification for a higher national level of protection	169
3.5.3	The role of international standards and the choice of a least trade restrictive measure achieving this level	173
4.	The changed status of Codex measures and the abolition of the Codex acceptance procedure	175
4.1	The obligations under the SPS Agreement and the TBT Agreement cover all documents adopted by the Codex Alimentarius Commission	176
4.2	The different objective and content of the obligations resulting from Codex measures in the context of the WTO agreements	177
4.3	Consequences related to the different membership of both institutions	178
4.4	The role of the 'explicit consent' under the acceptance procedure and under the SPS Agreement and the TBT Agreement	178
4.5	Consequences of the abolition of the Codex acceptance procedure	182
5.	The inter-institutional relationship of the WTO – Codex Alimentarius Commission	182
5.1	The dispute settlement mechanism	184
5.1.1	Lack of judicial review to ensure legitimacy of the Codex procedures	185
5.1.2	WTO panels and the interpretation of Codex measures	188
5.1.2.1	The scope of the obligation to use customary rules of interpretation	190
5.1.2.2	The competence to seek information from 'outside' sources	192
5.2	The SPS Committee: Monitoring the process of international harmonisation	195
6.	Conclusions	197

Chapter V
The legitimacy of the Codex Alimentarius, the standard-setting procedure and the institutional framework — 201

1.	Introduction	201
2.	Questions of legitimacy related to the institutional structure	203
2.1	The normative competence of the Codex Alimentarius Commission	204
2.2	Delegation of tasks to Codex Committees: a question of decentralisation	206
2.2.1	Concerns related to the position of the Codex Committees	207
2.2.1.1	Codex Committees and the initiation of new work	207
2.2.1.2	The definition of inter-Committee relationship	209
2.2.2	Instruments to supervise and co-ordinate the activities of Codex Committees	211
2.2.2.1	The 'Criteria for Establishing Work Priorities', the 'Medium-Term Plan' and their application by the Codex Alimentarius Commission and the Executive Committee	211
2.2.2.2	The definition of inter-relationship between Codex Committees	214
2.2.3	Amendments of the 'Procedures for the Elaboration of Codex Standards and Related Texts' and the new function of the Executive Committee	215
2.3	The mandate of expert bodies	217
2.3.1	Relation risk assessor – risk manager	218
2.3.2	Working procedures to ensure independence of the expert bodies	221
2.4	The role of other international institutions in the elaboration procedure	224
3.	Procedural legitimacy	228
3.1	Consensus	230
3.1.1	Final decision-making and the rules regulating consensus	231
3.1.2	Consensus-building: the role of Codex Committees	236
3.1.3	Managing the procedure	237
3.2	Participation	241
3.2.1	Participation possibilities of developing countries	242
3.2.1.1	Obstacles to a *de facto* effective participation of developing countries	243
3.2.1.2	Ways to stimulate the participation of developing countries	245
3.2.1.2.1	Co-chairing and hosting of Codex Committee meetings in developing countries	246
3.2.1.2.2	The Trust Fund for the participation of developing countries and countries in transition in the work of the Codex Alimentarius Commission	247
3.2.2	Analysing the intergovernmental character of participation rights	248
3.2.3	The participation of industry INGOs *vis-à-vis* public interest INGOs	250
3.3	Transparency	254

3.3.1	Transparency and the standard-setting procedure	255
3.3.2	Transparency and the Procedural Manual	256
3.3.3	Openness of the meetings of the involved bodies	257
3.3.4	Access to documents	259
3.3.5	Problems related to translations	259
4.	Substantive legitimacy	260
5.	Conclusions	263

Conclusions 269
1.	General conclusions	269
2.	Main concerns, recommendations and suggestions	277
2.1	Consequences resulting from the fragmentation of law	277
2.2	Towards a more structural approach to ensure the legitimacy of the Codex Alimentarius Commission, its standard-setting procedure and its standards	279

Annexes

Annex I:	Codex International Individual Standard for Gouda, Codex Stan C-5-1966	285
Annex II:	Codex Standard for Chocolate and Chocolate Products, Codex Stan 87-1981 (Rev. 1-2003)	289
Annex III:	General Standard for the Labelling of Prepackaged Foods, Codex Stan 1-1985 (Rev. 1-1991)	301
Annex IV:	Code of Hygienic Practice for the Transport of Food in Bulk and Semi-Packed Food, CAC/RCP 47-2001	309
Annex V:	Guidelines for Food Import Control Systems, CAC/ GL 47-2003	315

Bibliography 325

Table of Cases 339

Index 341

Summary 345

Summary in Dutch 348

Curriculum vitae 352

LIST OF ABBREVIATIONS

Acute RfDs	Acute Reference Doses
ADIs	Acceptable Daily Intakes
ALARA	As Low As Reasonably Achievable
BSE	Bovine Spongiform Encephalopathy
CAC	Codex Alimentarius Commission
CCCF	Codex Committee on Contaminants in Foods
CCFA	Codex Committee on Food Additives
CCFAC	Codex Committee on Food Additives and Contaminants
CCFH	Codex Committee on Food Hygiene
CCFICS	Codex Committee on Food Export and Import Inspection and Certification Systems
CCFL	Codex Committee on Food Labelling
CCFO	Codex Committee on Fats and Oils
CCGP	Codex Committee on General Principles
CCMAS	Codex Committee on Methods of Analysis and Sampling
CCMMP	Codex Committee on Milk and Milk Products
CCPFV	Codex Committee on Processed Fruit and Vegetables
CCPR	Codex Committee on Pesticides Residues
CCRVDF	Codex Committee on Residues of Veterinary Drugs in Food
CVPM	Committee for Veterinary Medicinal Products
DDT	Dicholoridipheryltrichloroethane
DG	Directorate General
DSB	Dispute Settlement Body
DSU	Dispute Settlement Understanding
EBDC	Ethylene bisdithiocarbamate
EC	European Community
ECJ	European Court of Justice
EEC	European Economic Communities
EFSA	European Food Safety Authority
EMRL	Extraneous Maximum Residue Limit
EPA	Environmental Protection Agency
FAO	Food and Agriculture Organisation of the United Nations
GAP	Good Agricultural Practice in the Use of Pesticides
GATT	General Agreement on Tariffs and Trade
GCPF	Global Crop Protection Federation
GIFAP	Groupement International des Associations Nationales des Fabricants de Produits Agrochimiques
GMP	Good Manufacturing Practice
GVP	Good Veterinary Practice
HACCP	Hazard Analysis and Critical Control Point

IACFO	International Association of Consumer Food Organisations
IAEA	International Atomic Energy Agency
IARC	International Agency for Research on Cancer
IDF	International Dairy Federation
IGO	International Governmental Organisation
IIA	International Institute for Agriculture
IMF	International Monetary Fund
INGO	International Non-Governmental Organisation
IPCS	International Programme on Chemical Safety
IPPC	International Plant Protection Convention
JECFA	Joint FAO/WHO Expert Committee on Food Additives
JEMRA	Joint FAO/WHO Expert Meetings on Microbiological Risk Assessment
JMPR	Joint FAO/WHO Meeting on Pesticide Residues
LIC	Low-income countries
LMIC	Lower middle-income countries
MLs	Maximum Levels
MRLPs	Maximum residue levels for pesticide residues
MRLs	Maximum residue levels
MRLVDs	Maximum residue levels for veterinary drugs
NCCCs	National Codex consultative committees
NGO	Non-Governmental Organisation
OECD	Organisation for Economic Co-operation and Development
OIE	Office International des Epizooties (World Organisation for Animal Health)
PRA	Pest Risk Assessment
RCP	Recommended Code of Practice
SCAN	Scientific Committee for Animal Nutrition
SCF	Scientific Committee for Food
SEA	Single European Act
SPS Agreement	Agreement on the Application of Sanitary and Phytosanitary Measures
TBT Agreement	Agreement on Technical Barriers to Trade
UMIC	Upper middle-income countries
UN	United Nations
UNECE	United Nations Economic Commission for Europe
UNEP	United Nations Environmental Programme
WHO	World Health Organisation
WTO	World Trade Organisation

INTRODUCTION

1. THE RISE OF THE CODEX ALIMENTARIUS COMMISSION AND ITS STANDARDS

This book aims to define the status of Codex standards and other measures adopted by the Codex Alimentarius Commission under the WTO agreements and within the EC legal order.[1] It seeks to analyse the legal consequences of the increased use of Codex measures in these legal systems, and to examine the legitimacy of the institutional framework of the Codex Alimentarius Commission, its standard-setting procedure and the consequences of the disappearance of the Codex acceptance procedure.

The Codex Alimentarius Commission was established in 1963 in order to meet the increasing need for international standards which harmonised national food requirements and to follow up on the already existing regional initiatives (the Codex Alimentarius Europaeus and the Latin American Food Code) that were created in view of the expanding international food market after the Second World War. The Codex Alimentarius Commission was positioned under the auspices of the Food and Agriculture Organisation (FAO) and the World Health Organisation (WHO), both recognised for their expertise in the area of food issues. In its role as the responsible institution for the execution of the Joint FAO/WHO Food Standards Programme, the Codex Alimentarius Commission has adopted a large number of international measures over the years.

During the first 10-15 years of its existence, the Codex Alimentarius Commission regularly received the attention of legal scholars, who were particularly interested in the nature of Codex standards that deviated from the classical sources of international law.[2] However, probably due to the lack of binding force of Codex measures, this attention diminished over the years.

---

[1] The term 'standards' in the title of this book also covers other measures that have been adopted by the Codex Alimentarius Commission to assist in the harmonisation of food requirements, such as recommended codes of practice and guidelines. Throughout this book, the ensemble of Codex standards, recommended codes of practice and guidelines shall be referred to as 'Codex measures'. The reason for not using the term 'measures' in the title of the book is the fact that this term – in other contexts – is sometimes used to refer to political decisions, as well as for internal use and may be misleading with regard to the aim of this research.

[2] J.P. Dobbert, 'Le Codex Alimentarius vers une nouvelle méthode de réglementation internationale', 15 *Annuaire Français de Droit International* (1969) pp. 677-717. J.P. Dobbert, 'Decisions of International Organizations-Effectiveness in Member States. Some Aspects of the Law and Practice of FAO', in S.M. Schwebel (ed.), *The Effectiveness of International Decisions - Papers of a conference of The American Society of International Law and the Proceedings of the conference* (Leyden, Sijthoff 1971). S. Shubber, 'The Codex Alimentarius Commission under International Law', 21 *International and*

This lack of interest has changed considerably since the entry into force of the WTO agreements in 1995.[3] During the last decade, the Codex Alimentarius Commission has received renewed attention from scholars, as well as from governmental and non-governmental organisations.[4] Two of the annexed agreements to the WTO Agreement, the Agreement on the Application of Sanitary and Phytosanitary Measures (hereinafter the *SPS Agreement*) and the Agreement on Technical Barriers to Trade (hereinafter the *TBT Agreement*) make reference to international standards. These two agreements have legally-binding force as part of an integral package of agreements for all states that have become Members of the WTO. Furthermore, disputes relating to the WTO agreements are resolved by panels and by the Appellate Body. The reports of the panels, as may be amended by the Appellate Body, are adopted by a reverse consensus by the Dispute Settlement Body of the WTO. The reference to international standards in both the *SPS Agreement* and the *TBT Agreement* has raised questions as to the exact legal nature of the standards.[5] Mentioned explicitly in the *SPS Agreement* and recognised as relevant in the context of the *TBT Agreement* by the Panel in *EC-Sardines*, the standards adopted by the Codex Alimentarius Commission are at the heart of this debate.

Comparative Law Quarterly (1972) pp. 631-655. C.H. Alexandrowicz, *The Law-Making Functions of the Specialised Agencies of the United Nations* (Sydney, Angus and Robertson 1973). D.L. Leive, *International Regulatory Regimes* (Lexington, Lexington Books 1976).

[3] See on the WTO agreements, P. Van den Bossche, *The Law and Policy of the World Trade Organization. Text, Cases and Materials* (Cambridge, Cambridge University Press 2005), J.H. Jackson, *The World Trading System. Law and Policy of International Economic Relations*, 2nd edn. (Cambridge, Massachusetts, MIT Press 2000), M. Trebilcock and R. Howse, *The Regulation of International Trade*, 2nd edn. (London, Routlegde 1999).

[4] For instance, the reaction of the European Parliament on the decision of the WTO Panel in *EC-Hormones*, EP resolution, OJ 1997, C222/53. N. Avery, et al., *Cracking the Codex* (London: National Food Alliance, 1993). S. Suppan, *Governance in the Codex Alimentarius Commission* (Consumers International 2005). J. Bizet, 'Sécurité alimentaire: le Codex Alimentarius', 450 *Les Rapports du Sénat* (1999-2000). D.G. Victor, 'Risk Management and the World Trading System: Regulating International Trade Distortions Caused by National Sanitary and Phytosanitary Policies', *Incorporating Science, Economics, and Sociology in Developing Sanitary and Phytosanitary Standards in International Trade: Proceedings of a Conference* (National Academies Press, 2000), pp. 118-169. G. Bossis, *La sécurité sanitaire des aliments en droit international et communautaire. Rapports croisés et perspectives d'harmonisation* (Bruxelles, Bruylant 2005). T. Makatsch, *Gesundheitsschutz im Recht der Welthandelsorganisation (WTO). Die WTO und das SPS-Übereinkommen im lichte von Wissenschaftlichkeit, Verrechtlichung und Harmonisierung* (Berlin, Duncker & Humblot 2004). G. Sander, 'Gesundheitsschutz in der WTO - eine neue Bedeutung des Codex Alimentarius im Lebensmittelrecht?', 3 *Zeitschrift für Europarechtliche Studien* (2000). F. Veggeland and S.O. Borgen, 'Negotiating International Food Standards: The World Trade Organization's Impact on the Codex Alimentarius Commission', 18 *Governance: An International Journal of Policy, Administration, and Institutions* (2005) pp. 675-708.

[5] For example, G. Bossis, *supra* n. 4, pp. 121-122. R. Muñoz, 'La Communauté entre les mains des normes internationales: les conséquences de la décision Sardines au sein de l'OMC', 4 *Revue du Droit de l'Union Européenne* (2003), p. 463. D. Roberts, 'Preliminary Assessment of the Effects of the WTO Agreement on Sanitary and Phytosanitary Trade Regulations', 1 *Journal of International Economic Law* (1998), p. 379.

In addition, European Community food legislation refers to and increasingly applies Codex standards and other Codex measures. In 2002, the European Community adopted Regulation 178/2002, better known as the General Food Law, which includes an obligation for both member states and Community institutions to take international standards into account.[6] In addition, an increasing number of sectoral Community food legislation refers to Codex measures.[7]

This increased use of Codex measures within the framework of both the WTO agreements and the EC legislation has made the Codex standards considerably more important. This raises questions as to both the place of these measures in international law and in EU law, *and* the legitimacy of the Codex Alimentarius Commission as the responsible institution for the adoption of these measures, the legitimacy of its standard-setting procedure, and the legitimacy of its measures as an instrument for harmonisation.

2. INTERNATIONAL GOVERNANCE AND FRAGMENTATION OF INTERNATIONAL LAW; QUESTIONS AS TO THE LEGAL NATURE OF CODEX STANDARDS

There is no doubt that globalisation has augmented the inter-dependence between states. Since the Second World War, international relations between states have increasingly been undertaken within the framework of international organisations, which has resulted in a proliferation of international organisations[8] and of dispute settlement mechanisms.[9] One of the features of international law-making activities has been an increasing specialisation of dealing with area-issues, also referred to as 'functional differentiation'.[10] From its inception, the UN system is based upon the elements of decentralisation and co-ordination[11] and on the assumption that the

---

[6] European Parliament and Council Regulation 178/2002/EC laying down the general principles and requirements of food law, establishing the European Food Safety Authority and laying down procedures in matters of food safety [2002] L 31/1, E. Vos, 'Risicobeheersing door de EU: het nieuwe beleid op het gebied van voedselveiligheid', in G. van Calster and E. Vos (eds.), *Risico en voorzorg in de rechtsmaatschappij* (Antwerpen, Intersentia 2005).

[7] See, for a detailed discussion on the increasing number of references to Codex measures, Chapter III.

[8] N.M. Blokker and H.G. Schermers (eds.), *Proliferation of International Organizations: legal issues* (The Hague, Kluwer Law International 2001).

[9] Th. Buergenthal, 'Proliferation of International Courts and Tribunals: Is It Good or Bad?', 14 *Leiden Journal of International Law* (2001) pp. 267-275. P.-M. Dupuy, 'The danger of fragmentation or unification of the international legal system and the International Court of Justice', 31 *New York University Journal of International Law and Politics* (1999) pp. 791-807. J.I. Charney, 'Is International Law Threatened by Multiple International Tribunals?', 63 *Receuil des Cours* (1998). C.P.R. Romano, 'The proliferation of international judicial bodies: the pieces of the puzzle', 31 *New York University Journal of International Law and Politics* (1999) pp. 709-751.

[10] M. Koskenniemi, Fragmentation of international law: difficulties arising from the diversification and expansion of international law, in Report of the International Law Commission on its 58th Session, A/CN.4/L.682, p. 11.

[11] The concept of decentralisation refers to the preference that activities be undertaken by more specialised bodies, instead of a centralised body, which the League of Nations had been. This shift

distinction between the areas of competence of its different 'specialised agencies' is clear-cut.[12] In this matter, the International Court of Justice has explicitly recognised that the system of UN institutions is based upon the principle of speciality, that is characterised by the relationship that exists between the United Nations, invested with powers of a general scope, and the 'various autonomous and complementary organisations, invested with sectoral powers'.[13] Activities undertaken in the framework of different international organisations (within and outside the UN system) have led to distinct 'treaty regimes'. However, with the increasing inter-dependence of different fields, such as environmental protection, economic development, human rights, etc., it became increasingly apparent that the competences of the various international institutions were beginning to overlap. These developments have led to what is often called the fragmentation of international law.[14] The consequences are that regulation on the same issues are adopted by different institutions (fragmentation of primary rules) and that they can be invoked before different dispute settlement mechanisms (fragmentation of secondary rules).[15] This 'unorganised system',[16] in which parts interact with one another, is full of 'intra-systematic tensions, contradictions and frictions'.[17] The fragmentation of primary rules inevita-

---

away from a centralistic approach was inspired by the 'Bruce Report' of 1939 (Rapport du Comité Spécial, 'Le développement de la collaboration internationale dans le domaine économique et social', A.23.1939, Genève, le 23 août 1939). The Bruce Committee was charged with the examination of appropriate institutional measures which would allow for extended activities of the League of Nations related to economic and social aspects of the international society.

[12] C. Tietje, 'Global Governance and Inter-Agency Co-operation in International Economic Law', 36 *Journal of World Trade* (2002) pp. 501-515: 'As can be seen by looking at the historical development and the judgment of the ICJ, the "United Nations system" is based upon the idea that a clear-cut distinction between the competencies and areas of activities of different international organizations is possible. However, the process of globalization that can be seen, particularly in the economic area, challenges this perspective.'

This functional differentiation is visible in the traditional separation between Bretton Woods institutions (the World Bank, the International Monetary Fund (IMF) and the GATT, now the WTO) and the traditional United Nations (UN) institutions during the early years of their existence: the Bretton Woods institutions dealt with the world's economic problems and the UN institutions dealt with world's political problems (See J. Pauwelyn, 'Bridging fragmentation and unity: international law as a universe of inter-connected islands', 25 *Michigan Journal of International Law* (2003), p. 903). It is also visible among other UN specialised agencies or programmes (such as the FAO, the WHO, the United Nations Environmental Programme (UNEP), etc.).

[13] Legality of the Use by a State of Nuclear Weapons in Armed Conflict, 8 July 1996, ICJ, Advisory Opinion, Preliminary Objections, I.C.J. Reports 1996, para. 26.

[14] M. Koskenniemi, *supra* n. 10, p. 11. Fragmentation is also referred to in less negative terms such as 'international legal pluralism', see W.W. Burke-White, 'International Legal Pluralism', 25 *Michigan Journal of International Law* (2003) pp. 963-979.

[15] See for this distinction between fragmentation of primary rules (or substantive law) and secondary rules (rules that aim to ensure the observance of the primary rules), G. Hafner, 'Pros and Cons ensuing from Fragmentation of International Law', 25 *Michigan Journal of International Law* (2003), pp. 856-858.

[16] K. Zemanek, 'The Legal Foundations of the International System: General Course on Public International Law', 62 *Recueil des Cours: Collected Courses of the Hague Academy of International Law* (1997), p. 266.

[17] G. Hafner, *supra* n. 15, p. 850.

bly leads to contradictory obligations for states, and the question arises as to which rule to apply.[18] This situation is worsened by the presence of regulations adopted by regional institutions which often address the relevant concerns more specifically. The fragmentation of secondary rules is characterised by the fact that each dispute settlement mechanism is, first and foremost, concerned with the application of the substantive law of its own system, and will, consequently, solve a dispute in the context of its regime, which is not necessarily in line with the perspective of other regimes.[19] Fragmentation in these cases concerns, first and foremost, the interpretation of norms.[20] What needs to be avoided is the formation of 'self-contained regimes', which do not take general international law or substantive law adopted by other institutions into account.[21] In order to respond to this risk, it has, for example, been suggested that a clearer distinction between the jurisdiction of the relevant dispute settlement body and the applicable law be applied.[22] However, the fragmentation is more complex than just an unorganised system of treaty regimes. Other sources of law that cannot be classified as one of the classical sources of international law, such as standard-setting by public or private bodies, increasingly gain 'global validity'.[23]

These concerns, which result from the fragmentation of international law, are at the heart of this research into the status of Codex measures under the WTO agreements and within the EC legal order, and their consequences. The WTO, the Codex Alimentarius Commission and the European Community (EC) all deal with the harmonisation of national food measures, which may lead to tension, contradictions and friction. They do not have an overarching body that can supervise any incoherence between their rules. The Codex Alimentarius Commission is part of the UN system; the WTO and the EC, on the other hand, are not. Furthermore, although the Codex Alimentarius Commission is an intergovernmental body, its standards cannot be defined as being a 'classical' source of international law. This causes particular problems with regard to their legal status. Hence, what is the consequence of the references to international standards in the WTO agreements for the status of Codex measures? If they have acquired increased status, of what does this status consist? Does the reference create friction or contradictions between the WTO regime and the regime of the Codex Alimentarius Commission? What is the meaning of Codex measures in the EC legal order?

---

[18] Ibid., pp. 856-857.
[19] Ibid., p. 857-858.
[20] P.-M. Dupuy, *supra* n. 9, p. 792.
[21] J. Pauwelyn, *supra* n. 12, p. 904. See, on the concept of 'self-contained' regimes, B. Simma, 'Self-Contained Regimes', 16 *Netherlands Yearbook of International Law* (1985) pp. 112-136. Given the fact that there is no regime which fully excludes general international law, Koskenniemi, in the report of the Study Group of the International Law Commission, has suggested that the terminology 'self-contained' regime be replaced with 'special' regime; see M. Koskenniemi, *supra* n. 10, p. 82.
[22] J. Pauwelyn, *supra* n. 12, p. 910 and 915. W.W. Burke-White, *supra* n. 14, p. 964.
[23] A. Fischer-Lescano and G. Teubner, 'Regime-collisions: the vain search for unity in the fragmentation of global law', 25 *Michigan Journal of International Law* (2003), p. 1011.

3. LEGITIMACY AND THE CODEX ALIMENTARIUS COMMISSION, ITS STANDARD-SETTING PROCEDURE AND ITS STANDARDS

The increasing importance of Codex standards in the international and European legal orders makes their content and the way in which they are adopted more important. This leads to the question of the legitimacy of the Codex Alimentarius Commission, often referred to as a gentlemen's club of which little information was available, as the responsible body to prepare and adopt the international food standards.[24]

Legitimacy is a pre-requisite for an effective functioning international organisation, and an international organisation which is perceived as legitimate creates a strong incentive for states to comply with its norms.[25] Legitimacy has been defined by Hurd as the 'normative belief by an actor that a rule or institution ought to be obeyed', and has been qualified as a 'relational between actor and institution, defined by the actor's perception of the institution'.[26] Thus, both to exist and to operate, international organisations need the recognition, particularly the recognition of states, as a legitimate authority.[27]

Until recently, the question of the legitimacy of international organisations was hardly raised, mainly because the authority of the organisation was 'self-imposed' by its member states: for their operation, international organisations were largely dependent upon the consent of their member states.[28] Furthermore, it was mainly the perception of the member states that determined the presence of the legitimacy of international organisations. However, these two conditions have changed over the years. The authority of international organisations increasingly exceeds the con-

---

[24] See, for instance, Resolution on the Commission Green Paper on the general principles of food law in the European Union, A4-0009/98, para. 74, the reaction of the European Parliament on the decision of the WTO Panel in EC-Hormones, EP resolution, OJ 1997, C222/53, the question by Marianne Thyssen, E-2929/96, 8 November 1996. Furthermore, the FAO and WHO, as parent organisations of the Codex Alimentarius Commission, have commanded an evaluation of the Joint FAO/WHO Food Standards Programme to improve its operation. See also, concerns expressed by academic scholars, H. Horn, J.H.H. Weiler, 'European Communities – Trade Description of Sardines: Textualism and its Discontent', in H. Horn and P. Mavroidis (eds.), *The WTO Case Law of 2002* (Cambridge, Cambridge University Press 2005), p. 255. T. Makatsch, *supra* n. 4. R. Muñoz, *supra* n. 5, pp. 457-484. R. Romi, 'Codex Alimentarius: de l'ambivalence à l'ambiguité', *Revue Juridique de l'Environnement* (2001) pp. 201-213. S.A. Shapiro, 'International Trade Agreements, Regulatory Protection, and Public Accountability', 54 *Administrative Law Review* (2002) pp. 451-453. D. Bodansky, 'The Legitimacy of International Governance: A Coming Challenge for International Environmental Law?', 93 *American Journal of International Law* (1999) pp. 596-624.

[25] Th.M. Franck, *The Power of Legitimacy among Nations* (New York, Oxford University Press 1990), p. 24.

[26] I. Hurd, 'Legitimacy and Authority in International Politics', 53 *International Organization* (1999), p. 381.

[27] J.M. Coicaud, 'Conclusion. International Organizations, the evolution of international politics, and legitimacy', in J.M. Coicaud and V. Heiskanen (eds.), *The Legitimacy of international organizations* (Tokyo, United Nations University Press 2001), p. 523. D. Beetham, and Ch. Lord, *Legitimacy and the European Union* (London, Longham 1998), p. 11.

[28] D. Bodansky, *supra* n. 24, p. 597.

sent of their member states.²⁹ Moreover, the activities and decisions of international organisations increasingly limit the national sovereignty that states possess over their own citizens by imposing obligations on their regulatory choices. The European Union is a clear example in which these two developments have taken place.³⁰

This tendency can also be detected with regard to the Codex Alimentarius Commission and its measures. It is therefore not surprising that the Codex Alimentarius Commission has received renewed attention and that concerns of legitimacy have been raised.³¹ These concerns relate to the Codex standard-setting procedure, such as the dominant position of developed Codex Members and industrial-interest groups within the Codex standard-setting procedure, and the lack of transparency on procedural rules.³² They also include concern about the distribution of powers among the different subsidiary Codex bodies and other bodies involved in the standard-setting procedure.³³

---

²⁹ B. Reinalda and B. Verbeek (eds.), *Decision Making Within International Organizations* (London, Routledge 2004): 'All in all, it can be argued that over the last 15 years these developments have contributed in principle to an increase of policy autonomy for international organizations, albeit in different degrees for different organizations.'

³⁰ See, on the legitimacy of the European Union, D. Beetham and Christopher Lord, *supra* n. 27.

³¹ For instance, the reaction of the European Parliament on the decision of the WTO Panel in EC-Hormones, EP resolution, OJ 1997, C222/53. N. Avery, et al., *supra* n. 4. S. Suppan, *supra* n. 4. J. Bizet, *supra* n. 4, D.G. Victor, *supra* n. 4, pp. 118-169. G. Bossis, *supra* n. 4. T. Makatsch, *supra* n. 4. G. Sander, *supra* n. 4. F. Veggeland and S.O. Borgen, *supra* n. 4, pp. 675-708.

³² Numerous concerns expressed by Codex Members during sessions of the Codex Alimentarius Commission and Codex Committees, Report of the 22ⁿᵈ Session of the Joint FAO/WHO Codex Alimentarius Commission, Geneva, 23-28 June 1997, ALINORM 97/37, paras. 45 and 93, Report of the 21ˢᵗ Session of the Joint FAO/WHO Codex Alimentarius Commission, Rome, 3-8 July 1995, ALINORM 95/37, para. 46, Report of the 20ᵗʰ Session of the Codex Committee on General Principles, Paris, 3-7 May 2004, ALINORM 04/27/33A, paras. 133-134, Report of the 11ᵗʰ Session of the Codex Committee on General Principles, ALINORM 95/33, 1994, para. 35, Report of the 11ᵗʰ Session of the Codex Committee on General Principles, ALINORM 95/33, 1994, para. 54. Report of the 16ᵗʰ Session of the Codex Committee on General Principles, ALINORM 01/33A, 2001, para. 103 and further. Report of the 19ᵗʰ Session of the Codex Committee on General Principles, ALINORM 04/27/33, 2003, para. 13 and further, and 42 and 43. See also Report of the evaluation of the Codex Alimentarius and other FAO and WHO food standards work, ALINORM 03/25/3, paras. 193-194. See also, concerns expressed by academic scholars, N. Avery, et. al., *supra* n. 4, D.L. Leive, *supra* n. 2, pp. 435-436. D.G. Victor, *Effective Multilateral Regulation of Industrial Activity: Institutions for Policing and Adjusting Binding and Nonbinding Legal Commitments*, Ph.D. Thesis (Massachusetts, Harvard University, Institute of Technology 1997), pp. 198-201. D. McCrea, 'A View from Consumers', in N. Rees and D. Watson (eds.), *International Standards for Food Safety* (Gaithersburg, Maryland, Aspen Publishers 2000), p. 155. S. Suppan, *supra* n. 4. G.E. Spencer, et al.; 'Effects of Codex and GATT', 9 *Food Control* (1998), p. 179. L. Salter, *Mandated Science, Science and Scientists in the Making of Standards* (Dordrecht, Kluwer Academic Publishers 1988), pp. 70-71 and p. 74. L. Rosman, 'Public participation in international pesticide regulation: when the Codex Commission decides, who will listen?', 12 *Virginia Environmental Law Journal* (1993), p. 346. G. Sander, *supra* n. 4.

³³ See, for instance, the opinion expressed by a team of consultants examining the structure and mandate of Codex Committees, Report of the 55ᵗʰ Session of the Executive Committee, Rome 9-11 February 2005, ALINORM 05/28/3, para. 27, the opinion of the evaluation team responsible for the Joint FAO/WHO Evaluation of the Codex Alimentarius Commission, Report of the evaluation of the

This book distinguishes three dimensions of legitimacy inspired by the distinction made by Steffek, which will be referred to as: institutional legitimacy, procedural legitimacy, and substantive legitimacy.[34] This division enables the analysis of different types of concerns regarding the legitimacy of the Codex Alimentarius Commission and its standards. It will assist in defining whether the recent efforts of the Codex Alimentarius Commission and its parent organisations to ameliorate its institutional and procedural aspects adequately address the causes of the concerns.

Institutional legitimacy: the Codex Alimentarius Commission and the attribution of powers

The first dimension relates to the function and the attributed powers of the international organisation and its subsidiary organs. This dimension of legitimacy is closely related to what Steffek calls 'the dimension of legitimacy and the scope of international governance'.[35] It touches upon the fundamental question related to the necessity of the organisation itself; the legitimacy of the decision that certain activities can only be undertaken by an international organisation.[36] This dimension is closely related to questions of legality, which form an important basis for the legitimacy of international organisations.[37] As the competence of both international organisations and their organs to act depends upon the attribution of their powers by governments, their competence is limited to the acts that are necessary to perform the functions of the organisation as laid down in its constitutional documents.[38] Furthermore, constitutional documents, rules of procedures and other internal rules are adopted by members to regulate the actions of both the organs and the subsid-

---

Codex Alimentarius and other FAO and WHO food standards work, ALINORM 03/25/3, para. 116, Consultant's report, Review of the Working Procedures of the Joint FAO/WHO Meeting on Pesticide Residues (JMPR), 2002, p. 38, Committee of Experts on Tobacco Industry Documents, Tobacco Company Strategies to Undermine Tobacco Control Activities at the World Health Organization, July 2000, p. 1, and concerns expressed by Codex Members throughout meetings of Codex Committees and the Codex Alimentarius Commission, such as, for instance, Report of the 17th Session of the Codex Committee on General Principles, ALINORM 03/33, 2002, para. 94.

[34] J. Steffek, *The Power of Rational Discourse and the Legitimacy of International Governance*, EUI Working Paper RSC 2000/46 (San Domenico, European University Institute 2000), pp. 20-27.

[35] Ibid., p. 21. Beetham and Lord also emphasise this aspect of legitimacy and distinguish two sources of legitimacy deficit related to the performance of the European Union, D. Beetham and Ch. Lord, *supra* n. 27, pp. 23-24: 'There are two possible sources of legitimacy deficit here. One is the existence of fundamental disagreement about the character and scope of the tasks the EU should undertake: the definition of the ends and purposes it should serve. A second concerns the ability of its institutional matrix to deliver effective policy in the arenas it undertakes, to meet some basic criteria of effective decision making, and to demonstrate a capacity for correction and renewal in the event of "failure".'

[36] J.-M. Coicaud, *supra* n. 27, p. 524. According to Coicaud, a major condition for being perceived as a legitimate authority is the ability of the international organisation to overcome the limitations of states by enhancing international co-operation.

[37] D. Beetham, Ch. Lord, *supra* n. 27, pp. 11-12.

[38] H.G. Schermers and N.M. Blokker, *International Institutional Law*, 4th edn. (The Hague, Martinus Nijhoff Publishers 2003), p. 155.

iary bodies of the organisation. Legitimacy is enhanced by the possibility that the members have to verify that the organs and bodies act in accordance with these rules, hence granting legitimacy through legality.[39]

The dimension of institutional legitimacy does not solely address the mandate of the Codex Alimentarius Commission itself, but also focuses on the distribution of powers between the different organs and international institutions involved, and on the delegation of powers to the subsidiary bodies of the international organisation and the relation between these different bodies.[40] The duplication of powers, the lack of a clear separation of powers between the bodies, and the exceeding of the attributed powers may easily lead to disagreement and jeopardise the legitimacy of the international organisation at hand. In other words, questions of legitimacy addressed in the first dimension relate to the institutional structure of the international organisation. The issues under discussion are the mandates of the organs and subsidiary bodies involved in the Codex standard-setting procedure and the rules which establish the relationship between the bodies.

Procedural legitimacy: the Codex standard-setting procedure

The second dimension of legitimacy relates to the process of decision-making. Procedural legitimacy refers to the belief on the part of the member countries that the procedures for establishing decisions are defined in such a way that they ensure an outcome which is perceived as right. Most areas of international law are still characterised by a state-oriented rule-making process. The transfer of rule-making authority to international institutions is limited by the major role that member countries still have in the rule-making process.[41] As long as the process ensures the adoption of international norms that are perceived to be in the self-interests of the member countries, they accept the normative authority of the international institution. Aspects that reflect this include decision-making by consensus or unanimity, and the consent of the member country as a condition for the binding force of a norm. In other words, the consent of states is generally perceived as a legitimate method for establishing binding norms and enhances the legitimacy of the international institution in its competence as normative authority. This means that a legitimate procedure ensures the inclusion of the interests of states and their citizens. This state-oriented process rests upon the assumption that these interests are best mediated through the representatives of states. This means, according to Beetham and Lord, that, from the point of view of the citizens, the relevant model of legiti-

---

[39] See on legality as the basis for legitimacy, A.J. Hoekema, *Legitimiteit door legaliteit, over het recht van de overheid* (Nijmegen, Ars Aequi Libri 1991).

[40] See, on questions of legality related to the powers and functions of organs of international organisations, F. Morgenstern, 'Legality in international organizations', 48 *British Yearbook of International Law* (1976-1977), p. 246.

[41] D. Bodansky, *supra* n. 24, p. 604. K. Lenaerts and M. Desomer, 'New models of constitution-making in Europe: the quest for legitimacy', 39 *Common Market Law Review* (2002), p. 1225.

macy for international organisations is an indirect model.[42] Clearly, it is recognised that the state-oriented process results in a democracy-deficit, of which the European Union is a leading example.[43] However, this research is based upon the opinion that, at this present moment and in the context of the Codex Alimentarius Commission, democracy at inter-state level merits the main attention. As held by Hey, 'attaining greater legitimacy at the interstate level is pre-requisite for enhancing legitimacy in the inter-actions between states and individuals and groups in society at the international level of decision-making'.[44] Seeking to address the democratic deficit at this moment may even prove to be counter-productive and impede the establishment of inter-state democracy. Thus, this research primarily concentrates on 'inter-state democracy', also referred to as horizontal democracy.[45] As defined by Wouters, De Meester and Ryngaert, it refers to the equal rights of all states to participate in the decision-making procedure.[46]

Substantive legitimacy: Codex measures as instruments for harmonisation

Substantive legitimacy relates to the outcome of decision-making procedures within international organisations: the legitimacy of the measures adopted by international organisations as such. Although the legitimacy of the outcome is partly determined by the perceived legitimacy of the decision-making procedures, the dimension of substantive legitimacy concentrates on the measures adopted at the end of the procedure. The acceptance of these measures as legitimate depends upon their conformity with broadly shared values and interests.[47] Three elements identified by Chayes and Chayes that may obstruct compliance of states with international rules are:

1. ambiguity and indeterminacy of the language;
2. limitations on the capacity of states to carry out their undertakings; and
3. the temporal dimension of the social and economic changes contemplated by regulatory treaties.[48]

---

[42] D. Beetham, Ch. Lord, *supra* n. 27, p. 11. Hey refers to this as horizontal legitimacy as it is the perception of states that count in contrast to the concept of vertical legitimacy, which also concerns the perception of individuals and groups of in society. E. Hey, 'Sustainable development, normative development and the legitimacy of decision-making', 34 *Netherlands Yearbook of International Law* (2003), p. 13.

[43] J. Wouters, et al., 'Democracy and international law', 24 *Netherlands Yearbook of International Law* (2003), p. 178. D. Bodansky, *supra* n. 24, p. 618. E. Stein, 'International integration and democracy: no love at first sight', 95 *American Journal of International Law* (2001) pp. 489-534.

[44] E. Hey, *supra* n. 42, p. 7. See also, J. Brunnée, 'COPing with Consent: Law-Making Under Multilateral Environmental Agreements', 15 *Leiden Journal of International Law* (2002), pp. 7-15.

[45] J. Wouters, et al., *supra* n. 43, p. 177.

[46] Ibid.

[47] J.H.H. Weiler, *The Constitution of Europe. "Do the new clothes have an emperor?" and other essays on European integration* (Cambridge, Cambridge University Press 1999), p. 80. R. Howse, 'The legitimacy of the World Trade Organization', in J.M. Coicaud and V. Heiskanen (eds.), *The Legitimacy of international organizations* (Tokyo, United Nations University Press 2001), p. 363.

[48] A. Chayes and A.H. Chayes, 'On compliance', 47 *International Organization* (2001), p. 188.

The aim of this research is not to consider each Codex measure individually and to examine whether it responds to the values and interests of Codex Members. Instead, it focuses upon the techniques used in Codex measures that advance the harmonisation of national food requirements. The reason behind this is twofold. First, harmonisation was the incentive to establish the Joint FAO/WHO Food Standards Programme and to assign the Codex Alimentarius Commission with the responsibility of preparing and adopting international food standards. Harmonisation was also the incentive to include references to international standards in the *SPS Agreement* and the *TBT Agreement*. Second, the use of a wrong harmonisation technique may seriously obstruct the compliance of the adopted measures, and, for this reason, the reconsideration of the harmonisation techniques used is a core element in promoting the legitimacy of Codex standards. In this context, it has to be noted that the instrument of harmonisation is a far-reaching means of international co-operation. It requires more transfer of sovereignty by states than other trade-facilitating means, such as negative integration, the concept of equivalence and the principle of mutual recognition. In contrast to the EC legal order, in which positive integration is firmly established,[49] within the context of international trade law, the role of positive integration is not always evident.[50] The great diversity of circumstances that reign in different countries, such as agricultural production practices, temperature, the dietary habits of the population, the market structure and even the perceived risk by a given society raises the question of whether harmonisation is an appropriate tool to promote trade liberalisation.[51] After all, harmonisation means a real threat to the autonomy of states.[52] On the other hand, it would be premature to reject harmonisation as means of international co-operation for these reasons. In fact, harmonisation is by no means a homogeneous instrument and does not necessarily mean uniformity of standards or measures.[53] Harmonisation can be defined as an approximation of regulatory measures, or making regulatory measures more similar in certain respects.[54] The regulatory approach taken by the European Community provides us with clear examples of the fact that harmonisation can take different forms.[55] Provisions can be addressed to different actors and can result in

---

[49] G. Majone, *International Economic Integration, National Autonomy, Transnational Democracy: An Impossible Trinity?*, EUI Working Paper RSC 2002/48 (San Domenico, European University Institute 2002), pp. 4-12.

[50] Some authors hold that the role of harmonisation in the WTO agreements is a minor one. See for instance, G. Majone, *supra* n. 49, p. 5.

[51] J.N. Bhagwati and R.E. Hudec (eds.), *Fair Trade and Harmonization* (Cambridge, MA, MIT Press 1996).

[52] G. Majone, *supra* n. 49, p. 5.

[53] See on the meaning of harmonisation within the EC legal order, D. Vignes, 'The Harmonisation of National Legislation and the EEC' 15 *European Law Review* (1990), p. 361. P.J. Slot, 'Harmonisation', 21 *European Law Review* (1996), p. 379.

[54] A.O. Sykes, 'The (limited) role of regulatory harmonization in international goods and services markets', 2 *Journal of International Economic Law* (1999), p. 51.

[55] The question of harmonisation has been dealt with extensively over the years in the context of the discussion over the distribution of authority between the Community and its member states and solutions have been sought in different types and different degrees of harmonisation. D. Vignes, *supra*

different degrees of retained sovereignty. The different techniques of harmonisation used in the EC legal order, which aim for different degrees of harmonisation, can be recognised by the following elements: the exhaustion of the issues covered by harmonisation measures, a general safeguard clause allowing member states to maintain or to adopt national measures for justified reasons,[56] the inclusion of safeguard clauses in the harmonisation measures themselves and the formulation of the harmonisation provisions themselves (for instance, by using the term 'should' instead of 'shall').

The dimension of substantive legitimacy will concentrate on questions relating to the harmonisation techniques incorporated in Codex standards in the light of the changes resulting from the adoption of the WTO agreements.

4. THE STRUCTURE OF RESEARCH

The book therefore starts with a thorough legal analysis of the institutional framework of the Codex Alimentarius Commission (Chapter I). It examines the legal basis for the different competences of the Codex Alimentarius Commission, its relations with its parent organisations, the FAO and the WHO, and the structure of subsidiary Codex bodies and other bodies operational in the Joint FAO/WHO Food Standards Programme. In addition, the approach taken by the Codex Alimentarius Commission to execute its prime mandate is investigated: the adoption of a collection of measures that assist in the harmonisation of food requirements (Chapter II). Here, account is taken of the Codex acceptance procedure, a procedure that was abolished in July 2005 and thus no longer exists. However, as it formed an essential element in the harmonisation policy of the Codex Alimentarius Commission, it will be examined. Subsequently, the status that has been attributed to Codex measures in other legal systems which promote the harmonisation of food regulations will be analysed: the EC legal system (Chapter III), and the WTO system (Chapter IV), and the consequences for the relations between these different institutions. Finally, the findings of this research are reconsidered in the light of the new role of the Codex Alimentarius Commission that results from the references to its work in the WTO agreements and in secondary EC food regulation. In this perspective, it analyses the institutional framework of the Codex Alimentarius Commission, its procedures and its standards to reveal and comprehend the major concerns of legitimacy (Chapter V). In conclusion, the findings of this research will be summarised, and, based upon the belief that guaranteeing legitimacy is a continuous process,[57] some suggestions for the improvement of the legitimacy of the Codex Alimentarius Commission will be advanced.

---

n. 53, pp. 358-374. P.J. Slot, *supra* n. 53, pp. 378-397. R. Barents and L.J. Brinkhorst, *Grondlijnen van Europees Recht*, 9th edn. (Deventer, Kluwer 1999), pp. 401-411.

[56] For instance, Arts. 95.4 and 95.5 EC.

[57] K. Lenaerts and M. Desomer, *supra* n. 41, pp. 1217-1253.

Chapter I
THE CODEX ALIMENTARIUS COMMISSION: THE INSTITUTIONAL FRAMEWORK

1. INTRODUCTION

This chapter examines the institutional framework that has been installed in order to enhance international co-operation by means of international food standards (Joint FAO/WHO Food Standards Programme). It concentrates on the position of the Codex Alimentarius Commission therein, and its use of mandated competencies that have allowed it to develop into a solid institution.

The Codex Alimentarius Commission was not established under the most convenient circumstances for it to develop into a strong institution. Although the necessity of launching a food standard programme to overcome differences between national food regulations through international co-operation was recognised, the operation of the Codex Alimentarius Commission was set up amidst the already existing food standard-setting bodies. In 1962, at the time of its establishment, several other initiatives related to food standard-setting existed, such as the initiatives undertaken by the United Nations Economic Commission for Europe (UNECE) on fruit and fruit juices, by the Food and Agriculture Organisation of the United Nations (FAO) on fish, by the International Dairy Federation (IDF) on milk and milk products, and by the Organisation for Economic Co-operation and Development (OECD). Hence, even at its start, it faced duplication with other international institutions in its mandated work. Furthermore, the Codex Alimentarius Commission was established as a subsidiary body to two international organisations: the FAO and the World Health Organisation of the United Nations (WHO). Consequently, it depends (and has always depended), for the assignment of its competences, on the delegation of powers by its parent organisations.

In order to appreciate the institutional challenges and opportunities that the Codex Alimentarius Commission has faced over the years, this chapter analyses the legal aspects that form its basis. It starts with the origin of the Joint FAO/WHO Food Standards Programme. Then, in the next section, it will examine the institutional aspects of the Codex Alimentarius Commission, its subsidiary bodies and the other joint subsidiary bodies of the FAO and the WHO which are involved in the Joint FAO/WHO Food Standards Programme.

2. THE ORIGIN OF THE JOINT FAO/WHO FOOD STANDARDS PROGRAMME

Numerous new and advanced ways of transportation in the 19th Century made an extension in foods markets possible, from local to trans-national markets. Furthermore, several severe hunger crises in the 19th Century underscored the dependency on food from other countries to overcome temporal lacks of self-sufficiency. The increased trans-national transportations of, and commerce in, food products furthered the need for international co-operation. Even by the beginning of the 20th Century, it was recognised that the international trade of food products could be jeopardised by national food laws which constituted trade barriers.[1] As a consequence, international co-operation increasingly included issues related to the distribution, commerce and consumption of food, and eventually led to the decision to establish a Joint FAO/WHO Food Standards Programme.

As from the beginning of the 20th Century, food issues were dealt with at an international level. Early attempts failed because of a lack of commitment on the part of governments to adapt their food legislation.[2] However, the initiatives which took place during the 1930s were more successful. Under the auspices of the International Institute for Agriculture (IIA), the predecessor of the FAO, several conventions were adopted in order to ensure fair trade practices, an element that would subsequently come under the mandate of the Codex Alimentarius Commission.[3] Examples of this include the Brussels Convention for the marking of eggs of 10 December 1931, and the Rome Convention on the unification of the sampling and analysis methods for cheese of 26 April 1934.[4] Furthermore, on the initiative of the International Dairy Federation (IDF), several proposals for conventions were discussed during conferences of the IIA and were presented for adoption in 1939: these included, the regulation of the production and trade of condensed milk, the regulation on the production and trade in fat and low-fat milk, the unification of the methods of analysis of condensed milk and milk-powder, and the regulation on the production and trade in 'soft' cheese. Although the outbreak of the Second World War prevented their final adoption, these documents served as the basis for successive international co-operation.

After the Second World War, the need for the co-ordination of national food laws was reflected in the establishment of several regional and international *fora*.[5] The regional *fora* comprised the initiative of a group of Latin American countries to

---

[1] A. Randell, 'International Food Standards: The Work of Codex', in N. Rees and D. Watson (eds.), *International Standards for Food Safety* (Gaithersburg, Maryland, Aspen Publishers 2000), p. 3.

[2] With regard to these items, governments were hesitant to commit and adapt their food legislation accordingly. F.C. Lu; 'The Joint FAO/WHO Food Standards Programme and the Codex Alimentarius', 24 *WHO Chronicle* (1970), p. 198.

[3] The International Institute of Agriculture was established at a diplomatic conference convened by the King of Italy on June 7, 1905, by an international treaty. The institute was dissolved in agreement with the FAO in 1945. See also, O. Ribbelink, *Opvolging van internationale organisaties. Van Volkenbond - Verenigde Naties tot ALALC - ALADI* (The Hague, T.M.C. Asser Institute 1988), p. 121. See also, H.F.W.M van Haastert, *Het Internationaal Landbouw Instituut (I.I.A.) en de Organisatie voor Voedsel en Landbouw (F.A.O.)*, ('s-Hertogenbosch, N.V. Zuid-Nederlandsche Drukkerij 1947), p. 52.

[4] The Sampling and Analysis Methods were used to verify the correct naming of cheeses, as many cheese names were protected either through the place of origin or as a result of an acquired reputation for quality.

[5] D.L. Leive, *International Regulatory Regimes* (Lexington, Lexington Books 1976), p. 377.

establish a Latin American Food Code, and the 1958 initiative which created a Council of the Codex Alimentarius Europaeus to develop a European Food Code. At international level, it was mainly under the auspices of the FAO that initiatives were launched, including the creation of a Committee of government experts on the Code of Principles concerning Milk and Milk Products in 1957. Furthermore, joint initiatives between the FAO and the WHO led to the establishment of the Joint FAO/WHO Expert Committee on Food Additives in 1955, and to the organisation of a joint meeting on pesticides in 1961.[6]

Three of these developments are considered to form the basis of the establishment of the Joint FAO/WHO Food Standards Programme[7] and to have been influential in its definition:[8]

- the Committee of government experts on the Code of Principles concerning Milk and Milk Products;
- the Joint FAO/WHO Expert Committee on Food Additives and the Joint FAO/WHO meeting on pesticides; and
- the Codex Alimentarius Europaeus.

The Council of the Codex Alimentarius Europaeus was created under the auspices of the International Commission on Agricultural Industries and the Permanent Bureau of Analytical Chemistry.[9] The idea of the creation of a Codex Alimentarius Europaeus was based upon the Codex Alimentarius Austriacus, a system of food regulations used in the Austro-Hungarian Monarchy as from 1897.[10] The establishment of the Codex Alimentarius Europaeus was intended to update and to extend the principles of this system.[11] In order to promote wider participation, the Council of the Codex Alimentarius Europaeus proposed that the FAO should take over the work that had commenced. This request led to the adoption of Resolution 12/61 by the FAO Conference in 1961, which emphasised the rapidly growing importance of internationally accepted food standards as a means of protecting both consumers and producers in all states, and proposed to pay attention to the possibilities of a joint FAO/WHO Food Standards Programme.[12] Consequently, a Joint FAO/WHO Conference on food standards was organised to discuss these possibilities, which

---

[6] *The First Ten Years of the World Health Organization* (Geneva, WHO 1958), pp. 425-427, *The Second Ten Years of the World Health Organization* (Geneva, WHO 1968), pp. 157-158, and pp. 225-226.

[7] J.P. Dobbert, 'Le Codex Alimentarius vers une nouvelle méthode de réglementation internationale', 15 *Annuaire Français de Droit International* (1969), p. 679, D.L. Leive, *supra* n. 5, p. 377.

[8] D.L. Leive, *supra* n. 5, pp. 380-381.

[9] F.C. Lu, *supra* n. 2, p. 199.

[10] Ibid. See also, R. van Havere, 'Codex Alimentarius', in R. Kruithof, *Levensmiddelenrecht* (Brussels, Ced-Samson 1979), p. 22.

[11] F.C. Lu, *supra* n. 2, p. 199. R. van Havere, *supra* n. 10, p. 22.

[12] Resolution 12/61 of the FAO Conference of November 1961. See S. Shubber, 'The Codex Alimentarius Commission under International Law', 21 *International and Comparative Law Quarterly* (1972), p. 631.

resulted in the decision to establish the Joint FAO/WHO food standard programme with the objective of simplifying and co-ordinating the work on international food standards previously undertaken by various international non-governmental and governmental organisations.[13] As the main tool to achieve these objectives, the development of the Codex Alimentarius was agreed upon: namely, the publication of internationally-adopted food standards presented in a unified form.

The fact that the Council of the Codex Alimentarius Europaeus sought to incorporate its work in a world-wide food standards programme explains the strong European orientation of the Programme in its initial years.[14]

In addition, the Committee of government experts on the Code of Principles concerning Milk and Milk Products left its marks on the organisation of the Joint FAO/WHO Food Standards Programme. The methods of work developed during the early years of the Codex Alimentarius Commission drew upon the experience obtained in the Committee of government experts on the Code of Principles concerning Milk and Milk Products.[15] Elements of the methods of work of the Committee of government experts that were inserted into the methods of work of the Joint FAO/WHO Food Standards Programme were amongst others:

– the intergovernmental character of the institution;
– the preparation of draft standards by outside bodies (such as the function of the International Dairy Federation, an industry INGO, that had proposed the establishment of the Committee on government experts); and
– the loose-leaf form of the finalised standards to be published.[16]

For their part, both the Joint FAO/WHO Expert Committee on Food Additives and the Joint FAO/WHO Expert Meeting on Pesticides were incorporated as expert bodies into the institutional framework responsible for the Joint FAO/WHO Food Standards Programme.

3. A JOINT FAO/WHO INSTITUTIONAL FRAMEWORK

The establishment of the Joint FAO/WHO Food Standards Programme and the development of the *Codex Alimentarius* as core element of the Programme was decided upon by the decision-making bodies of both the FAO (in 1961) and the WHO (in 1963).[17] Together, both organisations have established the institutional frame-

---

[13] Report of the Joint FAO/WHO Conference on Food Standards, Geneva 1-5 October 1962, ALINORM 62/8.
[14] D.L. Leive, *supra* n. 5, pp. 380-381.
[15] Ibid., p. 381.
[16] Ibid. These elements shall be further discussed in this chapter, Chapter II and Chapter V.
[17] Resolution No. 12/61 of the FAO Conference, 'Codex Alimentarius', adopted at the 11th Session of the FAO Conference, Report of the 11th Session of the FAO Conference, Rome 4-24 November 1961, para. 263. Resolution WHA 16.42 'Joint FAO/WHO Programme on Food Standards (Codex

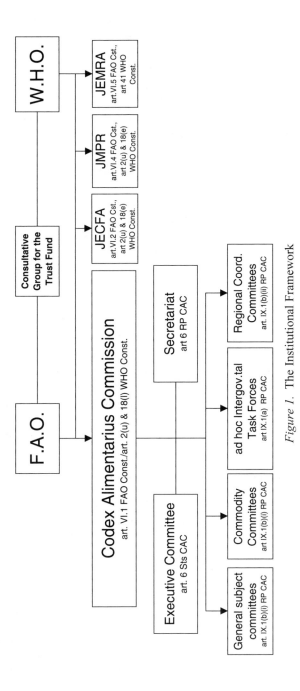

Figure 1. The Institutional Framework

work responsible for the execution of the Joint FAO/WHO Food Standards Programme that functions under their authority. As illustrated in Figure 1 of this chapter, the institutional framework can be divided into the Joint FAO/WHO Codex Alimentarius Commission and the subsidiary bodies which fall under its authority, on the one hand, and the other Joint FAO/WHO bodies, such as the Joint FAO/WHO expert bodies or expert meetings, and the Joint FAO/WHO Consultative Group for enhanced participation of developing countries to the Codex Alimentarius, on the other. The scientific expert bodies and the Joint FAO/WHO Consultative Group function independently from the Codex Alimentarius Commission and fall under the direct responsibility of the FAO and the WHO.

3.1 The Joint FAO/WHO Codex Alimentarius Commission

The Joint FAO/WHO Codex Alimentarius Commission was appointed as principal organ responsible of the implementation of the Joint FAO/WHO Food Standards Programme.[18] This section provides a legal analysis of the Codex Alimentarius Commission, its powers, membership, budgetary aspects and rules of procedure as regards its meetings.

3.1.1 *Powers and internal rules*

In 1963, the Joint FAO/WHO Codex Alimentarius Commission was established on the basis of Article VI of the FAO Constitution and Article 18(1) in combination with Article 2(u) of the WHO Constitution, as the principal responsible body to enact the Joint FAO/WHO Food Standards Programme.[19] At the very least, its position should be viewed as exceptional, as it was established as subsidiary body born out of two separate specialised agencies of the United Nations, the FAO and the WHO. This characteristic is reflected in various institutional elements of the

---

Alimentarius)', adopted at the 16th Session of the World Health Assembly, Geneva, 7-23 May 1963, Off. Rec. World Health Organization, 124, 74.

[18] See, for instance, Resolution WHA 16.42 'Joint FAO/WHO Programme on Food Standards (Codex Alimentarius)', adopted at the 16th Session of the World Health Assembly, Geneva, 7-23 May 1963, Off. Rec. World Health Organization, 124, 74, Art. 1 of the Statutes of the Codex Alimentarius Commission.

[19] Art. 18(1) of the WHO Constitution grants the WHO the competence to establish any institution it considers desirable to achieve its objectives. Art. 2(u) provides for the development of food standards. Art. VI.1 of the FAO Constitution states that 'The Conference or Council may establish commissions, the membership of which shall be open to all Member Nations and Associate Members, or regional commissions open to all Member Nations and Associate Members whose territories are situated wholly or in part in one or more regions, to advise on the formulation and implementation of policy and to coordinate the implementation of policy. The Conference or Council may also establish, in conjunction with other intergovernmental organisations, joint commissions open to all Member Nations and Associate Members of the Organization and of the other organizations concerned, or joint regional commissions open to Member Nations and Associate Members of the Organization and of the other organizations concerned, whose territories are situated wholly or in part in the region.' (emphasis added). See also, S. Shubber, *supra* n. 12, pp. 633-634.

Codex Alimentarius Commission. First and foremost, being a subsidiary organ means that, for the assignment of its competences, the Codex Alimentarius Commission is dependent on the delegation of powers by its 'parent organisations'. Nevertheless, being the subsidiary organ of *two* international organisations means that, in most cases, the Codex Alimentarius Commission has the advantage of 'inheriting' the powers and the rules of the 'parent' organisation which has the broadest powers. In the following section, the legal status, the powers and the legal order of the Codex Alimentarius Commission will be examined.

3.1.1.1 The subsidiary character of the Codex Alimentarius Commission

Several elements determine whether an intergovernmental institution can be classified as an international governmental organisation (IGO), such as whether the institution has been established by a treaty or other instrument governed by international law, and whether the institution has its own international legal personality.[20] In the recognition of the institution as a subject of international law, it is important that the authority of an international organisation is attributed directly by states. The international agreement that establishes the international organisation functions as a constitutional document which aims to provide the organisation (either implicitly or expressly) with a basis for the legal personality which enables the organisation to act as a subject of international law. At the same time, the scope of its authority to act as a subject of international law is determined by the states which consent to the attribution of powers as contained in the international agreement.[21]

---

[20] Draft Art. 2 of the draft articles on responsibility of international organizations, International Law Commission, report on the work of its 55th Session, 5 May-6 June and 7 July-8 August 2003, Official Records General Assembly, 58th Session, Supplement No. 10 (A/58/10), Chapter IV, Responsibility of International Organizations, p. 33 and pp. 38-41, <http://www.un.org/law/ilc>. First report on responsibility of international organizations by G. Gaja, Special Rapporteur, International Law Commission, 55th Session, 5 May-6 June and 7 July-8 August 2003, (A/CN.4/532). See also, H.G. Schermers and N.M. Blokker, *International Institutional Law*, 4th edn. (The Hague, Martinus Nijhoff Publishers 2003), pp. 21-22. See for other definitions of international organisation, R. Bernhardt (ed.), *Encyclopedia of public international law* (Max Planck Institute for Comparative Public Law and International law, Amsterdam, North Holland 1995), p. 1289: 'The term international organizations denotes an association of States established by and based upon a treaty, which pursues common aims and which has its own special organs to fulfil particular functions within the organization.' See also, M. Diez de Velasco Vallejo, *Les organisations internationales*, Collection Droit International (Paris, Economica 2002), p. 10, D.W. Bowett, et al., *Bowett's law of international institutions,* 5th edn. (London, Sweet & Maxwell 2001), C.A. Colliard and L. Dubouis, *Institutions Internationales*, 10th edn. (Paris, Dalloz 1995), p. 173, M.N. Shaw, *International Law*, 5th edn. (Cambridge, Cambridge University Press 2003), pp. 1193-1195. P. Malanczuk, *Akehurst's modern introduction to international law*, 7th edn. (Routledge, London 1997), p. 92. See also, for instance, on the definition of non-governmental organizations, ECOSOC Resolution 288(x) of 27 February 1950: 'Any international organization which is not established by intergovernmental agreement shall be considered as a non-governmental organization for the purposes of these arrangements.'

[21] Also referred to as the doctrine of attributed competences. See, for instance, H.G. Schermers and N.M. Blokker, *supra* n. 20, p. 155.

The establishment of the Codex Alimentarius Commission took place through the adoption of a resolution by the decision-making organs of two international organisations: the FAO Conference and the WHO Health Assembly. The decision-making organs of both organisations acted on the legal basis which allowed them to establish subsidiary bodies. Consequently, the Codex Alimentarius Commission cannot be considered to be a new and autonomous international organisation which has its own legal personality.[22] Instead, its authority is derived from the attributed powers of its parent organisations and not directly by states. For this reason, the Codex Alimentarius Commission has to be classified as a subsidiary organ of both the FAO and the WHO. This subsidiary position of the Codex Alimentarius Commission is reflected in several elements. For instance, the Directors-General of both the FAO and the WHO have far-reaching powers with regard to the establishment of the Commission's Agenda.[23] Furthermore, despite the fact that the Codex Alimentarius Commission can establish and amend its own Rules of Procedure, the entry into force of the Rules of Procedure and the amendments thereof require the approval of the Directors-General of both the FAO and the WHO.[24] Likewise, the FAO and WHO governing bodies have to approve the strategic vision and the Medium-Term Plan of the Commission, which serves as the basis for determining the priorities of Codex when deciding to develop new standards. One of the most essential consequences of this subsidiary position is the financial dependency of the Codex on both organisations, and the fact that the Secretariat of the Codex Alimentarius Commission is part of the staff of the FAO. As a subsidiary body, the Codex Alimentarius Commission does not have the competence to enter into agreements with international organisations in order to further co-operation, but depends upon the rules and policies of its parent organisations.[25] Further proof of its subsidiary position can be found in the fact that the current evaluation of the Codex Alimentarius Commission is undertaken by both parent organisations, and that important alterations to the operation of the Codex Alimentarius Commission which

---

[22] According to the Yearbook of International organizations, the Codex Alimentarius Commission is not classified as a conventional international governmental organization, YIO 2004/05, Vol. IA, at p. 1111 and Vol. IB at p. 2911 (Appendix 2). See for a discussion on the consequences of the adoption of a 'Codex treaty', A. Herwig, 'Legal and institutional aspects in the negotiation of the Codex Alimentarius Convention', 2 *Zeitschrift fur das gesamte Lebensmittelrecht* (2001). Although the constitutional character of such a treaty transforming the Codex Alimentarius Commission into a new international organisation is not explicitly addressed, some ideas raised are interesting. See on subsidiary bodies (including the Codex Alimentarius Commission) as de facto international organisations, Ch. Martini, 'States' Control over New International Organization', 6 *Global Jurist Advances* (2006), pp. 1-25. At <http://www.bepress.com/gj/advances/vol6/iss3/art4>.

[23] Rules VII.1 and VII.6 of the Rules of Procedure, Procedural Manual, 16th edn., 2006. The Procedural Manual can be found on the website of the Codex Alimentarius Commission: <www.codexalimentarius.net>. See, for further details, Section 3.1.3 of this chapter.

[24] Art. 8, Statutes, Procedural Manual, 16th edn., 2006. The firstly adopted rules of procedure by the Commission were sent to the FAO Committee on Constitutional and Legal Matters by the Director-General before final adoption could take place.

[25] See, for instance, Rule IX.5 of the Rules of Procedure, Procedural Manual, 16th edn., 2006.

involve financial consequences must be approved by the competent bodies of both the parent organisations.

3.1.1.2 Delegation of powers as basis for Codex Alimentarius Commission's authority

As a result, the authority of the Codex Alimentarius Commission depends upon its parent organisations. The question arises as to which powers have been delegated to the Codex Alimentarius Commission and to what extent the FAO and the WHO maintain responsibility over the actions taken by the Codex Alimentarius Commission.

The explicit powers delegated to the Codex Alimentarius Commission are laid down in Articles 1 and 7 of its Statutes and in Rules XI and XII.1 of its Rules of Procedure.[26] Article 1 of the Statutes expressly provides the Codex Alimentarius Commission with consultative powers as regards the implementation of the Joint FAO/WHO Food Standards Programme. Article 7 of the Statutes and Rule XI of the Rules of Procedure state that the Codex Alimentarius Commission may establish the subsidiary bodies necessary for the preparation of draft food standards.[27] In addition, Rule XII.1 provides that the Codex Alimentarius Commission may establish and amend its own standard-setting procedure. The adoption and amendment of the procedure is not subject to the approval of its parent organisations.

Moreover, careful reading of Article 1 of the Statutes in conjunction with the Rules of Procedures illustrates that the powers of the Codex Alimentarius Commission are not limited to these expressly stated competencies.[28] Practice resulting from the operation of the Codex Alimentarius Commission over the years confirms this. Article 1 states that:

> 'The Codex Alimentarius Commission shall ... be responsible for making proposals to, and shall be consulted by the Directors-General of the Food and Agriculture Organization (FAO) and the World Health Organization (WHO) on all matters pertaining to the implementation of the Joint FAO/WHO Food Standards Programme:
> (a) protecting the health of the consumers and ensuring fair practices in the food trade,
> (b) promoting coordination of all food standards work undertaken by international governmental and non governmental organizations,
> (c) determining priorities and initiating and guiding the preparation of draft standards through and with the aid of appropriate organizations;
> (d) finalizing standards elaborated under (c) above and, after acceptance by governments, publishing them in a Codex Alimentarius either as regional or world wide

---

[26] Procedural Manual, 16th edn., 2006. Both 'constitutional' documents have been adopted by the decision-making organs of the FAO and the WHO and can only be amended upon approval by these same organs. See, for more detailed discussion, Section 3.1.1.3 of this chapter. On the powers of the Codex Alimentarius Commission, See also, J.P. Dobbert, *supra* n. 7, p. 680, S. Shubber, *supra* n. 12, pp. 634-648, D.L. Leive, *supra* n. 5, pp. 384-386.

[27] See also, Section 3.4 of this chapter.

[28] S. Shubber, *supra* n. 12, pp. 634-635.

standards, together with international standards already finalized by other bodies under (b) above, wherever this is practicable;

(e) amending publishing standards, after appropriate survey in the light of developments.'

At first sight, the enumeration under Article 1 seems to reflect only the content of the Joint FAO/WHO Food Standards Programme and remains imprecise as to the consequential powers for the Codex Alimentarius Commission as the responsible organ for the implementation of the Programme. However, it has been held that the enumeration under Article 1 implicitly contains some of the powers of the Codex Alimentarius Commission, in particular, when it comes to sub-paragraphs (b) to (d).[29] Over the years, practice confirms that these elements are, in effect, the autonomous powers of the Codex Alimentarius Commission. This practice as developed by the Codex Alimentarius Commission has not been the subject of objections raised by either the FAO or the WHO.[30] Two important powers are worth mentioning:

1. the authority to promote co-ordination of all food standards work at an international level;
2. normative powers through standard-setting.

As discussed above under Section 2 of this chapter, the urge to co-ordinate international food standard-setting activities was one of the prime reasons for the adoption the Joint FAO/WHO Food Standards Programme. The promotion of the co-ordination of all food standards work undertaken by other international governmental and non-governmental organisations is included as one of the objectives of the Joint FAO/WHO Food Standards Programme, enumerated in Article 1 of the Statutes.[31] However, Article 1 of the Statutes does not appoint an organ with the task of carrying out the co-ordination of international standard-setting activities. As responsible organ for the implementation of the Joint FAO/WHO Food Standards Programme, it would be logical for the Codex Alimentarius Commission to carry out this task.[32] This is confirmed by the fact that the Codex Alimentarius Commission is entitled to establish its own standard-setting procedure.[33] This competence includes the adop-

---

[29] J.P. Dobbert, *supra* n. 7, p. 680, S. Shubber, *supra* n. 12, p. 635. See on the doctrine of 'implied powers' of international organisations in general, for instance J.E. Alvarez, *International Organizations as Law-makers* (Oxford, Oxford University Press 2005), pp. 92-95.

[30] See for the situation in 1976, D.L. Leive, *supra* n. 5, p. 384. Again, in 2003, the management board of the FAO and the Director-General of the WHO in their reactions to the report of the Joint FAO/WHO Evaluation implicitly reaffirm the standard-setting powers of the Codex Alimentarius Commission, see FAO Management Response and the Report of the Director-General (WHO) to the WHO Executive Board, Evaluation of the Codex Alimentarius Commission and other FAO and WHO food standards work, December 2002, ALINORM 03/25/3-Add. 1.

[31] Art. 1 under (b) of the Statutes, Procedural Manual, 16th edn., 2006, p. 3.

[32] S. Shubber, *supra* n. 12, p. 635.

[33] Rule XII.1 of the Rules of Procedure, Procedural Manual, 16th edn., 2006, p. 16.

tion of provisions on the function of international organisations within the standard-setting procedure and does not require the prior approval of its parent organisations. In practice, the Codex Alimentarius Commission uses this competence to promote co-ordination by seeking close co-operation with other international (non-) governmental organisations in the context of the standard-setting procedure.[34] Consequently, work submitted by other international (non-) governmental organisations is subject to the final approval of the Codex Alimentarius Commission.[35]

Sub-paragraphs (c) to (e) concentrate on the elaboration, preparation, adoption and publication of food standards and other documents to be included in the Codex Alimentarius. It clearly results from the Rules of Procedure that the Codex Alimentarius Commission is the recognised responsible organ for this task. Rule XI.1 gives the Codex Alimentarius Commission the right to establish the subsidiary bodies which are necessary 'for the accomplishment of its work in the finalization of draft standards'. Furthermore, as mentioned above, the Codex Alimentarius Commission may adopt its own rules on the standard-setting procedure.[36] Likewise, the function of the Codex Alimentarius Commission as a decision-making body in the standard-setting procedure is recognised by Rule XII.2, which states that the Commission 'shall make every effort to reach agreement on the adoption or amendment of standards by consensus'. As neither the publication of the standards, nor the submission of the standards to its members for acceptance has been subject to a decision of the FAO or the WHO, this makes the Commission the final decision-maker with regard to the standards and other decisions contained in the Codex Alimentarius. Hence, contrary to the literal wording of Article 1 of the Statutes, its decisions to adopt and publish standards are *not* submitted to its parent organisations for approval. This provides the Codex Alimentarius Commission with independent normative powers with regard to the standard-setting procedure, as it deals directly with its members without the intervention of its parent organisations. Furthermore, as the organ responsible for the preparation, adoption and publication of standards that are often of a trans-scientific nature, the task of the Commission includes what is often referred to as risk management.[37] Risk management is a process of weighing several policy alternatives in order to reduce risks. Although based upon risk assessment, the process of risk management is distinct from the latter. Risk assessment can be described as a scientific evaluation in order to identify and characterise hazards and to characterise the resulting risks. The Codex Alimentarius Commis-

---

[34] It has adopted the 'Guidelines on the co-operation with other international organisations' for this purpose. See, for further discussion, Section 3.4 of this chapter and Section 2.4 of Chapter V.

[35] See, for further discussion on the standard-setting procedure, Section 4 of Chapter II.

[36] Rule XII.1 Rules of Procedure, Procedural Manual, 16th edn., 2006, p. 16.

[37] See also, the Working Principles for Risk Analysis for Application in the Framework of the Codex Alimentarius, Procedural Manual, 16th edn., 2006, p. 103. The term 'risk management' is used in the context of 'risk analysis', a process of decision-making to address trans-scientific issues, which consists of risk assessment, risk management and risk communication.

sion does not undertake risk assessments itself, but depends on scientific bodies that are situated outside its own institutional structure.[38]

It can thus be concluded from the above, that, despite the subsidiary position of the Codex Alimentarius Commission, its parent organisations have, in fact, delegated it with some significant powers.[39]

3.1.1.3 The Procedural Manual

The internal rules of the Codex Alimentarius Commission and its subsidiary bodies are laid down in the Procedural Manual. As mentioned above in Section 3.1.1.1, the Procedural Manual contains two legal documents that require a decision from the competent organs of the FAO and the WHO in order to be amended, namely, the Statutes, and the Rules of Procedures of the Codex Alimentarius Commission.

With regard to the Statutes, only the FAO Conference and the World Health Assembly can amend them. However, the Codex Alimentarius Commission may make recommendations for amendments by a two-thirds majority on condition that a quorum of the members is present at the moment of voting.[40] This quorum consists of half the total number of Codex Members.

The Statutes only provide general provisions on the functioning of the Codex Alimentarius Commission. In order to allow the Codex Alimentarius Commission to establish more detailed rules for the regulation of its functioning, the Statutes explicitly mention the possibility of adopting and amending its own Rules of Procedure.[41] However, amendments of, and additions to, the rules of procedure require a quorum of Codex Members, and a two-thirds majority-vote in order to be adopted, as well as the approval of the Directors-General of the FAO and the WHO in order to enter into force.[42]

In addition to the Statutes and the Rules of Procedure, the Codex Alimentarius Commission has adopted an extensive collection of internal rules laid down in the Procedural Manual. These internal rules regulate the functioning of the institutional framework of the Codex and its subsidiary bodies as well as the standard-setting procedure in great detail. They include 'General Principles of the Codex Alimentarius', 'Guidelines for Codex Committees and Task Forces', 'Principles for the Participation of International Non-governmental Organisations in the Work of Codex', 'Criteria for Work Priorities and for Establishing Subsidiary Bodies', and 'Guidelines for the Inclusion of Specific Provisions in Codex Standards'. Most of the provisions contained in the Procedural Manual are binding upon the Codex Alimentarius Commission, its subsidiary bodies and upon all Codex Members to the extent that they act as part of these bodies. For instance, the 'Guidelines for

---

[38] See, for more details on the expert bodies, Section 3.5 of this chapter.
[39] See, for more details on its standard-setting power, Section 4 of Chapter II.
[40] Rule VI.7 Rules of Procedure, Procedural Manual, 16th edn., 2006, p. 11.
[41] Art. 8 of the Statutes, Procedural Manual, 16th edn., 2006, p. 4.
[42] Ibid. Rules XV and XVI of the Rules of Procedure, Procedural Manual, 16th edn., 2006, p. 18.

Codex Committees and Task Forces' are binding on host countries that are responsible for the organisation of subsidiary committees.[43]

3.1.2 Composition

3.1.2.1 Members of the Codex Alimentarius Commission

The membership of the Codex Alimentarius Commission is open to all countries and its decisions are taken by the representatives of the governmental organs of these countries. Thus, like its parent organisations, it can be classified as an intergovernmental body which has a universal character. However, the universal character of the Codex Alimentarius Commission faces one restriction: countries eligible for membership to the Codex Alimentarius Commission have to be members or associate members of at least one of its parent organisations.[44] In practice, this restriction to the universal nature of the Codex Alimentarius Commission can hardly be called a restriction, as most of the countries are members of the FAO or the WHO. The membership of the Codex Alimentarius Commission has grown from 30 participating members at its first session in 1963, to the current number of 171 member countries and 1 regional economic integration organisation: hence, the universal character of the Codex Alimentarius Commission. In contrast with the early years of the Codex Alimentarius Commission, when its membership was dominated by European countries, membership today comprises nations from all regions of the world.

The admission of new members occurs by notifying the Directors-General of the FAO and the WHO.[45] The granting of membership to countries is not subject to voting or any other means of acceptance by the Codex Alimentarius Commission. The fact that a state is a member of either one of the parent organisations and the notification procedure are sufficient for it to become a member. The reason behind this is that the relevant decision-making with regard to the acceptance of new members has already taken place in one of the parent organisations in accordance with their rules. The possibility of withdrawing from the Codex Alimentarius Commission membership is not mentioned in either the Statutes or the Rules of Procedure. Therefore, it is not clear whether unilateral withdrawal by a Codex Member is recognised.[46] Until 2003, membership was only open to countries; regional

---

[43] The host countries are the Codex Members that have been appointed to host and organise the meetings of a Codex Committee or Task Force; see, for further details, Section 3.4.4 of this chapter.

[44] Art. 2 of the Statutes, Procedural Manual, 16th edn., 2006, p. 3 and Rule I.1 of the Rules of Procedure, Procedural Manual, 16th edn., 2006, p. 6. Associate members are these territories that do not have independence, but are part of another member. For example, Puerto Rico used to be an associate member to the WHO. This possibility within the WHO aims to ensure universality through its membership. However, nowadays this option is rarely used as its significance is reduced due to the reduction in the number of such territories. See, for instance, H.G. Schermers and N.M. Blokker, *supra* n. 20, p. 123.

[45] Art. 2 of the Statutes, Procedural Manual, 16th edn., 2006, p. 3 and Rule I.1 of the Rules of Procedure, Procedural Manual, 16th edn., 2006, p. 6.

[46] The WHO does not contain a provision in its constitution on withdrawal. Prior to 1985, 10 countries withdrew their membership to the WHO. The WHO did not recognise their withdrawal as it

organisations were excluded from membership. Since 1991, proposals for an amendment to the Rules of Procedure to allow regional organisations to become members of the Codex Alimentarius Commission have been subject to discussion.[47] During the Codex Alimentarius Commission meeting in 2003, the proposal for this amendment was finally adopted.[48] As it had already been a full member of the FAO since 1991, this amendment allowed the European Community to become member to the Codex Alimentarius Commission.[49]

Members have the right to participate in the meetings of the Codex Alimentarius Commission and in most of the subsidiary Codex bodies.[50] Furthermore, it is only Codex Members that have the right to vote. Each member has one vote,[51] and each member has the right to request a vote. Except for the adoption or amendment of standards and other related Codex measures, the Codex Alimentarius Commission takes its decisions by voting.[52] The voting procedure occurs by a roll-call vote,[53] which takes place as follows: the names of the member countries are called out in alphabetical order and the representative of that member country indicates whether it is in favour of the proposal, against it or whether it wishes to abstain.[54] The first member to be called is selected by drawing lots.[55] Each vote is then reported. The adoption or amendment of standards and other related Codex measures is decided upon by consensus, and a vote takes place only if all efforts to reach consensus have failed.[56]

Furthermore, Codex Members have the right to insist on having their views or reservations included in the report of the subsidiary bodies.[57]

---

was of the view that withdrawal was impossible as long as the constitution did not explicitly provide for it. See H.G. Schermers and N.M. Blokker, *supra* n. 20, pp. 97-98. Previously, unilateral withdrawal by a Codex Member would probably not have been very controversial. However, due to the changes in the resulting obligations from the Codex decisions triggered by the adoption of the WTO agreements (as will be discussed in Chapter III), withdrawal from the Codex by its Members may become a subject of discussion.

[47] See Rule I.3 and Rule II of the Rules of Procedure, Procedural Manual, 16th edn., 2006, p. 6.

[48] See Report of the 26th Session of the Codex Alimentarius Commission, Rome, 30 June-7 July 2003, ALINORM 03/41, paras. 19-24.

[49] See, on the membership of the EC to the FAO, for instance, W. van de Voorde, 'De E.E.G. als lid van een internationale instelling: het geval F.A.O.', 45 *Studia Diplomatica* (1992) pp. 49-73.

[50] The exception to this right is participation in the Executive Committee as membership of this Committee is limited and participation in regional co-ordinating Committees, as membership of these committees is limited to the Codex Members of the relevant region. See below for further discussion in Section 3.2, and Section 3.4.1.3 respectively of this chapter.

[51] Rule VIII.1 of the Rules of Procedure, Procedural Manual, 16th edn., 2006, p. 12. The Commission of the EC representing its member states is an exception as it has as many votes as EC member states present at the moment of voting.

[52] Rule VIII.2 of the Rules of Procedure, Procedural Manual, 16th edn., 2006, p. 12.

[53] Rule VIII.4 of the Rules of Procedure, Procedural Manual, 16th edn., 2006, p. 12.

[54] Rule XII.7(a) of the General Rules of the FAO.

[55] Ibid.

[56] Rule XII.2 of the Rules of Procedure, Procedural Manual, 16th edn., 2006, p. 16. See also, for more details on the elaboration procedure, Section 4.2 of Chapter II.

[57] Rule X.1 of the Rules of Procedure, Procedural Manual, 16th edn., 2006, p. 14 and the 'Guidelines on the conduct of meetings of Codex Committees and ad hoc Intergovernmental Task Forces', Procedural Manual, 16th edn., 2006, pp. 53-54.

Regional organisations that are members of the Codex Alimentarius Commission have similar rights. However, their rights depend on the number of member states belonging to the regional organisation that are present at the sessions. Several conditions are laid down in the Rules of Procedure to prevent the membership rights of the regional organisations and their members from being exercised simultaneously. For instance, the member organisation or one of its members has to inform the Codex Alimentarius Commission about the division of competences between the regional organisation and its members.[58] Upon the request of any Codex Member, the regional organisation or any of its members has to give information on this division of competences.[59] The member organisation has the right to vote and has a number of votes which reflects the number of its members entitled to vote.[60] However, the member organisation may not vote for positions subject to election in the Codex Alimentarius Commission or its subsidiary bodies.[61]

3.1.2.2 Observers of the Codex Alimentarius Commission

Observer status can be granted to states, international organisations and non-governmental organisations.[62] Three categories of observers can be distinguished. The first category is that of observer countries that are members of either the FAO or the WHO, but have not yet become members of the Codex Alimentarius Commission. The second category consists of observers that are members of the United Nations, but are not members of the FAO or of the WHO. And the last category consists of international non-governmental organisations (INGOs) and international governmental organisations (IGOs).

The Codex Alimentarius Commission has a large number of registered observers: in total, it has 217 observers: 62 IGOs (of which 16 are UN-related organisations) and 157 INGOs.[63] Currently, it does not have observer states.

---

[58] Rule II.5 of the Rules of Procedure, Procedural Manual, 16th edn., 2006, p. 7.

[59] Rule II.6 of the Rules of Procedure, Procedural Manual, 16th edn., 2006, p. 7.

[60] As all its member states are Codex Members, as regards the EC this rule means in practice that the Commission (as representative of the EC) has the amount of votes that equals the amount of its Members present at the session at the time of the vote.

[61] Rule II.4 of the Rules of Procedure, Procedural Manual, 16th edn., 2006, p. 6. See also, Section 3.1.3 of this Chapter.

[62] See Art. 3 and 4 of the Statutes on observer status of states. Observer status of international organisations in the work of the Commission is falling under the responsibility of its parent organisations and is regulated through the rules of FAO and WHO (Rule XVII of the FAO General Rules and Art. 18(h) of the WHO Constitution and Rule 48 and 49 of the General Rules of Procedure of the World Health Assembly), as well as by the applicable regulations of FAO and WHO on relations with international organisations; such relations shall be handled by the Director-General of FAO or of WHO as appropriate (see Rule IX.5 of the Rules of Procedure, Procedural Manual, 16th edn., 2006, p. 13). There is no distinction between granting observer status of FAO/WHO members and those nations which are not members of either one of the parent organisations.

[63] Website of the Codex Alimentarius Commission <www.codexalimentarius.net>, checked on the 10th of June 2007. Most of the INGOs consists of industry-based groups. See also S. Suppan, *Governance in the Codex Alimentarius Commission* (Consumers International 2005), p. 17.

Observers have the right to participate in the sessions of the Codex Alimentarius Commission and may participate in the discussions when invited to participate by the chairperson.[64] They have the right to receive all working documents and discussion papers in advance of the sessions and to submit memoranda to the sessions of the Codex Alimentarius Commission and to the sessions of its subsidiary bodies.[65] However, unlike Codex Members, observers do not have the right to vote. Similarly, observers (except for observer countries) do not have the right to have their positions formally recorded in the reports. However, the Codex Alimentarius Commission and its subsidiary bodies have followed a rather flexible approach with regard to these rules.[66]

3.1.3 Meetings of the Codex Alimentarius Commission

Sessions of the Codex Alimentarius Commission are held every year and normally last one week.[67] Until 2003, the frequency of the meetings held was every other year.[68] Its first session was held in 1963, and from 1963 till 2005, it met 28 times.

---

[64] As regards observer states, Rule IX.1 and 2 of the Rules of Procedure of the Codex Alimentarius Commission (Procedural Manual, 16th edn., 2006) and provision C.2 of the 'Statement of principles relating to the granting of observer status to nations'. As regards international governmental organisations, Rule IX.4 of the Rules of Procedure of the Codex Alimentarius Commission and Rule XVII.1 and XVII.2 of the General Rules of the FAO and Rule 48 of the General Rules of Procedure of the World Health Assembly. As regards international non-governmental organisations as observers, Rule IX.4 of the Rules of Procedure of the Codex Alimentarius Commission and Rule XVII.3 of the General Rules of the FAO, provision 19(a) of the FAO Policy concerning relations with international non-governmental organisations, Rule 49 of the General Rules of Procedure of the World Health Assembly and Provision 6.1(i) of the 'Principles governing relations between the World Health Organization and non-governmental organizations'.

[65] As regards observer states, Rule IX.1 of the Rules of Procedure of the Codex Alimentarius Commission (Procedural Manual, 16th edn., 2006) and provision C.3, respectively C.4 of the FAO 'Statement of principles relating the granting of observer status to nations'. As regards international governmental organisations as observers, Rule XVII.1 and XVII.2 of the General Rules of the FAO and Rule 48 of the General Rules of the World Health Assembly. As regards international non-governmental organisations as observers, Provision 5.1(a), (b) and (e) of the 'Principles concerning the participation of international non-governmental organizations in the work of the Codex Alimentarius Commission' (Procedural Manual, 16th edn., 2006), Rule XVII.3 of the General Rules of the FAO, provision 19(c) and (d) of the FAO Policy concerning relations with international non-governmental organisations and Provision 6.1(ii) and (iii) of the 'Principles governing relations between the World Health Organization and non-governmental organizations'.

[66] Report of the 16th Session of the Joint FAO/WHO Codex Alimentarius Commission, Geneva, 1-12 July 1985, ALINORM 85/47, para. 84.

[67] There are Codex Alimentarius Commission meetings that have a shorter duration, such as the Extraordinary 25th Session of the Codex Alimentarius Commission which concentrated specifically on the evaluation process of the Codex Alimentarius.

[68] The reason behind having meetings of the Codex Alimentarius Commission more frequently is to speed up the standard-setting procedure. When this change was proposed some opposition from the developing countries, as participation to all Codex Committee meetings and annual meetings of the Codex Alimentarius Commission cannot be ensured by these countries and may result in difficulties to keep track of the number of standards and other documents to be adopted. There was however no opposition during the Codex Alimentarius Commission meeting of 2003 to hold its next meeting in the

Its meetings take place on a rotating basis either at the Headquarters of the FAO in Rome or are hosted by the WHO in Geneva.[69]

The sessions of the Codex Alimentarius Commission are chaired by a chairperson, who is elected from among the representatives and advisers of the delegations of the members of the Commission, with the consent of the head of the delegation concerned.[70] The chairperson is assisted by three vice-chairpersons who are also elected from among delegates of the member countries. The candidates for the vacancies are nominated, and the election of chairpersons and vice-chairpersons takes place by a role-call vote with a secret ballot. The chairperson and vice-chairpersons hold office from the end of the Codex Alimentarius Commission's session at which they have been elected until the end of the following session. They can be re-elected twice, provided that by the end of their second term of office they have not serve for a period longer than two years. The chairperson presides over the meetings of the Commission and exercises all the other functions necessary for the facilitation of the work of the Commission. In the absence of the chairperson, a vice-chairperson takes over the functions of the chairperson.[71] In the event that neither the chairperson nor the vice-chairpersons can serve in their capacity, the Commission may elect a staff member of the FAO or the WHO to replace him or her temporarily.[72] The current Chairperson of the Codex Alimentarius Commission is Mr. Claude Mosha (United Republic of Tanzania) and the current Vice-chairpersons are Ms. Karen Hulebak (United States of America), Ms. Noraini Mohd. Othman (Malaysia) and Mr. Wim van Eck (the Netherlands), who were all elected during the session of the Codex Alimentarius Commission in 2005 and re-elected during its last session in 2006.[73]

A provisional agenda for each session of the Codex Alimentarius Commission is prepared by the Directors-General of the FAO and the WHO, after consultation with the chairperson of the Commission or with the Executive Committee.[74] Although the final version of the Agenda is adopted by the Codex Alimentarius Commission itself, no items included on the Agenda by the governing bodies or by the Directors-General of the FAO or the WHO can be deleted from it.[75]

---

following year 2004. See, for detailed discussion of the participation of developing countries, Section 3.2.1 of Chapter V.

[69] This alternation started in 1981. Before that year, meetings of the Codex Alimentarius Commission were hosted by the FAO in Rome.

[70] Rule III, Rules of Procedure of the Codex Alimentarius Commission, Procedural Manual, 16th edn., 2006, p. 8.

[71] It is, however, the practice of the Codex Alimentarius Commission to give opportunity to all Vice-chairpersons to preside a morning or afternoon session of the Codex Alimentarius Commission meeting.

[72] Rule III.3 of the Rules of Procedure, Procedural Manual, 16th edn., 2006, p. 8.

[73] Report of the 28th Session of the Codex Alimentarius Commission, Rome 4-9 July 2005, ALINORM 05/28/41, para. 242. Report of the 29th Session of the Codex Alimentarius Commission, Geneva 3-7 July 2006, ALINORM 06/29/41, para. 227.

[74] Rule VII.1 of the Rules of Procedure, Procedural Manual, 16th edn., 2006, p. 11.

[75] Rule VII.6 of the Rules of Procedure, Procedural Manual, 16th edn., 2006, pp. 11-12.

3.1.4 *Budgetary issues*

The budget is administered by the FAO on behalf of both organisations.[76] The initial idea was to make the bodies of the Joint FAO/WHO Food Standards Programme financially independent from its parent organisations by setting up a Trust Fund formed from donations from the member countries.[77] The establishment of the Trust Fund allowed industry, albeit indirectly through the governments of Codex Members, to make contributions.[78] However, this was seen to obstruct some countries from joining the Codex Alimentarius Commission as they were unable to participate financially. In addition, it formed a somewhat uncertain basis for the continuous activities of the Codex Alimentarius Commission. For these reasons, the Trust Fund was replaced by an annual quota from both the FAO and the WHO in order to finance the Joint FAO/WHO food standards programme.[79]

Although, formally, they are equal partners in this co-operation, the financial influence of the FAO has been the dominant one, at least during the first 30 years of the existence of the Codex Alimentarius Commission. This is, for instance, reflected in the fact that the FAO paid 75% of the budget, whereas the WHO only committed itself to 25%. Recently, the responsible bodies of the WHO have taken decisions to make the Joint FAO/WHO Food Standards Programme a priority among its activities. Whereas the overall budget of the WHO has decreased slightly, its contribution to the Codex Alimentarius Commission has, however, significantly increased. The budget allocated to the Codex Alimentarius Commission by the WHO for the years 2003/2004 amounts to an increase of 19% as regards to the allocated budget of 2001/2002.[80] Furthermore, in May 2003, the World Health Assembly adopted a resolution on the Codex Alimentarius, in which it made a commitment to increase its own direct involvement.[81]

---

[76] Art. 9 of the Statutes, Procedural Manual, 16th edn., 2006, p. 4.

[77] Former Art. 8 of the Statutes as adopted by the 11th Session of the FAO Conference, Report of the 11th Session of the FAO Conference, Rome 4-24 November 1961, Appendix D. See also, Report of the 1st Session of the Codex Alimentarius Commission, Rome, 25 June-3 July 1963, ALINORM 63/12 Appendix B Rules of Procedure for the Commission. The idea of the trust fund required that no contribution was to exceed 20% of the overall budget or be less than 500 dollars per year. The contributions were to be agreed upon between the countries and the Directors-General to be determined on the basis of its interest in the international food trade.

[78] D.G. Victor, *Effective Multilateral Regulation of Industrial Activity: Institutions for Policing and Adjusting Binding and Nonbinding Legal Commitments*, Ph.D. Thesis (Massachusetts, Harvard University, Institute of Technology 1997), p. 183.

[79] Report of the 3rd Session of the Codex Alimentarius Commission, Rome, 19-28 October 1965, ALINORM 65/30, para. 12.

[80] See statement of WHO's Director-General Bruntland during the Extraordinary Meeting of the Codex Alimentarius Commission in February 2003.

[81] Resolution on the Joint FAO/WHO Evaluation of the work of the Codex Alimentarius Commission, 56th World Health Assembly, 19-28 May 2003, WHA56/2003/REC/1.

3.2 The Executive Committee

3.2.1 *Legal basis, powers and internal rules*

Article 6 of the Statutes of the Codex Alimentarius Commission requires the Commission to establish an Executive Committee. At the first session of the Codex Alimentarius Commission, the Rules of Procedure which established the Executive Committee were adopted.[82] Its first session took place right after the first session of the Codex Alimentarius Commission in July 1963. The main function of the Executive Committee is to act as the Executive organ of the Commission between the sessions.[83] This task consists of making proposals to the Codex Alimentarius Commission with regard to the general orientation and programme of work, the studying of special problems and providing assistance in the implementation of the programme. At its 20th Session in 1993, the Codex Alimentarius Commission decided to assign the Executive Committee with the responsibility of reviewing standards at Step 5 and advancing them to Step 6 in the standard-setting procedure.[84] In the past, the Executive Committee also had the competence to decide upon the initiation of new work.[85] In 2004, the Executive Committee obtained the responsibility for critically reviewing proposals to initiate the elaboration of new Codex measures and for monitoring the progress of the development of standards.[86] In contrast to earlier responsibilities, the authority of the Executive Committee is today limited to a consultative function, and it no longer has the power to take decisions with regard to the initiation of the standard-setting procedures or the advancing of draft measures from Step 5 to Step 6.[87]

---

[82] Report of the 1st Session of the Codex Alimentarius Commission, Rome, 25 June-3 July 1963, ALINORM 63/12, para. 4 in conjunction with Rule III of Appendix B.

[83] Art. 6 of the Statutes, Procedural Manual, 16th edn., 2006, p. 4.

[84] See Report of the 20th Session of the Joint FAO/WHO Codex Alimentarius Commission, Geneva, 28 June-7 July 1993 ALINORM 93/40, para. 93 and Report of the 10th Session of the Codex Committee on General Principles, Paris, 7-11 September 1992, ALINORM 93/33, 1992, para. 43. See, for more details on the standard-setting procedure, Section 4 of Chapter II. The standard-setting procedure of the Codex Alimentarius Commission consists of 8 steps, which will be explained in further detail in Section 4 of Chapter II.

[85] See Procedural Manual (12th edn.), 2001, pp. 19-20. In 1981, at the request of the FAO Conference the Codex Committee on General Principles carried out a step by step examination of the elaboration procedure and considered that the Executive Committee should be able to approve decisions of subsidiary bodies to start new work on the basis of the 'Criteria'. See Report of the 7th Session of the Codex Committee on General Principles, ALINORM 81/33, 1981, para. 30: 'The Committee carried out a step by step examination of the proposed Procedures for the Elaboration of Worldwide Codex standards in the light of government comments. The Committee considered that the Executive Committee should be able to exercise the prerogative of the Commission in approving decisions of the subsidiary bodies to commence work on the elaboration of worldwide standards in accordance with the "Criteria for the Establishment of work priorities and for the Establishment of Subsidiary Bodies"....'

[86] Report of the 27th Session of the Codex Alimentarius Commission, Geneva, 28 June-3 July 2004, ALINORM 04/27/41, para. 13 and Appendix II.

[87] See, for further details on the standard-setting procedure, Section 4 of Chapter II, and on its legitimacy, Sections 2.2.2, 2.2.3 and 3.1.3 of Chapter V.

3.2.2 Composition

According to the Statutes, the composition of the Committee has to ensure an adequate geographical representation of all the areas of the world covered by the Codex Members.[88] The Rules of Procedure specify that the Executive Committee be composed of the following members (currently 17 members): the chairperson and vice-chairpersons of the Codex Alimentarius Commission, the co-ordinators of the six Co-ordinating Committees, plus the members from Asia, Africa, Europe, Latin America and the Caribbean, the Near East, North America and the South West Pacific.[89] The chairperson and vice-chairpersons of the Codex Alimentarius Commission act as chairperson and vice-chairpersons of the Executive Committee.[90] The seven members from different geographical areas are elected by the Codex Alimentarius Commission by majority-voting with a secret ballot.[91] Their term of office starts at the end of the Codex Alimentarius Commission session in which they were elected and ends at the end of the second succeeding session of the Commission. They may be re-elected for one succeeding term. The seven members of different regions do not act as representatives of the region in question, but are elected on this basis simply to ensure geographical spread.[92] Membership may not include two members of the same country.[93] This rule creates complications as regards the region of North America, as this region is made up of only two countries. Consequently, if the chairperson or one of the vice-chairpersons comes from one country of the North American region, the regional member has to come from the other country.

Initially, internal rules on observers and their rights were not included. However, due to the increasing amount of advisors that some members of geographic regions brought along to the meetings of the Executive Committee, the Codex Alimentarius Commission adopted the following understanding to Rule IV.1 (currently Rule V) of the Rules of Procedure in 1989:

> 'Except for the Chairman and the three Vice-Chairmen, the six (read seven) further members of the Executive Committee elected by the Commission to represent the geographic locations are countries not individuals.
> The delegate of a Member may be accompanied by not more than two advisors from the same geographic location.

---

[88] Art. 6 of the Statutes, Procedural Manual, 16th edn., 2006, p. 4.
[89] Rule V.1 of the Rules of Procedure, Procedural Manual, 16th edn., 2006, p. 9.
[90] Rule V.5 of the Rules of Procedure, Procedural Manual, 16th edn., 2006, p. 10.
[91] Rule V in conjunction with Rule VIII.2 and Rule VIII.5 of the Rules of Procedure, Procedural Manual, 16th edn., 2006.
[92] The regions selected to guarantee geographical spread coincide more or less with the different regional Co-ordinating Codex Committees (see Section 3.4.1.3 of this chapter), except for the Codex Committee for North America and the South West Pacific, which region has two members in the Executive Committee.
[93] Rule V.1 of the Rules of Procedure, Procedural Manual, 16th edn., 2006, p. 9.

Regional Co-ordinators shall be invited to attend meetings of the Executive Committee as observers.
Only members or, with the permission of the Chairman, observers, may take part in the discussions.'[94]

3.2.3 *Meetings and budgetary issues*

Since its first meeting in July 1963, the Executive Committee has met 58 times. Normally, the Executive Committee meets once a year. Until 2003, the Codex Alimentarius Commission met once every two years, which meant that the Executive Committee met more frequently than the Codex Alimentarius Commission. Since the decision to hold the sessions of the Codex Alimentarius Commission annually, the Executive Committee meets immediately prior to every session of the Codex Alimentarius Commission.[95] However, according to Rule IV.4 of the Rules of Procedure the Directors-General of the FAO and the WHO may convene sessions of the Executive Committee as often as necessary. It normally meets in the same city in which the session of the Codex Alimentarius Commission is to take place that year, which is either Rome or Geneva.[96] The Executive Committee reports to the Codex Alimentarius Commission.[97]

The operating expenses of the Executive Committee are borne by the budget of the Joint FAO/WHO Food Standards Programme, which is jointly administered by the FAO and the WHO.[98]

3.3 The Codex Secretariat

In contrast to the inclusion of a requirement to establish an Executive Committee and the possibility of establishing subsidiary bodies where necessary, the Statutes of the Codex Alimentarius Commission do not foresee in the possibility of establishing a Secretariat which is responsible for administration. However, the Rules of Procedure do state that the Directors-General of the FAO and the WHO are to be requested to appoint a Secretary and other officials to perform all the duties that the work of the Commission may require.[99]

The Secretariat consists of a Secretary and 10 other officials (3 senior food standards officers, 5 food standards officers, 1 associate professional officer and 1 consultant). The Secretariat is part of the Food and Nutrition Division of the Economic

---

[94] Addition inserted by the author as at the time of the adoption of the understanding the Executive Committee still consisted of six regional members instead of the current number of seven. Report of the 18th Session of the Joint FAO/WHO Codex Alimentarius Commission, Geneva, 3-12 July 1989, ALINORM 89/40, para. 183.
[95] Rule V.6 of the Rules of Procedure, Procedural Manual, 16th edn., 2006, p. 10.
[96] See Section 3.1.3 of this chapter.
[97] Rule V.7 of the Rules of Procedure, Procedural Manual, 16th edn., 2006, p. 10.
[98] Art. 9 of the Statutes, Procedural Manual, 16th edn., 2006, p. 4. See, on the budget of the Joint FAO/WHO Food Standards Programme, also, Section 3.1.4 of this chapter.
[99] Rule III.5 of the Rules of Procedure, Procedural Manual, 16th edn., 2006, p. 8.

and Social Department of the FAO and is situated at the Headquarters of the FAO in Rome. The officers of the Secretariat are recruited by the FAO and fall under the responsibility of the Director-General of the FAO. In the past, this has posed problems, as their functions often did not solely consist of the working activities for the Codex Alimentarius Commission, but also contained tasks for the FAO.[100] At present, the Secretariat dedicates all its time to the work of the Codex Alimentarius Commission and the staff members are overloaded with work that results from preparing and attending all Codex Committee meetings and other relevant meetings.[101] Although the task of administration to support the work of the Codex Alimentarius Commission and its subsidiary bodies is an important issue, the capacity of the Secretariat is simply inadequate.[102] Furthermore, the Secretariat has no legal counsellors. When legal advice is required, the Codex Alimentarius Commission and the Secretariat regularly consult legal counsellors who are staff members of the FAO and of the WHO.[103] However, it has to be kept in mind that these consultations are part of their work, which means that their task is partly to ensure that the internal rules of the Codex Alimentarius Commission are in conformity with the rules of the FAO and the WHO. With regard to this construction, the dependency of the Codex Alimentarius Commission on its parent organisations for its administrative activities is clearly visible.

3.4 The Codex Committees and Task Forces

As mentioned above in Section 3.1.1.2, the Codex Alimentarius Commission has the power to establish subsidiary bodies. Article 7 of the Statutes allows the Codex Alimentarius Commission to establish subsidiary bodies, as it deems necessary, in order to accomplish its tasks and subject to the availability of funds. This competence has extensively been used by Codex Alimentarius Commission. At its first session in 1963, the Commission established 9 Codex Committees. Over the years, the number of sub-bodies has increased to the current number of 10 General Codex Committees, 11 Commodity Codex Committees (of which 4 are adjourned) and 3 Task Forces.[104] Given their major role in the standard-setting procedure, the proliferation of subsidiary bodies can be considered as giving content, *ratione materiae*, to the mandate of the Codex Alimentarius Commission.

This competence has also allowed the Codex Alimentarius Commission to strengthen its position as the promoter that co-ordinates international activities re-

---

[100] Report of the evaluation of the Codex Alimentarius and other FAO and WHO food standards work, ALINORM 03/25/3, para. 104.

[101] Ibid., paras. 101-102.

[102] Report of the evaluation of the Codex Alimentarius and other FAO and WHO food standards work, ALINORM 03/25/3 paras. 101-103. The evaluation report emphasised the need for more senior staff at the Secretariat.

[103] See, for instance, Report of the 27th Session of the Codex Alimentarius Commission, Geneva, 28 June-3 July 2004, ALINORM 04/27/41, Appendix X.

[104] See Report of the 29th Session of the Codex Alimentarius Commission, Geneva 3-7 July 2006, ALINORM 06/29/41, Appendix XII.

lated to food standard-setting. Throughout the years, it has built an institutional framework which allows it to co-operate with other international standard-setting bodies, and, in some cases, even replace, these outside bodies with Codex Committees.[105] The initial idea was that external expert bodies and international organisations would play a major role in the preparation of draft standards and would not be replaced by Codex *ad hoc* expert groups.[106] Codex *ad hoc* expert groups were only to be established in cases where no convenient outside body existed or could be established. However, during the 3rd Session, the Codex Alimentarius Commission decided to establish a Codex Committee for fish and fisheries products, which took over the tasks of the FAO Fisheries Division and the OECD, which had initially had the responsibility of preparing standards on fish and fisheries products. Likewise, co-operation was sought with the UNECE, which resulted in two groups: Joint UNECE/Codex Alimentarius Group of Experts on quick-frozen foods and the Joint UNECE/Codex Alimentarius group of experts on the standardisation of fruit juices. At first, these expert groups were established outside the framework of the Codex Alimentarius Commission, with a similar function to that of the Codex Committees in the process of standard-setting.[107] Later on, both these expert groups were abolished and their work was transferred to a Codex Committee, or a Codex Task Force established for the purpose of transferring the work completely within the institutional structure of the Codex.

3.4.1 *Different types of Committees*

As laid down in Rule XI of the Rules of Procedure, the Codex Alimentarius Commission can establish different types of subsidiary bodies.

---

[105] See also, D.G. Victor, *supra* n. 78, p. 184: 'Within a decade most of other, competing, standard-setting activities had either withered away or, in some instances, were integrated into the Codex Alimentarius system. Whereas in 1963 an outside observer would find a bewildering and rapidly growing array of efforts to harmonize food standards, by the early 1970s the food safety regime was largely unified loosely centred on the Codex Alimentarius Commission.'

[106] Report of the Joint FAO/WHO Conference on Food Standards, Geneva 1-5 October 1962, ALINORM 62/8, para. 28-30: 'In some cases, work is already being or can conveniently be undertaken by an inter-governmental organization of regional or sub-regional coverage which has its own methods of preparation and of finalization of standards at the government level. Examples are the Permanent Commission of the Latin American Food Code, the Organization for Economic Co-operation and Development (OECD) working in conjunction with the Economic Commission for Europe (UNECE) and the European Economic Community (EEC, the European Common Market). In such cases, the Conference recommended that the Commission make full use of the work carried out by these organizations. In allocating preparatory work on standards, full use should likewise be made of the wide technical knowledge and facilities offered by existing non-governmental specialist organizations and by the International Organization for Standardization. In agreement with these organizations, draft standards prepared by them would be made available to the Commission for finalizing at the governmental level ... Whenever it appears to the Commission that no appropriate outside international body already exists or can conveniently be set up, for example to handle the general part of the Codex..., preparatory work should be undertaken by an ad hoc expert group of representatives of national Codex Alimentarius Committees, wherever such bodies have been established, under the leadership of one of their number specifically appointed for this purpose by the Commission....'

[107] See Procedural Manual, 16th edn., 2006, p. 158, n. 40.

Paragraph 1 of Rule XI specifies that the Commission can create Codex Committees which are responsible for the preparation of draft standards (Rule XI 1(b)(i)) and Co-ordinating Committees for regions or groups of countries (Rule XI 1(b)(ii)). The Commission has the right to abolish these Codex Committees, or to declare them to be in a state of 'adjourned *sine die*'.[108] The advantage of the adjournment *sine die* of certain subsidiary committees, which have completed their tasks at a given moment, is to be able to reactivate them again whenever the developed standards which fall within their mandate require review or amendment.

In addition to these Codex Committees, the Codex Alimentarius Commission may establish any other subsidiary bodies that it deems necessary in the light of finalising standards (Rule XI 1(a)). The Commission has used Rule XI 1(a) to establish *ad hoc* Intergovernmental Task Forces, as can be seen in Figure 2 below.

3.4.1.1 General Codex Committees and Commodity Codex Committees

The first category of subsidiary bodies consists of the Codex Committees. This category is composed of two types of committees: General Committees and Commodity Committees. General Committees are responsible for the elaboration of 'horizontal' issues or issues that relate to more than one commodity or group of commodities, such as food additives, pesticide residues, or food labelling. Currently, there are 10 General Codex Committees. Examples of General Codex Committees include the Codex Committee on Food Additives and Contaminants (CCFA), the Codex Committee on Contaminants in Foods (CCCF), the Codex Committee on Pesticides Residues (CCPR), the Codex Committee on Methods of Analysis and Sampling (CCMAS), and the Codex Committee on Food Labelling (CCFL).[109] The work elaborated by these committees result either in the form of a general standard, a code of practice or a guideline, or form a part of a commodity standard.

In contrast, Commodity Committees are in charge of the elaboration of commodity standards, and deal with one specific commodity or group of commodities. At present, there are 11 Commodity Codex Committees, of which 6 are adjourned *sine die*.[110] Examples of active Commodity Codex Committees are the Codex Com-

---

[108] For instance, in 1999, the Codex Alimentarius Commission abolished the Codex Committee on Meat and Poultry Products, which was re-established by the 24th Session of the Codex Alimentarius Commission in 2001. See Report of the 24th Session of the Codex Alimentarius Commission, Geneva, 2-7 July 2001, ALINORM 01/41, paras. 9 and 215.

[109] See also, Figure 2 on p. 37. The Codex Committee on General Principles is considered as a general codex committee as well, although its function is somewhat different as it is limited to policy and legal issues arising in the context of the Joint FAO/WHO Food Standards Programme. During the 28th Session of the Codex Alimentarius Commission in 2005, there was a recommendation to divide the CCFAC into two separate codex committees. The Codex Committee on General Principles is to develop the terms of reference of each of the separate committees. See Report of the 28th Session of the Codex Alimentarius Commission, Rome 4-9 July 2005, ALINORM 05/28/41, para. 143. This led in 2006 of two new Codex Committees, the CCFA and the CCCF. Report of the 29th Session of the Codex Alimentarius Commission, Geneva 3-7 July 2006, ALINORM 06/29/41, para. 26.

[110] The 6 Codex Committees that have been adjourned *sine die* are: the Codex Committee on Cereals, Pulses and Legumes, the Codex Committee on Natural Mineral Waters, the Codex Committee on Sugars, The Codex Committee on Cocoa Products and Chocolate, the Codex Committee on Meat Hygiene and the Codex Committee on Vegetable Proteins.

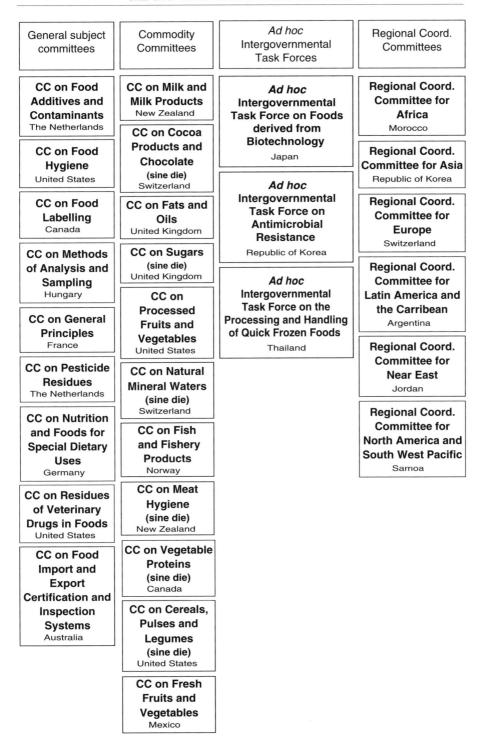

Figure 2. The subsidiary bodies of the Codex Alimentarius Commission and their host-countries

mittee on Fats and Oils (CCFO), the Codex Committee on Milk and Milk Products (CCMMP) and the Codex Committee on Processed Fruit and Vegetables (CCPFV).

3.4.1.2 *Ad Hoc* Intergovernmental Task Forces

The second category of subsidiary bodies consists of *ad hoc* Intergovernmental Task Forces. In 1999, the Codex Alimentarius Commission began establishing a 'new type' of subsidiary body in the form of these *ad hoc* bodies.[111] These bodies are responsible for the development of work within a limited time-frame. As a result, they are – by definition – of a temporary nature. The establishment of a task force is linked to a specific task, in contrast to a Codex Committee, which can determine its own tasks as long as they fall within its mandate. One of the consequences of this structure is that, if a task force fails to complete its task within the given period, it needs the approval of the Codex Alimentarius Commission to be able to complete its work. An illustration of this can be taken from the prior Task Force on Animal Feeding Practices, which had the task of developing a standard on Good Animal Feeding practices. As one of the provisions contained in the proposed standard was still subject to disagreement at the date upon which the task had to be completed (2003), the Codex Alimentarius Commission decided that the Task Force could convene one more meeting in order to complete its work.

On the other hand, a task force that has completed its task ceases to exist. In order to become operational again, it needs a decision from the Commission. For example, when the Task Force on Foods derived from Biotechnology had completed its work (2003), the Commission requested an appraisal of the need for further work,[112] and, as a consequence, the Task Force was re-established during the 27th Session of the Codex Alimentarius Commission in 2004.[113]

Currently, there are three Task Forces: the *Ad Hoc* Codex Intergovernmental Task Force on Food Derived from Biotechnology, the *Ad Hoc* Codex Intergovernmental Task Force on Antimicrobial Resistance and the *Ad Hoc* Codex Intergovernmental Task Force on the Processing and Handling of Quick Frozen Foods.[114]

3.4.1.3 Regional Co-ordinating Committees

Rule XI 1(b)(ii) provides the Commission with the possibility of establishing regional co-ordinating committees. Regional co-ordinating committees promote the mutual exchange of information between their members and discuss and promote

---

[111] L.R. Horton, 'Risk analysis and the law: international law, the World Trade Organization, Codex Alimentarius and national legislation', 18 *Food Additives and Contaminants* (2001), p. 1061.

[112] Report of the 26th Session of the Codex Alimentarius Commission, Rome, 30 June-7 July, 2003, ALINORM 03/41, para. 230.

[113] See Report of the 27th Session of the Codex Alimentarius Commission, Geneva, 28 June-3 July 2004, ALINORM 04/27/41, para. 89. Another example is the dissolution of the Task Force on Fruit and Vegetable Juices, that completed its task in 2005; see Report of the 28th Session of the Codex Alimentarius Commission, Rome 4-9 July 2005, ALINORM 05/28/41, para. 244.

[114] The moment of writing is December 2006.

interests that are inherently linked to most of the regional members. In some cases, Regional Codex Committees are responsible for preparing draft regional Codex standards.[115] There are six Regional Co-ordinating Committees: for Africa (1), Asia (2), Europe (3), Latin America and the Caribbean (4), the Near East (5), and for North America and the South West Pacific (6).[116]

3.4.2 *Powers and internal rules*

The definition of the responsibilities of the Codex Committees and Task Forces can be found in several documents included in the Procedural Manual: their Terms of Reference, the Guidelines for Codex Committees, and, to a certain extent, the Relations between Commodity Committees and General Committees. Codex Committees and Task Forces have an important function in the standard-setting procedure, as they are responsible for establishing work priorities and for preparing proposed draft standards and draft standards (which are submitted for approval by the Codex Alimentarius Commission at Step 5 (proposed draft standards) and Step 8 of the procedure (draft standards)).[117]

3.4.3 *Composition*

The members of the Codex Committees and Task Forces include members of the Codex Alimentarius Commission that have notified the Directors-General of their desire to participate as members of the subsidiary bodies.[118] Clearly, the membership of the Regional Codex Committees is restricted to the Codex Members that belong to the region in question.[119] Codex committees, in particular, the general

---

[115] See, on regional Codex standards, Section 2.2 of Chapter II and on the standard-setting procedure, Section 4 of Chapter II.

[116] Their terms of reference with respect to their region is:

'(a) defines the problems and needs of the region concerning food standards and food control;

(b) promotes within the Committee contacts for mutual exchange of information on proposed regulatory initiatives and problems arising from food control and stimulates the strengthening of food control infrastructures;

(c) recommends to the Commission the development of world wide standards for products of interest to the region, including products considered by the Committee to have an international market potential in the future;

(d) develops regional standards for food products moving exclusively or almost exclusively in intra regional trade;

(e) draws the attention of the Commission to any aspects of the Commission's work of particular significance to the region;

(f) promotes coordination of all regional food standards work undertaken by international governmental and non-governmental organizations within the region; exercises a general coordinating role for the region and such other functions as may be entrusted to it by the Commission;

(h) promotes the acceptance of Codex standards and maximum limits for residues by member countries.' (Procedural Manual, 16th edn., 2006, pp. 151-158)

[117] See, for more details on the Codex standard-setting procedure, Section 4 of Chapter II.

[118] Rule XI.2 Rules of Procedure, Procedural Manual, 16th edn., 2006, p. 14.

[119] Rule XI.3 Rules of Procedure, Procedural Manual, 16th edn., 2006, p. 15.

Codex committees, are often characterised by a high level of specialisation due to the expertise of the members in the area.[120] This is ensured by Rule XI.4 of the Rules of Procedure, which states that:

'Representatives of members of subsidiary bodies ... shall be specialists active in the fields of the respective subsidiary bodies.'[121]

Although the heads of the delegations are representatives of the government, they are often assisted by the specialised experts that they have selected.[122] Codex Committees and Task Forces are chaired by a representative that is appointed by the Codex Member responsible for the organisation and administration of the subsidiary body.[123] Each Regional Co-ordinating Codex Committee has a co-ordinator, who is appointed by the Regional Committee itself. As mentioned above in Section 3.2.2, co-ordinators have observer status in the Executive Committee on behalf of the region that they represent. Often, it is the chairperson who acts as co-ordinator.

Observers to the Codex Alimentarius Commission also have the right to participate as observers at the Codex Committees and the Task Forces.[124]

3.4.4 *Budgetary issues*

The possibility of establishing subsidiary bodies is subject to the availability of funds. As mentioned in Sections 3.1.1.1 and 3.1.4 above, the Codex Alimentarius Commission depends on its parent organisations for its financial resources. This would entail a serious restriction of its competence, if it meant that, even for the establishment of subsidiary bodies, the Codex Alimentarius Commission depended on the agreement of its parent organisations. However, Article 9 of the Statutes excludes all the operating expenses of a subsidiary body for which a member has accepted the chair from the budget of the Joint FAO/WHO Food Standards Programme. This provision opens the door to the establishment of subsidiary bodies, which are financially supported by the members: the construction of 'host countries'. This construction means that a particular Codex Member is responsible for the administration, organisation and financing of the subsidiary body in question. In this way, the Commission has been able to retain, to a large degree, its own discretion *vis-à-vis* its parent organisations with regard to the establishment of its subsidiary bodies.

---

[120] See also J.E. Alvarez, *supra* n. 29, pp. 245-246. As regards the Codex Committee on Pesticides Residues (CCPR), see L. Salter, *Mandated Science, Science and Scientists in the Making of Standards* (Dordrecht, Kluwer Academic Publishers 1988), p. 73: 'But unlike the Codex, CCPR members are often working scientists or technically trained individuals'.

[121] Procedural Manual, 16th edn., 2006, p. 15.

[122] See, for instance, the participation list of the CCFAC of 2005, Report of the 37th Session of the CCFAC, 25-29 April 2005, ALINORM 05/28/12, Appendix I. See also, D.L. Leive, *supra* n. 5, pp. 436-437. See also J.E. Alvarez, *supra* n. 29, pp. 245-246.

[123] See, for more details on this 'host-country construction', Section 3.4.4 of this chapter.

[124] Rule IX.1 and 4 of the Rules of Procedure, Procedural Manual, 16th edn., 2006, p. 13.

The assignment of a Codex Member as the host country for a Codex Committee is permanent. It only ends if, or when, the committee is dissolved. A recent example of this is the proposal to divide the Codex Committee on Food Additives and Contaminants into two individual Codex Committees.[125] Both of the two 'new' Codex Committees have been designated a host country.[126] The designation of the host country of a Task Force ends when the Task Force has finished its task.

The advantage of hosting a Codex committee is that it automatically grants the host country with the chair of the Codex committee in question, which is an advantage that is recognised by the Codex Members.[127] Furthermore, the host member is not only responsible financially, but also responsible for the administration and organisation.[128] Consequently, the host country has a major influence on the operation of the Codex Committee in question. This influence may, in some cases, give rise to disagreement between Codex Members when designating a host country for a certain Codex Committee, or when allocating a new task to the Codex Committee in question. For instance, discussion arose when the proposal was made to delegate the responsibility for working out the recommendations resulting from the evaluation report on the operation of the Codex Commission to the Codex Committee on General Principles, hosted by France. There were counter-proposals to establish a new task force, which could concentrate solely on the follow-up of the recommendations resulting from the evaluation in order to prevent the same member from being the host country.[129]

Despite this shortcoming, the construction of the 'host country' to ensure the operation of subsidiary bodies has worked well. The majority of the subsidiary bodies are financed by their host country.[130] This has allowed the Codex Alimentarius Commission to establish a large institutional capacity.

---

[125] Report of the 28th Session of the Codex Alimentarius Commission, Rome, 4-9 July 2005, ALINORM 05/28/41, para. 143.

[126] The Codex Committee on Food Additives is hosted by China and the Codex Committee on Contaminants in Food is hosted by the Netherlands, the former host of the Codex Committee on Food Additives and Contaminants.

[127] F.C. Lu, *supra* n. 2, p. 202.

[128] The host country established an international agreement with the FAO in order to be able to grant privileges and immunities to the participants to the relevant Codex Committee. See, for instance, the meeting of the CCPR in The Hague at 2-7 April 2001, 'Briefwisseling tussen de regering van het Koninkrijk der Nederlanden en de Voedsel- en Landbouworganisatie van de Verenigde Naties (FAO) houdende een verdrag betreffende de 33ste zitting van het Codex Comité reststoffen van pesticiden te 's-Gravenhage van 2 to 7 april 2001, Rome 5 October/22 December 2000', (2001) *Tractatenblad van het Koninkrijk der Nederlanden*, nr. 37.

[129] The result of this discussion can be found in Report of the 26th Session of the Codex Alimentarius Commission, Rome, 30 June-7 July, 2003, ALINORM 03-41, para. 169.

[130] An exception has been the Co-ordinating Committee for Africa. Its establishment was submitted for approval to the FAO and WHO as the Commission felt that its establishment had budgetary implications for both organisations. The Commission 'expressed the wish to have before it at its next session a report from the Directors-General on the administrative and financial implications of setting up a Co-ordinating Committee for Africa.' See Report of the 8th Session of the Codex Alimentarius Commission, Geneva, 30 June-9 July 1971, ALINORM 71/31, para. 82. S. Shubber, *supra* n. 12, p. 639. Likewise, the former Committee on Milk and Milk Products was directly sponsored by the FAO

3.5 Scientific expert bodies

As mentioned above in Section 3.1.1.2, the Codex Alimentarius Commission is responsible for the execution of risk management. For scientific evaluations, it relies upon risk assessment conducted by expert bodies which are external to the institutional framework of the Codex Alimentarius Commission. The institutional structure responsible for the execution of the Joint FAO/WHO Food Standards Programme is based upon a separation between risk assessors and risk managers within the standard-setting procedure. As specified in the 'Working Principles for Risk Analysis for Application in the Framework of the Codex Alimentarius':

> 'Within the framework of the Codex Alimentarius Commission and its procedures, the responsibility for providing advice on risk management lies with the Commission and its subsidiary bodies (risk managers), while the responsibility for risk assessment lies primarily with the Joint FAO/WHO expert bodies and consultations (risk assessors).'[131]

This separation is effectuated by the fact that expert bodies come under the direct responsibility of the FAO and the WHO.

There are several expert bodies or meetings which provide the Codex Alimentarius Commission with scientific assessments. The main three expert bodies or meetings consulted are the Joint Expert Committee on Food Additives (JECFA), the Joint Meeting for Pesticide Residues (JMPR), and the Joint FAO/WHO Expert Meetings on Microbiological Risk Assessment (JEMRA), which are all joint FAO/WHO initiatives. Scientific assessment is also provided by other initiatives which are consulted on a less regular basis than the previously discussed expert bodies or expert meetings. Such initiatives include, for instance, the Joint FAO/WHO/IAEA (International Atomic Energy Agency) Study Group on Food Irradiation, which was consulted in the context of the revision of the Codex general standard on Irradiated Foods. Expert conferences are also used as a scientific basis for the work undertaken by other subsidiary bodies of the Codex Alimentarius Commission. Examples include the Joint FAO/WHO Expert consultations on Food Safety Assessment for Foods derived from Biotechnology, which was frequently undertaken in order to assist the work of the Codex Task Force on Foods derived from Biotechnology. This section concentrates on the first three expert bodies and meetings, the JECFA, JMPR and the JEMRA.

3.5.1 *Mandate and internal rules*

The JECFA was established in 1955 on the basis of Article VI.2 of the FAO Constitution and Article 2(u) and Article 18(e) of the Constitution of the WHO as a result

---

and WHO. See F.C. Lu, *supra* n. 2, p. 203. However, its successor the Codex Committee on Milk and Milk Products is hosted by New Zealand.

[131] Consideration 3, Procedural Manual, 16th edn., 2006, p. 103.

of the recommendations of the first Joint FAO/WHO Conference on Food Additives which took place earlier that same year.[132] The JECFA is an established joint Committee. Its purpose is 'to consider chemical, toxicological and other aspects of additives, contaminants, natural toxins, and residues of veterinary drugs in food related to safety for human consumption and to report thereon'.[133]

The JMPR was established in 1963 on the basis of Article VI.4 of the FAO Constitution an Articles 2(u) and 18(e) of the Constitution of the WHO. It consists of the FAO Expert Panel on Residues in Food and the Environment (at the time it was established, it was called the Working Party of Experts on Pesticides Residues) and the WHO Core Assessment Group (previously called the WHO Expert Group on Pesticide Residues).[134] Unlike the JECFA, the JMPR is not a joint expert committee but a joint meeting of the FAO expert panel and the WHO scientific group. The purpose of the JMPR is to advise on all aspects regarding pesticides residues in food and the environment.[135] The FAO Panel on Pesticide Residues in Food and the Environment is responsible, among other things, for reviewing pesticide use patterns, data on the chemistry and composition of pesticides, methods of analysis for pesticide residues and for the estimation of maximum residue levels and the WHO Core Assessment Group concentrates on the estimation of Acceptable Daily Intakes (ADIs) and Acute Reference Doses (Acute RfDs).[136]

The JEMRA is a joint initiative which resulted from a request made by the Codex Alimentarius Commission to the FAO and the WHO to launch a series of expert meetings to address microbiological aspects of food safety.[137] The first joint FAO/WHO consultation, which was held in 1999, recommended the establishment of a vehicle for the provision of scientific advice on microbiological food safety risk assessment.[138] Unlike the JECFA, the JEMRA is not an established expert body. Likewise, in contrast to the JMPR, it is not a joint meeting of two scientific committees. The JEMRA is a joint initiative to organise meetings based upon the authority of the Director-Generals of the FAO and the WHO to convene expert meetings.[139]

---

[132] Introduction to the Guide to specification for General notices. To be found on: <www://app3.fao.org/jecfa/additive_specs/docs/t0368e/t0368e01.htm>.

[133] Directory of the FAO Statutory Bodies and Panels of Experts, FAO, 2002, available at the website of the FAO: <www.fao.org/DOCREP/MEETING/005/Y7713e/Y7713e04.htm>.

[134] See FAO/WHO, 'Legal arrangements for the provision of scientific advice', Discussion Paper 2a, FAO/WHO Electronic Forum on the provision of scientific advice to Codex Alimentarius and member countries, 1 October-14 November 2003, available at: <www.fao.org/es/esn/proscad/index_en.stm>.

[135] Directory of the FAO Statutory Bodies and Panels of Experts, FAO, 2002, available at the website of the FAO: <www.fao.org/DOCREP/MEETING/005/Y7713e/Y7713e04.htm>.

[136] 'Submission and evaluation of pesticide resiues data for the estimation of maximum residue levels in food and feed', (1st edn.), FAO, Rome 2002, p. 2. See also, L. Salter, *supra* n. 120, p. 76.

[137] Report of the 22nd Session of the Codex Alimentarius Commission, Geneva 23-28 July 1997, ALINORM 97/37, para. 139.

[138] Report of the Joint FAO/WHO Expert Consultation, 'Risk Assessment of Micro-biological Hazards in Foods', Geneva, 15-19 March 1999, p. 5.

[139] Art. VI.5 of the FAO Constitution and Art. 41 of the WHO Constitution.

As they are based upon Article VI.2 and VI.4 of the FAO Constitution, the function of the expert bodies and meetings is of an advisory nature.[140] Their advice may be requested by the FAO or the WHO, their member countries or by the Codex Alimentarius Commission and its subsidiary bodies.[141] Consequently, their work is not limited to giving scientific advice to the Codex Alimentarius Commission. The re-evaluation of data may be requested by the sponsors of the data in question, or may be initiated by the expert bodies themselves.[142] Requests from the Codex Alimentarius Commission and its subsidiary bodies constitute the majority of the expert bodies' work.[143] The internal rules of the expert bodies are found in the 'Regulations for Expert Advisory Panels and Committees' of the WHO and 'Principles and Procedures which should govern Conventions and Agreements concluded under Article XIV and XV of the Constitution, and Commissions and Committees established under Article VI of the Constitution' as established by the FAO Conference. Furthermore, the operation of the FAO and WHO expert groups are regulated by individual procedural guidelines.[144] These guidelines have been developed either by the FAO (with regard to FAO expert panels) or by the WHO (with regard to the WHO scientific groups).

3.5.2 Composition and meetings

The expert bodies and the expert meetings consist of experts who participate in their individual capacity as independent experts and who do not represent their own government or their institution.[145] All the expert bodies are composed of experts and are not open to observers or Codex Members.[146] There are two types of experts

---

[140] Art. VI.2 states that the '… Conference, Council or the Director-General on the authority of the Conference may establish committees and working parties to study and report …' (emphasis added). Art. VI.4 states that 'the Director-General may establish … panels of experts with the view to developing consultations …' (emphasis added).

[141] See, for instance, FAO procedural guidelines for the Joint FAO/WHO Expert Committee on Food Additives, WHO Procedural guidelines for the Joint FAO/WHO Expert Committee on Food Additives, Geneva, January 2001, WHO procedural guidelines for the Joint FAO/WHO Meeting on Pesticide Residues.

[142] The sponsors of data are in most cases the industry or governmental bodies that have submitted the data to the expert bodies to enable them to conduct a proper risk assessment.

[143] See also, FAO/WHO, 'Scientific advice provided by FAO/WHO', Discussion Paper 1a, FAO/WHO Electronic Forum on the provision of scientific advice to Codex Alimentarius and member countries, 1 October-14 November 2003, p. 2, available at: <www.fao.org/es/esn/proscad/index_en.stm>.

[144] For instance, FAO Joint Secretariat to JECFA, Residues of veterinary drugs in food. FAO procedural guidelines for the Joint FAO/WHO Expert Committee on Food Additives. 'Food Additives and contaminants. WHO procedural guidelines for the Joint FAO/WHO Expert Committee on Food Additives', Geneva, January 2001. Some responsibilities for the chairman and members of the FAO Panel of Experts on Pesticides Residues in Food and the Environment are found in the Manual on submission and evaluation of pesticide residues data for the estimation of maximum reside levels in food and feed, FAO, Rome 2002, (1st edn.).

[145] See FAO/WHO, preparatory meeting on the elaboration of a common framework for the functioning of joint FAO/WHO expert bodies and consultations, Rome, 3-4 September 2001, p. 6.

[146] Their membership is regulated according to the FAO rules with regard to expert committees, which are the strictest of the two parent organisations and which do not allow observers.

that participate at the expert bodies or meetings. The first type is the expert members. Members of the expert bodies and meetings act as 'decision-makers' during the risk assessment and, as such, are responsible for the conduct of the meetings.[147] The JECFA and the JMPR are each composed of a maximum of 15 expert members; a maximum of 7 selected by the WHO and a maximum of 8 selected by the FAO.[148] The number of expert members who participate in the JEMRA is approximately 16 experts.[149] The second type of experts consists of the experts who are responsible for the drafting of working papers (monographs). The working papers as prepared by the drafting experts are commented upon by the expert members and serve as a basis for the scientific discussion.[150] The 'drafting experts' are referred to by the FAO as consultants and by the WHO as temporary advisors. In both cases, they are part of the FAO or the WHO Secretariat during the time that they conduct their work as 'drafting experts'.[151] Furthermore, the sponsors of the available scientific data (the manufacturers of the pesticides, veterinary drugs or additives) are consulted during the meetings of the expert bodies.[152] As they have the responsibility of submitting all the published and unpublished data that are available on the substances, their presence is requested to provide background information on the data and to answer the questions of the experts.

The selection of the expert members, consultants and temporary advisors is undertaken by the FAO and the WHO separately, according to their own selection procedures. The FAO selects both the expert members and the consultants of the JECFA and the JMPR from a roster of experts.[153] To establish the roster of experts, the FAO launches a call for experts, and asks governments and institutions to iden-

---

[147] FAO procedural guidelines for the Joint FAO/WHO Expert Committee on Food Additives, Rome, September 2002, Section 5.3. WHO procedural guidelines for the Joint FAO/WHO Expert Committee on Food Additives, Geneva January 2001, Section 2. WHO Procedural guidelines for the Joint FAO/WHO Meeting on Pesticides Residues, Geneva, January 2001, Section 2.

[148] See FAO/WHO, preparatory meeting on the elaboration of a common framework for the functioning of joint FAO/WHO expert bodies and consultations, Rome, 3-4 September 2001, pp. 5-6.

[149] At the meeting in February 2004 the JEMRA consisted out of 16 expert members (Report of the Joint FAO/WHO Workshop on Enterobacter Sakazakii and other Micro-organisms in Powdered Infant Formula, Geneva, 2-5 February 2004. Also at the meeting in August 2002 16 expert members participated, see Report of the Joint FAO/WHO Consultation on Risk assessment of Campylobacter spp. in broiler chicken and Vibrio spp. in seafood, Bangkok, 5-9 August 2002, FAO Food and Nutrition Paper.

[150] FAO procedural guidelines for the Joint FAO/WHO Expert Committee on Food Additives, Rome, September 2002, Sections 5.1 & 5.2. WHO procedural guidelines for the Joint FAO/WHO Expert Committee on Food Additives, Geneva January 2001, Section 5.2. WHO Procedural guidelines for the Joint FAO/WHO Meeting on Pesticides Residues, Geneva, January 2001, Section 4.2.

[151] FAO procedural guidelines for the Joint FAO/WHO Expert Committee on Food Additives, Rome, September 2002, Section 2. WHO procedural guidelines for the Joint FAO/WHO Expert Committee on Food Additives, Geneva January 2001, Section 1. WHO Procedural guidelines for the Joint FAO/WHO Meeting on Pesticides Residues, Geneva, January 2001, Section 1.

[152] FAO procedural guidelines for the Joint FAO/WHO Expert Committee on Food Additives, Rome, September 2002, Section 6. WHO Procedural guidelines for the Joint FAO/WHO Meeting on Pesticides Residues, Geneva, January 2001, Section 6.

[153] FAO procedural guidelines for the Joint FAO/WHO Expert Committee on Food Additives, Rome, September 2002, Section 3.

tify the scientists whom they consider to be appropriate, and draws from its own experience in working with scientists in previous committees or from reviews of scientific publications.[154] When appointing experts to the roster of experts, it strives to achieve a balance between academic and regulatory experience and to ensure international representation in terms of diversity of knowledge and approaches.[155] Experts are appointed for a maximum of 4 years, which can be renewed. The members and consultants are appointed by the Director-General of the FAO in consultation with the FAO Secretariat of the JECFA.[156] The expert members and consultants who participate in the meetings of the JECFA are invited by the FAO Secretariat of the JECFA. The expert members selected by the WHO have to be members of a WHO Expert Advisory Panel (in the case of the JECFA, this is the WHO Food Safety Panel). The experts are appointed to be members of a WHO Expert Advisory Panel by the Director-General of the WHO on the basis of similar criteria as taken into account by the FAO. The WHO temporary advisors are directly selected from a 'participating institution' and do not need to be a member of a WHO Expert Advisory Panel.[157]

The experts (both the expert members as well as the drafting experts) who participate in the JEMRA meetings are selected from a joint FAO/WHO Roster of Experts for Microbiological Risk Assessment.[158] They are formally appointed by the Directors-General once they have been selected from the Roster to attend a JEMRA meeting.[159]

The JECFA held its first meeting in 1957 and meets twice a year, firstly on issues relating to additives and contaminants and successively on issues of residues of veterinary drugs.[160] The JMPR held its first meeting in 1963. Since its first meeting, it has functioned on an *ad hoc* basis. The JMPR convenes for two weeks once a year. Like the JMPR, the JEMRA holds its meetings on an *ad hoc* basis. Its first meeting was held in 1999. Since then, it has met 11 times.[161]

---

[154] Ibid.

[155] Ibid.

[156] Ibid.

[157] See also, FAO/WHO, 'Legal arrangements for the provision of scientific advice', Discussion Paper 2a, FAO/WHO Electronic Forum on the provision of scientific advice to Codex Alimentarius and member countries, 1 October-14 November 2003, available at <www.fao.org/es/esn/proscad/index_en.stm>. A 'participating institution' includes relevant national agencies, national academic and research institutes.

[158] Information available at: <www.fao.org/es/esn/jemra/selection_en.stm>.

[159] See also, FAO/WHO, 'Selection of experts to avail of best available expertise', Discussion Paper 2c, FAO/WHO Electronic Forum on the provision of scientific advice to Codex Alimentarius and member countries, 1 October-14 November 2003, available at: <www.fao.org/es/esn/proscad/index_en.stm>, p. 2.

[160] FAO Joint Secretariat of the Joint FAO/WHO Expert Committee on Food Additives (JECFA), FAO/WHO Roster of experts for JECFA for meetings addressing Exposure assessment of food chemicals (2002-2006), 18 March 2003, p. 1, available at: <ftp://ftp.fao.org/es/esn/jecfa/roster_ex_asses 2002.pdf>.

[161] See information on the JEMRA available at: <www.who.int/foodsafety/micro/jemra/meetings/en/index.htm>.

The JECFA evaluations are used by several subsidiary bodies of the Codex Alimentarius Commission, such as the Codex Committee on Food Additives (CCFA) the Codex Committee on Contaminants in Foods (CCCF) or the Codex Committee on Residues of Veterinary Drugs in Food (CCRVDF) to propose certain standards. They are used to develop Codex standards and other related measures. The reports of the meetings of the JECFA are published on behalf of both the FAO and the WHO, in the WHO Technical Report Series.[162] The JMPR conducts most of its evaluations at the request of the CCPR. The reports of the JMPR meetings are published as part of the FAO Plant Production and Protection Paper Series.[163] The JEMRA mainly provides scientific consultation to the Codex Committee on Food Hygiene, another subsidiary body of the Codex Alimentarius Commission. The reports of the JEMRA are published in the Microbiological Risk Assessment Series.[164]

3.5.3 Administration and finance

The expert bodies and meetings are administered by the joint secretariat of the FAO and the WHO and consist of staff members of the FAO and the WHO.[165] Their responsibilities are to organise the meetings, to invite the participants, to ensure that adequate documentation is prepared, and to edit and publish the report.[166]

The expert committee is jointly sponsored by the WHO and the FAO both through their regular budgets as well as through extra-budgetary funding.[167] However, a

---

[162] See also, FAO/WHO, 'Legal arrangements for the provision of scientific advice', Discussion Paper 2a, FAO/WHO Electronic Forum on the provision of scientific advice to Codex Alimentarius and member countries, 1 October-14 November 2003, available at: <www.fao.org/es/esn/proscad/index_en.stm>.

[163] Ibid.

[164] See also, FAO/WHO, 'Legal arrangements for the provision of scientific advice', Discussion Paper 2a, FAO/WHO Electronic Forum on the provision of scientific advice to Codex Alimentarius and member countries, 1 October-14 November 2003, available at: <www.fao.org/es/esn/proscad/index_en.stm>.

[165] For instance, the International Programme on Chemical Safety (IPCS) of the WHO is responsible for serving the WHO Secretariat of the JECFA and of the JMPR (WHO procedural guidelines for the Joint FAO/WHO Expert Committee on Food Additives, Geneva January 2001, Section 4, WHO Procedural guidelines for the Joint FAO/WHO Meeting on Pesticides Residues, Geneva, January 2001, Section 4) and the Pesticides Management Unit of the Plant Production and Protection Division of the FAO is responsible for serving the FAO Secretariat of the JMPR, while the Food and Nutrition Division serves as FAO Secretariat of the JECFA.

[166] WHO procedural guidelines for the Joint FAO/WHO Expert Committee on Food Additives, Geneva January 2001, Section 4.1. WHO Procedural guidelines for the Joint FAO/WHO Meeting on Pesticides Residues, Geneva, January 2001, Section 4.1. FAO procedural guidelines for the Joint FAO/WHO Expert Committee on Food Additives, Rome, September 2002, Section 4.1.

[167] See also, FAO/WHO, 'Legal arrangements for the provision of scientific advice', Discussion Paper 2a, FAO/WHO Electronic Forum on the provision of scientific advice to Codex Alimentarius and member countries, 1 October-14 November 2003, available at: <www.fao.org/es/esn/proscad/index_en.stm>.

lack of financial means has often been raised as a cause for concern.[168] The FAO and the WHO Secretariats have encountered financial limitations when organising the meetings of the JECFA and the JMPR.[169] Furthermore, only the travel expenses of the invited experts are paid, the experts do not receive *honoraria*.[170]

3.6 The Joint FAO/WHO Consultative Group for the Trust Fund

The Joint FAO/WHO Consultative Group for the Trust Fund was established in 2002 to 'guide' the 'Trust Fund for the participation of developing countries and countries in transition in the work of the Codex Alimentarius Commission' (hereafter referred to as the Trust Fund).[171] This Trust Fund aims, amongst other things, to promote the participation of developing countries by financially supporting their visits to the meetings of the Codex Alimentarius Commission and its subsidiary committees, and has a term of 12 years.[172] The Consultative Group has the mandate to provide strategic guidance to the project, to establish the relevant guidelines and criteria as a framework for project activities, and to monitor the application of the guidelines and criteria, the coherence with other funding mechanisms and the overall progress of the project.[173] The Joint FAO/WHO Consultative Group for the Trust Fund falls under the direct responsibility of the Directors-General of both organisations and functions independently from the Codex Alimentarius Commission and its subsidiary bodies. The Consultative Group is a clear reflection of the co-operation that exists between the two secretariats of the FAO and the WHO. The Consultative Group consists of three senior staff members of the FAO and three senior staff members of the WHO.[174] The Secretary of the Codex Alimentarius Commission participates in the meetings, but only in an advisory capacity and only

---

[168] Report of the evaluation of the Codex Alimentarius and other FAO and WHO food standards work, ALINORM 03/25/3, para. 192. See also, L. Salter, *supra* n. 120, p. 77.

[169] See also, FAO/WHO, 'Legal arrangements for the provision of scientific advice', Discussion Paper 2a, FAO/WHO Electronic Forum on the provision of scientific advice to Codex Alimentarius and member countries, 1 October-14 November 2003, available at: <www.fao.org/es/esn/proscad/index_en.stm>.

[170] FAO procedural guidelines for the Joint FAO/WHO Expert Committee on Food Additives, Rome, September 2002, Section 3. WHO procedural guidelines for the Joint FAO/WHO Expert Committee on Food Additives, Section 3. See also, FAO/WHO, 'Selection of experts to avail of best available expertise', Discussion Paper 2c, FAO/WHO Electronic Forum on the provision of scientific advice to Codex Alimentarius and member countries, 1 October-14 November 2003, available at: <www.fao.org/es/esn/proscad/index_en.stm>.

[171] The Trust Fund was launched during the 25th Session of the Codex Alimentarius Commission in February 2003 and has been operational since 1 March 2004. See, for more details on the Trust Fund, Section 3.2.1.2.2 of Chapter V.

[172] The minimum threshold that triggers the operation of the trust fund is an amount of $500,000 and the overall target for fundraising over 12 years is $40 million.

[173] Provision 11 of the Terms of Reference of the FAO/WHO Consultative Group; see 'Progress report of the FAO/WHO Project and Fund for enhanced participation in Codex', FAO and WHO, January 2003, ALINORM 03/25/4, Annex 1.

[174] Ibid., Provision 5.

upon specific invitation.[175] Decisions are taken by consensus.[176] The Consultative Group is chaired by a member who has been appointed from among the members.[177] The chair rotates between the two organisations on an annual basis.[178] The Consultative Group had its first meeting in December 2002[179] and meets at least twice a year at the location of the organisation holding the chair.[180] Daily management of the Trust Fund is undertaken by the Food Safety Department of the WHO.[181] This means that the financial regulations and procedural rules of the WHO apply to the daily management of the Trust Fund and that the Trust Fund will be incorporated in the financial report and the audited financial statements of the WHO.[182]

4. CONCLUSIONS

The launch of the Joint FAO/WHO Food Standards Programme in 1963 was the result of an increased recognition of the need for international co-operation in the field of food standards and the necessity for the co-ordination of several vertical and regional initiatives already taking place. Not surprisingly, in view of the mandate of the FAO and the WHO, their expertise and their previous activities of co-operation in the field of food quality and health related issues, this programme was launched under the auspices of both organisations. The core element in the execution of the Joint FAO/WHO Food Standards Programme is the creation of the Codex Alimentarius: a collection which consisted of uniformly-defined food standards with the purpose of giving assistance to the harmonisation of food requirements. For this purpose, the competent decision-making organs of both organisations established the Joint FAO/WHO Codex Alimentarius Commission.

The ability of the Codex Alimentarius Commission to give effect to this mandate depends largely on its institutional competencies. The authority of the Codex Alimentarius Commission is, however, restricted by the fact that it is derived from the attributed powers of its 'parent' organisations: the FAO and the WHO. Two elements that are essential for any international institution to develop into an effective institution are financial resources and the expertise of the administrative staff. Both the budgetary authority and the management of the Codex Secretariat's personnel are in the hands of its parent organisations, in particular, the FAO.

---

[175] Ibid., Provision 7.
[176] Ibid., Provision 8.
[177] Ibid., Provision 6.
[178] Ibid.
[179] 'Progress report of the FAO/WHO Project and Fund for enhanced participation in Codex', FAO and WHO, January 2003, ALINORM 03/25/4, p. 3.
[180] Provision 8 of the Terms of Reference of the FAO/WHO Consultative Group), See "Progress report of the FAO/WHO Project and Fund for enhanced participation in Codex", FAO and WHO, January 2003, ALINORM 03/25/4, Annex 1.
[181] Ibid., Provision 4.
[182] Ibid.

Despite these constraints related to its subsidiary position *vis-à-vis* its 'parent' organisations the Codex Alimentarius Commission has also been attributed with important powers. These powers have allowed the Codex Alimentarius Commission to develop into a strong standard-setting body through institution-building activities and by setting up different types of subsidiary bodies. It has overcome its financial dependency on its parent organisations by relying on a so-called 'host country' construction, in which Codex Members take the financial and administrative responsibility. This construction to establish subsidiary bodies has enabled the Codex Alimentarius Commission to give content, *ratione materiae*, to its mandate. Furthermore, independent normative powers and the task of acting as the promoter of co-ordinating international food standard activities, as laid down in its mandate, has provided the Codex Alimentarius Commission with a strong basis to position itself *vis-à-vis* other international organisations which act as competitors in this area.

Besides the Codex Alimentarius Commission and its subsidiary bodies, other bodies have been attributed with tasks that fall within the scope of the Joint FAO/WHO Food Standards Programme: the expert bodies and the Consultative Group for the Trust Fund. These bodies fall under the direct responsibility of the FAO and the WHO. The institutional relationship between the scientific expert bodies, on the one hand, and the Codex Alimentarius Commission, on the other, is based upon the separation of powers. The Codex Alimentarius Commission is responsible for risk management and relies, for scientific expertise, on the risk assessments conducted by the expert bodies.

Chapter II
THE CODEX ALIMENTARIUS: HARMONISATION THROUGH STANDARD-SETTING

1. INTRODUCTION

The previous chapter examined the institutional framework responsible for the establishment of the Codex Alimentarius and concluded that institution-building activities based upon its mandated competencies have positioned the Codex Alimentarius Commission as a strong international standard-setting body. However, its success in the long run will depend on its ability to carry out these activities, which enhances international co-operation. Or in other words, its success will be measured by whether it can maintain its justified position as an international institution. To this end, it has been entrusted with the normative powers to adopt and publish food standards that jointly constitute the Codex Alimentarius. The aim of the establishment of the Codex Alimentarius, as formulated in the 'General Principles of the Codex Alimentarius' is to assist in harmonising food requirements that ensure the protection of consumers' health and fair trade practices in foods in order to facilitate trade.[1]

Over the years, the Codex Alimentarius Commission has faced several complications in harmonising the diverse national circumstances and interests. Questions with regard to the need to promote the uniform application of Codex measures have arisen as well as concerns about a harmonisation process that is too exhaustive to respond to the wide diversity of circumstances, such as the difference in levels of development, production methods, eating habits and cultural perception.[2] It has sought ways to balance these sometimes contradictory interests, and has been able to develop an extensive system of different types of harmonisation measures.

This chapter examines the ways in which the Codex Alimentarius Commission has sought to advance the harmonisation process through its normative powers and its power to establish its own standard-setting procedure independently from the approval of its parent organisations: the Food and Agriculture Organisation (FAO) and the World Health Organisation (WHO). It also pays specific attention to the features of the Codex Alimentarius which function as instruments of harmonisation,

---

[1] Art. 1 of the General Principles of the Codex Alimentarius, Procedural Manual, 16th edn., 2006, p. 30.

[2] For instance, during the 1960s, there were extensive discussions on the nature of Codex standards both at the Joint FAO/WHO Conference preceding the establishment of the Joint FAO/WHO Food Standards Programme (Report of the Joint FAO/WHO Conference on Food Standards, Geneva, 1-5 October 1962, ALINORM 62/8, paras. 7-13), as well as at the Codex Alimentarius Commission. The discussion related to whether the standards should be trading standards, also referred to as 'recipe standards' or a minimum adaptable standard.

such as the scope of the measures and the specificity of their content, the formulation of the measures, and the legal binding force of the measures. The combination of these different features provides an indication of the harmonisation process which the Codex Alimentarius Commission aims to achieve.

2. THE STRUCTURE OF THE CODEX ALIMENTARIUS

One of the main purposes of the Joint FAO/WHO Food Standards Programme is the publication of the Codex Alimentarius, which is defined as a collection of internationally elaborated and adopted food standards.[3] Besides standards, the Codex Alimentarius is also composed of other measures designed to complement the collection of standards, such as codes of hygienic practice and guidelines on methods of sampling and analysis.[4]

The structure of the Codex Alimentarius is based upon two approaches which aspire to work towards its completion. During the process of developing the Codex Alimentarius over the years, these approaches have been altered. With regard to the first approach, in 1962, the Joint FAO/WHO Food Standards Conference which established the Joint FAO/WHO Food Standards Programme recommended working towards the completion of the Codex Alimentarius through the adoption of commodity-based standards.[5] This meant that the standards were to cover issues relating to one particular food product or commodity. To this end, the intention was to establish 'a collection of internationally adopted food standards presented in a unified form'.[6] All standards were to be prepared in accordance with a specific format in order to obtain this uniform presentation of the standards. This approach is still reflected in the 'General Principles of the Codex Alimentarius' and in the 'Format for Codex Commodity Standards', as laid down in the Procedural Manual.[7]

---

[3] Principle 1 of the General Principles of the Codex Alimentarius, Procedural Manual, 16th edn., 2006, p. 30.

[4] Ibid.

[5] Report of the Joint FAO/WHO Conference on Food Standards, Geneva, 1-5 October 1962, ALINORM 62/8, p. 8: 'The Conference recommended that the Codex in due course cover all the principal foods and their components in international trade. The type of standard to be included on the same long-term view should aim at covering all facets of the problem, especially: definition, composition, quality, designation, labelling, sampling, analyses and hygiene. These facets should be studied in their scientific, technical, economic, administrative and legal aspects in order to ensure that the products to which they apply are in all respects suitable for consumption from both the hygienic and commercial point of view, and are correctly described.'

[6] Report of the Joint FAO/WHO Conference on Food Standards, Geneva, 1-5 October 1962, ALINORM 62/8, p. 5.

[7] Art. 1 of the General Principles states: 'The Codex Alimentarius is a collection of internationally adopted food standards presented in a uniform manner.' (General Principles of the Codex Alimentarius, Procedural Manual, 16th edn., 2006, p. 30). The Format mentions that a Commodity Standard should contain the following provisions which are only required to be included if they are appropriate to the commodity standard in question: name of the standard, scope, description, essential composition and quality factors, food additives, contaminants, hygiene, weights and measures, labelling, and methods of analysis and sampling (Format for Codex Commodity Standards, Procedural Manual, 16th edn., 2006, pp. 91-94).

However, over the years, the function of general standards (standards that deal with general issues that relate to all food products or to a group of food products) has become more important *vis-à-vis* the commodity standards.

Furthermore, the initial idea was that the so-called regional standards, which are elaborated, adopted and applied by the Codex Members of one particular region, were to be an important component of the Codex Alimentarius. However, the role of the regional standards has been reduced.

2.1 From a vertical to a horizontal approach

During the 1960s, the standard-setting was clearly commodity-based, meaning that 'horizontal' issues, such as food additives, hygiene and labelling, were only negotiated to the extent relevant for the commodity standard in question. The first adopted standards were commodity standards.[8] In the field of cheese products, standards can even be referred to as 'variety' standards, as they focus on only one particular variety of the same commodity or sub-commodity group.[9] This can be explained by the fact that, during the first years of the Codex Alimentarius Commission, the already developed work of the Committee of government experts on the Code of Principles concerning Milk and Milk Products was submitted to the Codex Alimentarius Commission for adoption.[10] Examples include the standard on Gouda, on Emmentaler, on Camembert, and on Brie.

Commodity standards have to be prepared in conformity with the 'Format for Codex Commodity Standards' (hereinafter referred to as the 'Format') as contained in the Procedural Manual of the Codex Alimentarius Commission. The Format serves as a guide and enumerates elements which, where appropriate, have to be included in the commodity standard in question:

1. the name of the standard;
2. its scope;
3. a description;
4. the essential composition and quality factors;
5. food additives;

---

[8] The only important and early exception to this 'vertical approach' was the recommended code of practice on hygiene, adopted in 1969.

[9] The use of commodity in this context is based on the classification system of the FAO (FAO commodity codes), which distinguishes between different commodity groups. 'Products from Live Animals' is such a commodity group and contains sub-commodity groups, such as the one related to cow milk and sheep milk. The classification does, however, not differentiate the different varieties of cheese, such as Edam, Gouda or Emmentaler.

[10] Since 1958, the Committee of government experts on the Code of Principles concerning Milk and Milk Products has been working on the adoption of food standards in the area of milk products. It had adopted 16 individual standards on different types of cheese. See J.P. Dobbert, 'Le Codex Alimentarius vers une nouvelle méthode de réglementation internationale', 15 *Annuaire Français de Droit International* (1969), pp. 701-706. See also, Section 2 of Chapter I on the historical background of the Joint FAO/WHO Food Standards Programme.

6. contaminants;
7. hygiene;
8. weights and measures;
9. labelling; and
10. methods of analysis and sampling.

To give an illustration, the 'Codex International Individual Standard for Gouda' contains provisions on the kind of milk, on the authorised additives, on the principal characteristics of the cheese, such as the shape, dimensions, weights, rind and holes, and on the method of manufacture.[11] For instance, the standard regulates that the size of the holes of gouda cheese vary from a pin's head to a pea,[12] that the rind of gouda is dry or coated either with wax, a suspension of plastic, or a film of vegetable oil,[13] and that only cheese conforming to the characteristics indicated by the standard may be labelled as 'Gouda'.[14]

This commodity-based approach, also referred to as the 'vertical approach' of developing the Codex Alimentarius, was recognised as having several shortcomings. Commodity standards are often very detailed in content, but remain difficult to implement and apply in member countries, and are consequently not always considered to be appropriate as harmonisation instrument. Furthermore, it is considered to be a slow process towards the facilitation of harmonisation, as, with the adoption of each standard, only one commodity on the world market is covered.[15]

It was also recognised that more general standards needed to be prepared, as commodity standards might not cover all concerns. For instance, the 1981 'Codex Standard for Edible Fats and Oils not Covered by Individual Standards' illustrates the need to adopt provisions on aspects that the individual commodity standards on fats and oils had left uncovered.[16] Over the years, the deliberation of provisions on general issues, such as hygiene, food additives and labelling, increasingly took place in the form of 'independent' standards, instead of being part of the preparation of commodity standards. This resulted in the adoption of several general standards. Logically, the format of these standards deviates from the 'Format' for the preparation of commodity standards, as, in most cases, they concentrate on only one of the enumerated elements. Early examples include the adoption of the 1981 'General Standard for the Labelling of Food Additives when Sold as Such',[17] and the adoption of the 1983 'General Standard for Irradiated Foods'.[18] They are both a clear

---

[11] Codex Stan C-5-1966. See Annex I of this book.
[12] Provision 4.6.3 of Codex Stan C-5-1966
[13] Provision 4.4.2 of the Codex Stan C-5-1966.
[14] Provision 8 of the Codex Stan C-5-1966.
[15] J.P. Chiaradia-Bousquet, *Legislation governing food control and quality certification*, FAO Legislative Study 54 (Rome, FAO 1995), p. 13.
[16] On early discussions on the introduction of general standards, see D.L. Leive, *International Regulatory Regimes* (Lexington, Lexington Books 1976), pp. 404-409.
[17] Codex Stan 107-1981.
[18] Codex Stan 106-1983.

indication of an increased focus on general concerns. The General Standard for Irradiated Foods, for instance, concentrates on the irradiation process and contains general requirements for the use of irradiation, such as the different types of irradiation that may be used, the absorbed dose to be used, and the facilities and control of the process to be observed. Likewise, the adoption of the 1985 'General Standard for the Labelling of, and Claims for, Pre-packaged Foods for Special Dietary Uses' also illustrates this shift away from a purely 'vertical approach'.[19]

In 1985, the issue of food additives raised the important question of whether the Codex Committee on Food Additives should express opinions on food additives that were not covered by commodity standards.[20] As a response to this question, the Secretary of the Codex Alimentarius Commission wrote an important paper that paved the way towards a horizontal approach.[21] The major breakthrough in this development was the FAO/WHO Conference of 1991 on 'Food Standards, Chemicals in Food and Food Trade having implications for the Codex Alimentarius Commission', which emphasised the need for a 'horizontal' approach.[22] The horizontal approach emphasises that the elaboration of general issues is the most important way to complete the Codex Alimentarius.[23] However, the Codex Alimentarius Commission can still decide to elaborate commodity standards. This means that commodity standards are still adopted and do, in fact, still constitute an important part of standards that are adopted. During the extraordinary meeting of the Codex Alimentarius Commission in February 2003, which was dedicated to the evaluation

---

[19] This change equally was reflected in the adoption of more 'horizontal' guidelines and codes of practice, such as the 1979 General Guidelines on Claims and the 1987 General Principles on the addition of essential nutrients to foods.

[20] See K. Keefe, et al., 'The Codex General Standard for Food Additives-A Work in Progress', in N. Rees and D. Watson (eds.), *International Standards for Food Safety* (Gaithersburg, Maryland, Aspen Publishers 2000), p. 174. Should CCFA express opinions on food additives other than those included in Codex Commodity Standards? CX/FA 85/16. Rome: Codex Alimentarius Commission. See also, on the development of the mandate of the CCPR towards a more horizontal approach, L. Salter, *Mandated Science, Science and Scientists in the Making of Standards* (Dordrecht, Kluwer Academic Publishers 1988), p. 73, citing a CCPR working paper: 'Its first approach was largely commodity oriented and the first session was devoted to one group of commodities (cereals and cereal products) and the relevant pesticides. It soon became apparent that a pesticide oriented approach, covering all relevant commodities, would be more realistic, and the subsequent programme was, therefore, based on lists of pesticides selected according to agreed priorities.'

[21] G.O. Kermode, (1986). The future Direction of the Work of the Joint FAO/WHO Food Standards Programme. CX/EXEC 86/33/CRD 1. Rome: Codex Alimentarius Commission. See K. Keefe, et al., *supra* n. 20, p. 174.

[22] See Report of the 19th Session of the Joint FAO/WHO Codex Alimentarius Commission, Rome, 1-10 July 1991, ALINORM 91/40, App. IV: 'The Conference emphasized the importance of Committees focusing their efforts on the elaboration of horizontal provisions in Codex standards as related to consumer protection (i.e., health, safety) and facilitation of international trade. The importance of eliminating detail, where appropriate was also recognized as a major factor in simplifying standards and facilitating government acceptance of Codex standards.' See also, K. Keefe, et al., *supra* n. 20, p. 172.

[23] On the shift from vertical to a horizontal approach, See also, C.C. van der Tweel, *Een nieuw GATT-Verdrag en de Codex Alimentarius. Een bestuurskundig onderzoek naar internationale beleidsvorming en nationale beleidsruimte* (Leiden, Rijksuniversiteit Leiden 1993), p. 34.

process of the Codex Alimentarius Commission, there was again widespread agreement that the work of the Commission should concentrate on these general issues rather than on specific commodities.[24]

2.2 World-wide and regional Codex standards

Rule VIII.3 of the Rules of Procedure states that, when the majority of countries are in favour of the adoption of a standard and this majority represents a certain region or a group of countries, the standard shall be considered as being primarily intended for that region or group of countries.[25] During the first years of the Codex Alimentarius Commission, regional standards were expected to become an important part of the Codex Alimentarius. The main reason behind this was that the Joint FAO/WHO Food Standards Programme was partly a continuation of the *Codex Alimentarius Europaeus*, an earlier European initiative to develop a food code.[26] The former Council of the *Codex Alimentarius Europaeus* was incorporated into the institutional framework of the Codex Alimentarius Commission as the Regional Codex Committee for Europe.[27]

The justification for having regional standards, in addition to world-wide standards, was that the harmonisation of food standards at regional level would facilitate importation into that specific region. If no world-wide standard was feasible, a regional standard was considered to be at least a partial step towards harmonisation.[28] On the other hand, voices were raised in favour of reducing the role of regional standards.[29] It was held that the development of regional standards which covered products that were subject to trade that went beyond the region in question might, in fact, obstruct trade at international level. A proposal was made to amend the Rules of Procedure in order to limit the scope of the products covered by regional standards to solely those food products that were produced exclusively and con-

---

[24] See Report of the 25th Session of the Codex Alimentarius Commission (Extraordinary Meeting), Geneva, 13-15 July 2003, ALINORM 03/25/5. paras. 15-17 and Statement, para. 5 as included in Appendix II of the report: 'Within its mandate, the Commission emphasised that its first priority would be the development of standards having an impact on consumer health and safety.' Already in the Mid-Term Plan of the Codex Alimentarius Commission for the period 1998-2002, it was proposed that the Codex Alimentarius Commission reduces its work on vertical commodity specific standards by 2002 to those cases where there is a good reason for their development. See also, H.M. Wehr, 'Update on Issues before the Codex Alimentarius', 52 *Food and Drug Law Journal* (1997), p. 536.

[25] Procedural Manual, 16th edn., 2006, p. 12.

[26] D.L. Leive, *supra* n. 16, p. 409. See also, on the historical background of the Joint FAO/WHO Food Standards Programme, Section 2 of Chapter I.

[27] D.L. Leive, *supra* n. 16, p. 409. J.P. Dobbert, *supra* n. 10, p. 699.

[28] One of the arguments in favour of a wide scope of regional standards was that the harmonisation of food standards on a regional level would facilitate importation into the region, as importers would only need to comply with one standard. Report of the 6th Session of the Codex Alimentarius Commission, Geneva, 4-14 March 1969, ALINORM 69/67, para. 33.

[29] Report of the 5th Session of the Codex Alimentarius Commission, Rome, 20 February-1 March 1968, ALINORM 68/35, paras. 69 and 72.

sumed mainly within that region.[30] Despite this proposal, the Rules of Procedure have not been amended. However, the terms of reference of the Codex Co-ordinating Committees clearly reflect this restriction, as it states that the Committee 'develops regional standards for food products moving exclusively or almost exclusively in intra regional trade'.[31] In situations where the trade of the food products has an international market potential, the Codex Co-ordinating Committees recommend the preparation of world-wide standards. Proposals for the elaboration of a regional standard are made by the majority of the Codex Members from the region in question. However, the decision to elaborate a regional or world-wide standard is taken by the Codex Alimentarius Commission as a whole.[32] At this moment, there is only one regional standard: that of the European standard on fresh fungus 'chanterelle'.[33] Former regional standards have been either converted into world-wide standards or deleted. For example, the controversial adoption of the standard for natural mineral waters during the 22nd Session of the Codex Alimentarius Commission in 1997 was, in fact, a conversion from a European standard into a world-wide standard.[34] The European standard for Mayonnaise was withdrawn from the Codex Alimentarius during the 26th Session of the Codex Alimentarius Commission.[35]

Recently, there has been a discussion on whether to elaborate a regional or world-wide standard on ginseng. During the 26th Session of the Codex Alimentarius Commission in 2003, several Codex Members from Asia proposed that the elaboration of the Codex standard for ginseng, which had formerly been entrusted to the Codex Committee on Processed Fruit and Vegetables, be undertaken by the Codex Committee for Asia as a regional standard.[36] A number of delegations were against the elaboration of a regional standard, as 'ginseng was a commodity grown in countries outside the Asian region, and therefore, an international standard inclusive of all varieties was necessary'.[37] The decision on whether or not the standard will be a

---

[30] Report of the 5th Session of the Codex Alimentarius Commission, Rome, 20 February-1 March 1968, ALINORM 68/35, paras. 68-72. See also, D.L. Leive, *supra* n. 16, pp. 412-413.

[31] See, for instance, Provision (d) of the Terms of Reference of the FAO/WHO Co-ordinating Committee for Europe, Procedural Manual, 16th edn., 2006, p. 154.

[32] 'Uniform Procedure for the elaboration of Codex Standards and Related Texts', Part 3, Step 1 of the 'Procedures for the Elaboration of Codex Standards and Related Texts', Procedural Manual, 16th edn., 2006, p. 23.

[33] Codex Stan 40-1981.

[34] Adopted at the 12th Session of the Codex Alimentarius Commission in 1978 as European standard (Report of the 12th Session of the Codex Alimentarius Commission, Rome 17-28 April 1978, ALINORM 78/41, para. 176). In 1997, it was revised into a world-wide standard. The elaboration procedure of the world-wide standard on mineral waters led to controversial discussions due to the definition of mineral waters. The European region defines mineral waters as water that comes from mineral sources without necessarily being purified. However in the United States, mineral waters are considered to be purified drink-water. The final adoption resulted in a vote. See also, Section 3.1.1 of Chapter V.

[35] Report of the 26th Session of the Codex Alimentarius Commission, Rome, 30 June-7 July 2003, ALINORM 03/41, para. 225.

[36] Ibid., paras. 206-210.

[37] Ibid., para. 209.

regional standard or a world-wide standard was postponed to the moment the Codex Alimentarius Commission adopts the standard at Step 5.[38]

2.3 Other types of Codex measures

The 'General Principles of the Codex Alimentarius' of the Procedural Manual indicate that the Codex Alimentarius also includes other documents besides standards. It states that:

> 'The Codex Alimentarius is a collection of internationally adopted food standards presented in a uniform manner....The Codex Alimentarius also includes provisions of an advisory nature in the form of codes of practice, guidelines and other recommended measures intended to assist in achieving the purposes of the Codex Alimentarius ...'[39]

Although this authority of the Codex Alimentarius Commission is not explicitly foreseen in its Statutes, during the 5th Session of the Commission, it was agreed that Article 1(a) of the Statutes concerning the protection of the health of the consumers was a sufficient basis to develop codes of hygienic practice.[40] This conclusion was confirmed by the Legal Counsels of the FAO and the WHO, who stated that Article 1(a) in conjunction with Article 1(d) (the authority to finalise standards) gave the Codex Alimentarius Commission sufficient power to develop other advisory measures.[41] The General Principles were amended to express this interpretation.

Other types of measures that have been included in the Codex Alimentarius include: recommended codes of practice, guidelines and individual maximum residue levels (MRLs), which are published separately from Codex standards. Whereas Codex standards and MRLs are meant to have a mandatory status, it was decided that recommended codes of practice and guidelines would remain advisory.[42]

Recommended Codes of Practice concentrate mainly on hygienic practice, and contain requirements which aim to guarantee hygiene in the production, process-

---

[38] Report of the 27th Session of the Codex Alimentarius Commission, Geneva, 28 June-3 July 2004, ALINORM 04/27/41, para. 94. See, for more details on the standard-setting procedure of the Codex Alimentarius Commission, Section 4 of this chapter. The proposed draft standard on Ginseng is submitted to the 2007 Session of the Codex Alimentarius Commission for adoption at step 5.

[39] Provision 1 of the General Principles, Procedural Manual, 16th edn., 2006, p. 30. The General Principles were adopted during the 3rd Session of the Codex Alimentarius Commission in 1965 and amended in 1969 to include the possibility for the Codex Alimentarius Commission to adopt other measures (in 2005, the Principles were again amended to take out the provisions on the acceptance procedure). It contains many elements that were proposed by the Joint FAO/WHO Food Standards Conference in 1962 included in the annexed Guidelines for the Codex Alimentarius Commission. However, the additional documents such as codes of practice and guidelines were not initially foreseen.

[40] Report of the 5th Session of the Codex Alimentarius Commission, Rome, 20 February-1 March 1968, ALINORM 68/35, para. 46. Art. 1(a) of the Statutes state: 'Protecting the health of the consumers and ensuring fair practices in the food trade.' (Procedural Manual, 16th edn., 2006, p. 3).

[41] See Report of the 6th Session of the Codex Alimentarius Commission, Geneva, 4-14 March 1969, ALINORM 69/67. Appendix III, para. 1.

[42] See for more details on the legal status of Codex measures, Section 6 of this chapter.

ing, handling, packing, storage, transportation and distribution of food products.[43] The 'general' recommended code of hygienic practice is the Recommended International Code of Practice General Principles of Food Hygiene (RCP). This RCP contains the important Hazard Analysis and Critical Control Point System (HACCP), which is used in various countries as the basis for their regulation to ensure hygienic practice. The target group of recommended codes of practice is food manufacturers and food retailers, and the codes aim to ensure a uniform application of hygienic practices.

Guidelines deal with various issues. Most of the guidelines address the issue of control and inspections, and methods of analysis and sampling that are directly related to established hygiene requirements, MRLs, and requirements on the composition of foods in order to qualify for a trade description.[44] These guidelines are directed to Codex Members in order to advance the application of uniform conformity assessment procedures. Guidelines may also concentrate on issues such as the exchange of information between Codex Members and nutrient labelling claims.

Maximum permitted levels of food additives and contaminants are a component of the commodity standards, the general standards on additives, or the general standard on contaminants. In contrast, maximum residue levels for pesticide residues (MRLPs) and veterinary drugs (MRLVDs) are published separately. A maximum residue level is a value that expresses the maximum concentration of the substance in food products (expressed as mg/kg or µg/kg) considered by the Codex Alimentarius Commission as legally permitted to ensure safe foods.[45] These individual MRLs for pesticides residues and veterinary drugs were also intended to have a mandatory status once they were accepted by Codex Members.[46]

2.4 The relationship between the various measures

The shift from a vertical approach to a horizontal approach, as discussed above in Section 2.1, and the adoption of other measures, as discussed in Section 2.3, automatically led to the fact that the Codex Alimentarius became more than a mere *collection* of specific food standards as the establishers of the Joint FAO/WHO Food Standards programme had envisaged. Over the years, the Codex Alimentarius

---

[43] See, for an example of a recommended Code of Practice, Annex IV of this book.

[44] See, for an example of a guideline, Annex V of this book.

[45] See also, Definitions, Procedural Manual, 16th edn., 2006, pp. 42-43. The estimation of MRLs for pesticides, as undertaken by the Joint Meeting for Pesticides Residues (JMPR) is based on the assumption that practice complies with Good Agricultural Practice in the Use of Pesticides (GAP). These GAPs are based on different national authorised safe uses of pesticides which are used to examine a range of levels of pesticides applications up to the highest authorised use among these national authorised safe uses in order to define the smallest amount of pesticides practicably and reasonably achievable. Both types of MRLs are expressed as a quality of millegram per kilogram and reflect the maximum level of pesticides respectively veterinary drugs authorised for use during the production process and is based on the assumption that the use of pesticides or veterinary drugs is done in accordance with good agricultural practice alternatively good veterinary practice.

[46] See Section 6 of this chapter.

has developed into a complex system of food requirements. Today, the Codex Alimentarius consists of around 216 standards, 47 recommended codes of practice, 40 guidelines, maximum residue levels (MRLs) of pesticide residues for over 360 commodities and MRLs for 47 different veterinary drugs, and 7 other measures.

One of the consequences of this is that Codex measures overlap in scope, which may lead to conflict and duplication if they are not regulated well. This overlap exists both in the relationship between Codex standards and other Codex measures, as well as in the relationship between general standards and commodity standards. For instance, provisions on hygienic practice are laid down in commodity standards as well as in recommended codes of practice. Likewise, provisions on analysis and sampling methods are contained both in the commodity standards as well as in guidelines. During the 5th Session of the Codex Alimentarius Commission, it was indicated that provisions on hygienic practices (particularly those dealing with end-product specifications) that were intended to have a mandatory status could be included in the Codex standards.[47] Furthermore, it was emphasised that the Codex Alimentarius should contain an adequate cross-reference system.[48] Although not always expressed in links such as references, several practices of the different Codex Committees reflect the fact that these elements are considered as part of one and the same system. For instance, the role of the methods of analysis and sampling in relation to the MRLs as an enforcement instrument is emphasised by the fact that no MRLVDs go beyond Step 7 of the procedure if no appropriate methods of analysis and sampling can be defined.[49] Furthermore, the former acceptance procedure indicated that the non-acceptance of the section on the methods of analysis and sampling of a commodity standard means that the Codex standard itself is to be considered as having been accepted with specified deviations.[50]

The relationship between general standards and commodity standards is also regulated to avoid overlap and conflict.[51] Due to the shift from the 'vertical' to the 'horizontal' approach, preference is given to the adoption of general provisions. The co-existence of specific provisions additional or complementary to general provisions must be justifiable. For instance, with regard to the General Standard for the Labelling of Pre-packaged Foods, the Procedural Manual indicates that exemptions from, or additions to, the General Standard for the Labelling of Pre-packaged Foods, which are necessary for its interpretation with regard to the product concerned, should be fully justified, and should be restricted as much as possible. The

---

[47] Report of the 5th Session of the Codex Alimentarius Commission, Rome, 20 February-1 March 1968, ALINORM 68/35, para. 47. See provision on hygiene of the 'Format for Commodity Standards', Procedural Manual, 16th edn., 2006, p. 93.

[48] Report of the 5th Session of the Codex Alimentarius Commission, Rome, 20 February-1 March 1968, ALINORM 68/35, para. 49.

[49] Report of the 12th Session of the Codex Committee on Residues of Veterinary Drugs in Foods, Washington DC., 28-31 March 2000, ALINORM 01/31, para. 9.

[50] See 'Guidelines for the Acceptance Procedure for Codex Standards', para. 21, Procedural Manual, 14th edn., 2004, pp. 41-42.

[51] See, on early examination of the relationship between general standards and commodity standards, J.P. Dobbert, *supra* n. 10, p. 689, and D.L. Leive, *supra* n. 16, pp. 407-409.

'Format' and the document entitled 'Relations between Commodity Committees and General Committees' contain guidelines that specify the preference of working with reference to the general standards in the commodity standards.[52] For instance, the provision on the section on labelling contained in the 'Format' indicates that:

'... Provisions [on labelling in the commodity standard] should be included by reference to the General Standard for the Labelling of Prepackaged Foods ...'.[53]

Likewise, the general section of 'Relations between Commodity Committees and General Committees' states that:

'... Provisions of Codex General Standards, Codes or Guidelines shall only be incorporated into Codex Commodity Standards by reference unless there is a need for doing otherwise ...'.[54]

However, additional or even deviating provisions may be incorporated in commodity standards, when the circumstances of a particular food justify such inclusion.[55] In order to avoid conflict or duplication in these cases, the measures contained in the Codex Alimentarius consist of numerous provisions to avoid two provisions applying at the same time. For instance, Provision 4 of the 'General Standard for the Labelling of and claims for Pre-packaged Foods for Special Dietary Uses' states:

'The label of all pre-packaged Foods for Special Dietary Uses shall bear the information required by Sections 4.1 to 4.8 of this standard as applicable to the food being labelled, except to the extent otherwise expressly provided in a specific Codex standard.'[56]

In the event that neither the general standard nor the commodity standard covers such a provision, one may expect the relationship between general and commodity standards to be governed by the principle *lex specialis derogate legi generali* (the more specific provision has priority over the more general provision). It is more appropriate to apply this principle to the system of the Codex Alimentarius than to apply, for instance, the *lex posteriori* principle, according to which, the later adopted

---

[52] Procedural Manual, 16th edn., 2006.
[53] Procedural Manual, 16th edn., 2006, p. 94. Addition inserted by the author.
[54] Procedural Manual, 16th edn., 2006, p. 95.
[55] See, for instance, the general section of the 'Relations between Commodity Committees and General Committees', the section on Labelling of the 'Format for Commodity Standards', Section 5.4 of the 'General Standard for Labelling of and claims for prepackaged foods for special dietary uses', Codex Stan 146-1985.
[56] Codex Stan 146-1985. Another example is the General standard for bottled/packaged drinking water that excludes mineral waters as defined by the Codex standard on mineral waters from its scope. Likewise, the General standard on cheese states as regards its scope: 'Subject to the provisions of this Standard, standards for individual varieties of cheese, or groups of varieties of cheese, may contain provisions that are more specific than those in this Standard, and in these cases, those specific provisions shall apply.'

rule has priority over an earlier one.⁵⁷ The reason behind this is that the Codex Alimentarius is a system of norms that is under continuous development. The revision of provisions can be undertaken relatively easily, and, in fact, Codex measures are frequently revised. For this purpose, there is no reason to assume that the earlier adopted Codex measures would respond to a lesser extent to the intentions of the Codex Members than the more recently adopted Codex measures. Given this fact, the principle of *lex specialis* is simply more in line with the structure of the Codex Alimentarius and the governing rules contained in the Procedural Manual (as discussed above), which clearly establish the importance of the relation between general and more specific provisions.

3. THE SCOPE AND SPECIFICITY OF THE MEASURES CONTAINED IN THE CODEX ALIMENTARIUS

The scope and the specificity of the measures indicate the extent to which the harmonisation process of the Codex Alimentarius Commission is exhaustive. However, it has to be distinguished from the questions of to what extent the provisions are legally-binding upon the Codex Members, and whether the measures allow Codex Members to deviate.⁵⁸ These questions will be discussed in Sections 5 and 6 of this chapter.

This section addresses the following questions: What issues are covered by the scope of the Codex Alimentarius? To what extent is the content of the Codex Alimentarius exhaustive? For which issues do the Codex Members consider it appropriate to establish internationally-adopted provisions? How are varieties dealt with?

The General Principles give an indication on the scope of the areas and food products that are covered by the Codex Alimentarius. Principle 2 states that the Codex Alimentarius includes provisions on 'food hygiene, food additives, pesticide residues, contaminants, labelling and presentation, methods of analysis and sampling'.⁵⁹ It also specifies that the provisions apply to all foods, whether processed, semi-processed or raw, for the distribution to the consumer, as well as to the raw materials and other materials used in the production process.⁶⁰ Principle 1 of the

---

⁵⁷ The rationale between these two principles is different. The principle *lex specialis* aims to ensure that the most specific rule is applied in order to give effect to the intentions of the parties and to take into account the particularities of the case (A. Lindroos, 'Addressing Norm Conflicts in a Fragmented Legal System: The Doctrine of *Lex Specialis*', 74 *Nordic Journal of International Law* (2005), p. 36). The principle *lex posteriori* on the other hand, implies that it is the most recent rule that reflects the intention of the parties.

⁵⁸ On aspects of harmonisation at EC level, and, in particular, on the distinction between exhaustion versus harmonisation methods, see P.J. Slot, 'Harmonisation', 21 *European Law Review* (1996), pp. 388-389.

⁵⁹ Principle 2 of the General Principles of the Codex Alimentarius, Procedural Manual, 16th edn., 2006, p. 30.

⁶⁰ Ibid. See also, the definition of food in: 'Definitions for the Purposes of the Codex Alimentarius': 'any substance, whether processed, semi-processed or raw, which is intended for human consumption,

General Principles states that the aims of the provisions laid down in the food standards and other measures contained in the Codex Alimentarius are twofold:

A. to ensure the protection of consumer health; and
B. to ensure fair practices in food trade.

3.1 Ensuring the protection of health

A large part of the provisions contained in the standards and related documents of the Codex Alimentarius aim to protect consumer health from the risks which result from the consumption of food. During its Extraordinary Meeting in February 2003, the Codex Alimentarius Commission emphasised that its main priority was to develop standards related to the protection of consumer health and safety.[61] Food safety provisions are contained in commodity standards, general standards, MRLs, recommended codes of practice and guidelines. For instance, as mentioned above in Section 2.1, commodity standards are set up in accordance with the Format. They include (if relevant to the food products in question) a provision on the names and levels of food additives permitted, a provision on the level of contaminants, and, if considered necessary, any specific hygienic provision or provisions on the methods of analysis and sampling.[62] Furthermore, some general standards can be characterised as food safety standards as they mainly consist of food safety provisions. For instance, the 'General Standard for Food Additives' contains provisions on the permitted use of additives in foods and lists the permitted maximum level per food category.[63]

Food safety provisions are also contained in the recommended codes of practice that consist of hygiene requirements in order to reduce micro-biological risks. The aim of the development of food hygiene provisions is to deal with food control in a manner which is broader than the testing of end-products.[64] For example, Provision 1 of the 'Code of Hygienic Practice for Refrigerated Packaged Food with Extended Shelf Life' states that:

> 'the purpose of this code is to set out recommendations for processing, packaging, storage and distribution of refrigerated packaged foods with extended shelf-life. Its aim is preventing the outgrowth of pathogenic microorganisms and it is based on the principles of Hazard Analysis and Critical Control Point (HACCP).[65]

---

and includes drinks, chewing gum and any substance which has been used in the manufacture, preparation or treatment of food.' It does, however, exclude cosmetics or tobacco or substances used only as drugs.

[61] Report of the 25th Session of the Codex Alimentarius Commission (Extraordinary Meeting), Geneva, 13-15 July 2003, ALINORM 03/25/5, para. 24. See also, statement by the Director-General of the WHO, Report by the Director-General to the Executive Board, WHO 15 December 2002, EB111/29, para. 8.

[62] Format for commodity standards, Procedural Manual, 16th edn., 2006, pp. 91-94.

[63] Codex Stan 192-1995 (Rev. 6-2005).

[64] S. Hathaway, 'Management of food safety in international trade', 10 *Food Control* (1999), p. 248. G.E. Spencer, et al., 'Effects of Codex and GATT', 9 *Food Control* (1998), p. 179.

[65] CAC/RCP 46-(1999).

Furthermore, some Codex guidelines contain recommended methods of analysis and sampling in order to assist Codex Members in verifying compliance with MRLs or MLs. For instance, the 'Codex Guidelines for the Establishment of a Regulatory Programme for Control of Veterinary Drug Residues in Foods' defines the types and methods of sampling and the selection of samples in order to control effectively the amount of residues present and define whether the food products comply with the Codex maximum levels for veterinary drug residues (MRLVDs).[66] For example, it defines the minimum quantity per food product required for a laboratory sample and how to collect the sample of the food product. It also defines laboratory practices related to the residue control methods.

The adoption of these food safety provisions results from a standard-setting procedure in which a wide range of issues has been discussed and decided upon. These issues consist of considerations of both a scientific, as well as a political, nature: hence, the trans-scientific nature of the food safety provisions.[67] It is important to emphasise that the decision-making powers transferred to the international level under the auspices of the joint FAO/WHO bodies is composed of a risk assessment procedure as well as a risk management procedure. The standard-setting procedure is not just limited to defining risk assessment methods to assist Codex Members to determine their own level of protection. In point of fact, the Codex Members jointly determine the level of protection.

An example that illustrates the numerous issues of both a scientific and political nature covered by food safety provisions is the establishment of MRLs or MLs. This reflects the large scope of the issues that the harmonisation process of the Codex Alimentarius Commission aims for. Firstly, it has to be noted that the content is very specific. As mentioned above in Section 2.3, MRLs or MLs are expressed as a quality of mg per kg that reflects the maximum level of the substance allowed in foods.[68] Although expressed in a simple value, MRLs and MLs cover a wide range of considerations and implicit decisions. The calculation of MRLs and MLs is based upon Acceptable Daily Intakes (ADIs), i.e., the level of concentration of the substance in foods that is estimated not to present heath hazards on the basis of a calculated daily intake during a life-time. In order to be as representative as possible for the world's population, ADIs are based on 5 regional diets that reflect the different dietary habits in several regions in the world.

Moreover, both the establishment and the regulatory use of MRLs and MLs are closely linked with provisions that regulate the use of the additives, veterinary drugs or pesticides in question, or, in the case of contaminants, to the possibilities of the reduction of their presence.[69] For example, the maximum level of permitted con-

---

[66] CAC/GL 16-1993.

[67] L. Salter, *supra* n. 20, pp. 94-95.

[68] For instance, the permitted level of nisin in ripened cheese is 12.5 mg per kg cheese. See Provision 4 of the 'Codex General Standard for cheese', Codex Stan A-6-1978, Rev. 1-1999, amended 2003, p. 2.

[69] See, for an explanation on the relation between MRLs, ADIs and good practices, S. Hathaway, *supra* n. 64, p. 251.

taminants and toxins in foods, internationally determined, is based upon the so-called ALARA principle (As Low As Reasonably Achievable).[70] This means that the maximum level depends on the reasonability and economical feasibility of avoiding contaminants and toxins, the conditions of international trade, and the alternative solutions available to produce foods. The General Standard for Contaminants and Toxins in Foods lays down criteria to be taken into account when making decisions relating to the maximum acceptable levels of contaminants and toxins in foods.[71] Besides toxicological information, these criteria contain analytical data and intake data, fair trade considerations, such as existing or potential problems in international trade, and information about national regulations, in particular, on the data and considerations upon which these regulations are based.[72] The technological possibilities of preventing contaminants and toxins from entering the food chain and other alternative solutions are also considered.

Likewise, requirements on the use of food additives are connected with the desirability of their use. This is clearly expressed in Section 3.2 of the General Standard for Food Additives, which states that:

'the use of food additives is justified only when *such use has an advantage*, does not present a hazard to health of and does not mislead the consumer, and serves one or more of the technological functions set out by Codex and needs set out from (a) through (d) below, and only where these objectives cannot be achieved by other means which are economically and technologically practicable.'[73]

If such use is considered as 'desirable', the level of permitted residues is based upon the concept of Good Manufacturing Practice (GMP), which includes the condition that the quantity of the additive added to food should be limited to the lowest

---

[70] See also, E. Vos, C. Ni Ghiollarnath and F. Wendler, 'A review of institutional arrangements for European Union food safety regulation', pp. 53-54 (on file with author). The report discusses various elements that are subject to the harmonisation of approaches to risk assessment policy at EC level. The development of principles and instruments used for harmonisation purposes are often interlinked with those developed by the Codex Alimentarius Commission or the international expert bodies involved.

[71] Section 1.3.3 of Codex Stan 193-1995 (Rev. 1-1997).

[72] Ibid.

[73] Codex Stan 192-1995 (Rev. 3-2001). [Emphasis added] These technological functions and needs as established by Codex are:

'(a) to preserve the nutritional quality of the food; an intentional reduction in the nutritional quality of a food would be justified in the circumstances dealt with in sub-paragraph (b) and also in other circumstances where the food does not constitute a significant item in a normal diet;

(b) to provide necessary ingredients or constituents for foods manufactured for groups of consumers having special dietary needs;

(c) to enhance the keeping quality or stability of a food or to improve its organoleptic properties, provided that this does not change the nature, substance or quality of the food so as to deceive the consumer;

(d) to provide aids in the manufacture, processing, preparation, treatment, packing, transport or storage of food, provided that the additive is not used to disguise the effects of the use of faulty raw materials or of undesirable (including unhygienic) practices of techniques during the course of any of these activities.'

possible level necessary to accomplish its desired effect.[74] Likewise, the estimation of MRLs for pesticides and MRLs for veterinary drugs, as undertaken by the Joint Meeting for Pesticides Residues (JMPR) is based upon the assumption that practice complies with Good Agricultural Practice in the Use of Pesticides (GAP), respectively with Good Veterinary Practices (GVP). In the case of the use of pesticides, these GAPs are based upon different national authorised safe uses of pesticides. The examination of a range of levels of pesticide applications up to the highest authorised use among these national authorised safe uses aims to define the smallest amount of pesticides practicably and reasonably achievable.

3.2 Ensuring fair trade practices in foods

Other Codex provisions concentrate on consumer protection against false or misleading information or other fraudulent trade practices. They are included in general standards, commodity standards and guidelines. Although Codex measures are not exhaustive, their provisions often contain very detailed trade descriptions on the naming or labelling of food products.[75] For instance, provisions related to the naming of food products aim to assist the consumer in making an informed choice and to avoid misleading him or her with low quality products that are marketed under traditional names. It contains an international decision on which products may bear the traditional designation. Naming provisions included in commodity standards lay down the definition of the commodity in question. They determine the scope within which the food products can be named accordingly. For instance, the standard on chocolate specifically indicates the percentage of cocoa solids the product has to contain in order to be named 'chocolate'.[76] Likewise, the standard on yoghurt and sweetened yoghurt indicates the percentage of milkfat content and solid non-milkfat content that a dairy product has to contain in order to be named 'yoghurt'.[77] The large amount of naming requirements with regard to different cheese varieties is another good example. The provisions precisely state the principal characteristics and the specific method of production according to which the cheese has to respond in order to be labelled as the cheese variety in question.[78] Naming provisions can also be found in general standards. For example, Section 4.1 of the Codex General Standard for the Labelling of Pre-packaged Foods regulates, amongst

---

[74] Codex Stan 192-1995 (Rev. 3-2001), para. 3.3. MRLs for pesticides and veterinary drugs do not include any opinion on the desirability of the use and its purpose of the concerning pesticide or veterinary drug in question.

[75] See, for an interesting comparison between the Codex Guidelines on nutrient labelling and the relevant US regulations, Ch. Lewis and A. Randell, 'Nutrition labelling of foods: comparisons between US regulations and Codex guidelines', 7 *Food Control* (1996), pp. 285-293. The authors illustrate in what manner the US regulations deviate from the Codex Guidelines and to what extent the Guidelines allow for this. However, it has to be remarked that the Codex measures in question are guidelines, which are intended to remain advisory; see Sections 5 and 6 of this chapter for further discussion.

[76] Codex Standard for Chocolate (Codex Stan 87-1981), p. 3. See Annex II of this book.

[77] Codex Stan A-11(a)-1975, para. 2.1.1.

[78] See, for instance, Codex Stan C-34-1973 on Brie or Codex Stan C-9-1967 on Emmentaler.

others, the use of names as established in a Codex commodity standard in more general terms.[79] It specifies that, in cases where additional words or phrases are necessary to avoid misleading the consumer with regard to the nature and physical condition of the food, these additional words or phrases need to be included in conjunction with, or in close proximity to, the name of the food.[80] Examples of additional words laid down in the standard relate to the treatment that the food product in question has undergone, such as 'dried', 'concentrated', 'reconstituted', 'smoked', etc.[81]

Other labelling provisions to ensure fair practice in the food trade relate, amongst others, to the list of ingredients or nutrients,[82] net contents and drained weight,[83] and the date marking and the storage instructions for the product.[84]

3.3 Addressing the diversity of national circumstances

The variability of the national circumstances within the societies of the Codex Members is a challenge to the adoption of Codex measures that aim to assist the harmonisation process.[85] This section discusses some of the main complications due to these diverse national circumstances and how the Codex Alimentarius Commission responds to these complexities.

3.3.1 Scientific 'uncertainty' due to variability

One complication that arises with regard to the elaboration of food safety standards is that the success and acceptability of the outcome depends on the availability of data. Lack of sufficient data that represent the circumstances world-wide undermines the scientific basis of these provisions.

---

[79] Codex Stan 1-1985 (rev. 1-1991). See, for discussion on the work of the Codex Committee on Food Labelling in the beginning of the 1990s, R. van der Heide, 'The Codex Alimentarius on Food Labelling', *European Food Law Review* (1991), pp. 291-300. R. McKay, 'The Codex Food Labelling Committee - maintaining International Standards relevant to changing consumer demands', *European Food Law Review* (1992), pp. 70-80.

[80] Section 4.1.2 of Codex Stan 1-1985 (rev. 1-1991).

[81] Ibid.

[82] For instance, Section 4.2 of the Codex General Standard for the Labelling of Pre-packaged Foods (Codex Stan 1-1985 (rev. 1-1991)), Section 3.2 of Codex Guidelines on Nutrition Labelling (CAC/GL 2-1985 (Rev. 1-1993)), Section 7.2 of the Codex Standard for Chocolate (Codex Stan 87-1981), Section 8.3 of the Codex General Standard for edible fungi and fungus products (Codex Stan 38-1981).

[83] For instance, Section 4.3 of the Codex General Standard for the Labelling of Pre-packaged Foods (Codex Stan 1-1985 (rev. 1-1991)), Section 7.4 of the Codex Standard for Chocolate (Codex Stan 87-1981), Sections 4.3 and 5.3 of the Codex General Standard for the labelling of food additives when sold as such (Codex Stan 107-1981).

[84] For instance, Section 4.7 of the Codex General Standard for the Labelling of Pre-packaged Foods (Codex Stan 1-1985 (rev. 1-1991)), Section 7.3 of the Codex General Standard for Cheese (Codex Stan A-6-1978, Rev. 1-1999, Amended 2003).

[85] S. Hathaway, *supra* n. 64, p. 250.

With regard to the MRLs or MLs of veterinary drugs, pesticides residues, additives and contaminants, a proposal for the initiation of the elaboration procedure is only submitted if there is a commitment that all the relevant data will be made available for evaluation.[86] Whether the substance, which has not been evaluated previously, will be included on the agenda depends on whether the Joint Secretariat of the relevant expert body has received a positive indication that there will be one or more submitters of data, or that there will be data available from another source, such as a government organisation or published literature.[87] This means that the success of the standard-setting procedure, which includes the evaluation of pesticides, veterinary drugs and additives, heavily depends on the willingness of industry – as the sponsors of most of the data – to make that data available to expert bodies.[88]

It is frequently the case that the collection of data does not reflect a fair representation of the different conditions of the Codex Members and may thus lead to a failure to adopt a relevant and adequate Codex standard for all Codex Members.[89] Once an MRL appears on the agenda of the expert bodies, a call for data is launched to all Codex Members. However, developing countries often have problems in supplying data to the Joint Secretariat of the relevant expert body or bodies.[90] Although, in some cases, information on the allowance procedures can be found, there is hardly any information on GAPs and on conducted field trials, often due to the

---

[86] G.G. Moy and J.R. Wessel, 'Codex Standards for Pesticides Residues', in N. Rees and D. Watson (eds.), *International Standards for Food Safety* (Gaithersburg, Maryland, Aspen Publishers 2000), p. 215. 'to enter the Codex step procedure, a pesticide must first be placed on the CCPR priority list for MRL development. The criteria for the selection process are: 1. the pesticide is available for use as a commercial product, 2. its use gives rise to residues on a food commodity moving in international trade, 3. the presence of the residues may be matter of public health, 4 it has the potential to create problems in international trade. In addition to these criteria, there must be a commitment that all relevant data will be made available for evaluation.'

[87] See also, L. Salter, *supra* n. 20, p. 76.

[88] G.G. Moy and J.R. Wessel, *supra* n. 86, p. 215.

[89] See, for instance, the comment made by India during the 16th Session of the Codex Committee on General Principles (Report of the 16th Session of the Codex Committee on General Principles, Paris, 23-27 April 2001, ALINORM 01/33A, para. 76): 'The Delegation of India presented the document and stated that it had been prepared in the light of a number of decisions taken in the recent past by Codex Committees which demonstrated that scientific information from only a limited number of countries had been used as a basis for the decision-making and that draft standards had been advanced in some cases before completion of the risk assessment. The Delegation particularly addressed the discussions concerning Aflatoxins M1 in Milk and Lead in various foods as examples where global data, in particular data from developing countries, had not been taken into account. The paper contained proposals for specific guidelines that might be used to ensure that Codex standards were indeed based on global data.' See also, N. Rees, 'Exposure Assessment Supporting International Developments', in N. Rees and D. Watson (eds.), *International Standards for Food Safety* (Gaithersburg, Maryland, Aspen Publishers 2000), p. 112.

[90] Report of the 16th Session of the Codex Committee on General Principles, Paris, 23-27 April 2001, ALINORM 01/33A, para. 81. See also, R. Gonzalez, 'Enhancing the role of developing countries in developing scientific advice', Discussion Paper 3d, FAO/WHO Electronic Forum on the provision of scientific advice to Codex Alimentarius and member countries, 1 October-14 November 2003, available at: <www.fao.org/es/esn/proscad/index_en.stm>.

practices of small and medium-sized farms and due to the legal and administrative infrastructure of developing countries.[91]

A related problem for developing countries is the fact that Codex MRLs for pesticides are based upon information taken from the application practices of the developed countries, especially the US and the member states of the EU.[92] However, if, for instance, pesticides are withdrawn from the market in the states of these developed Codex Members, which means that no more data will be produced, the Codex MRLs are annulled. Notwithstanding this, these pesticides often still continue to be used in developing countries.[93]

Another complication resulting from the variability of national circumstances is that the collection of data from different countries may come from different sources which reflect diverse conditions or stages of the 'food chain'. To illustrate this, data-packages related to pesticides residues are, in principle, based upon the existing regulations on pesticide uses. In the past, the detection practices of the EU and the US (the most important donors of data) were different, as the US approach to detect the maximum residue levels concentrated on a relatively early stage of the food chain, the so-called 'farm-gate' MRLs, whereas the EU approach concentrated on the residue levels detectable at a later stage of the food chain, the so-called 'dinner-plate'.[94] In order to match the comparability of the data, the Codex has opted to use the US approach.[95]

In view of these shortcomings with regard to the collection of data, the Codex Alimentarius Commission has dedicated increasing importance to the role of Codex Codes of Practice in the area of contaminants. It is hoped that the implementation of the Codex Codes of Practice by Codex Members will result in more quantitative and 'uniform' data. One example of this can be found in the decision taken by the Codex Alimentarius Commission in 2003, to adopt temporarily the MRL for patulin in apple juice at 50 mg/litre while awaiting the results from the adopted code of practice on the prevention of contamination of apple juice with patulin.[96] There was discussion between Codex Members as to whether to permit a higher level than 50 mg/litre of patulin in apple juice or to reduce the level to 25 mg/litre.

---

[91] Interview with Dr. Wim van Eck. See also, L. Salter, *supra* n. 20, p. 83: '...Supervised field tests are not conducted systematically in every region or country. Field tests are conducted only if the regulatory authority in a specific country requires them. Since many countries do not require registration of pesticides, or local field tests for pesticide registration, such tests are often not done. Moreover, field tests are costly. Countries are often reluctant to undertake them, and manufacturers challenge any requirement for extensive field testing in several different countries.'

[92] Interview with Dr. Wim van Eck.

[93] Ibid.

[94] J.P. Frawley, 'Codex Alimentarius – Food Safety – Pesticides', 42 *Food, Drug and Cosmetic Law Journal* (1987) pp. 168-173. Interview with Dr. W. van Eck. See also, D.G. Victor, *Effective Multilateral Regulation of Industrial Activity: Institutions for Policing and Adjusting Binding and Nonbinding Legal Commitments*, Ph.D. Thesis (Massachusetts, Harvard University, Institute of Technology 1997), pp. 195-196.

[95] Interview with Dr Wim van Eck. See also, D.G. Victor, *supra* n. 94, p. 196.

[96] Report of the 26th Session of the Codex Alimentarius Commission, Rome, 30 June-7 July, 2003, ALINORM 03/41, paras. 43-44.

The reason for the temporary adoption of the MLR while awaiting the results of the implementation of the Code of Practice by Codex Members was to determine whether the lower level of 25 mg/litre, as requested by many countries, was actually reasonably achievable.[97]

In 2001, the Codex Alimentarius Commission addressed this concern in a more structural way and decided that it would not adopt a standard in cases of scientific uncertainty resulting from both a lack of scientific data as well as a lack of scientific evidence. It adopted a statement hereto:

> 'when there is evidence that a risk to human health exists but scientific data are insufficient or incomplete, the Commission should not proceed to elaborate a standard but should consider elaborating a related text, such as a code of practice, provided that such a text would be supported by the available scientific evidence.'[98]

In other words, in situations where the collection of data has resulted in insufficient information, the Codex Alimentarius Commission refrains from the adoption of a standard and leaves the competencies to its Members. There has been some discussion as to whether, and, indeed, to what extent, this statement can be referred to as the precautionary principle.[99] However, it has to be noted that the scope of this statement is still unclear. It is partly a codification of the practice in the area of contaminants, as mentioned above, to develop a code of practice in order to collect more data. Moreover, the purpose of collecting more data is mainly for the analysis of the possible achievement of a certain level of a specific contaminant in the food, rather than on the actual risks for human health. Nevertheless, the statement does, to a certain extent, reflect a preventive approach. During the absence of a Codex standard, due to insufficient data or lack of consensus, the adoption and the implementation of codes of practice will contribute to the reduction of risks.[100] Future application of this statement will illustrate whether its scope is more extensive than the above-mentioned practice in the area of contaminants.

---

[97] Ibid., para. 43: 'The Commission noted that the Committee on Food Additives and Contaminants had discussed the development of the proposed maximum level of 50 mg/kg of patulin with a view to establishing a lower level of 25 mg/kg in the future based on the application of the Code of Practice which was aimed at achieving lower patulin levels. The Commission supported the decision of the Committee to continue to collect data on the levels of patulin in apple juice and apple juice ingredients for other beverages with the aim of reconsidering a possible reduction of the maximum level once the code of practice had been implemented (after four years).'

[98] Report of the 24th Session of the Codex Alimentarius Commission, Geneva, 2-7 July 2001, ALINORM 01/41, para. 83. See, on the controversial way of the adoption of this statement, Section 3.1.2 of Chapter V.

[99] S. Poli, 'The European Community and the Adoption of International Food Standards within the Codex Alimentarius Commission', 10 *European Law Journal* (2004), p. 622.

[100] Report of the 15th Session of the Codex Committee on General Principles, Paris 10-14 April 2000, ALINORM 01/33, para. 52: '...The Delegation of Sweden indicated that precaution was reflected in the development of Codex codes of practices when the risk assessment of certain contaminants was not completed, but it was necessary to address public health problems through preventive action.'

3.3.2 *Diversity of national circumstances and 'non-scientific' factors*

Diverse national perceptions of how to deal, for example, with food production and distribution practices, and consumer protection complicate the adoption of Codex measures related to food safety as well as the adoption of fair trade practices.[101] One example related to provisions on fair practices in the food trade can be found in the discussion which occurred prior to the adoption of the Standard on Mineral Waters. The draft standard submitted for adoption was based upon a European Codex Standard on Mineral Waters and constituted a conversion of the regional standard into a world-wide standard. It contained the proposed definition of mineral waters, delimiting the scope of the standard, as:

> '… it is obtained directly from natural or drilled sources from underground water…it is packaged close to the point of emergence of the source with particular hygienic precautions…'.[102]

As, under this definition, bulk transport of water was not permitted before bottling, it was felt that this criterion was uniquely based upon the European traditions and did not take into account the difficulties of establishing a bottling plant close to the source in other regions of the world.[103] Another example is the discussions on the basis for the nutrient values to be declared as part of label information.[104] The discussion concentrated on whether nutrient values should be expressed per portion of 100g or per serving size, such as cups or tablespoons, a basis used in both the United States and the United Kingdom.[105]

Furthermore, with regard to food safety provisions, non-scientific considerations often vary according to the circumstances of the Codex Members and the consumers' perception of an acceptable level of risk. As explained above in Section 3.1, these non-scientific considerations form an essential part of the adoption of food safety provisions. For instance, the calculation of MRLs for pesticide residues is based on the definition of good agricultural practices (GAPs). However, the perception of what is good agricultural practice may vary from country to country, depending, for instance, on the weather and the particular pests that have to be dealt with.[106] Different perceptions of risk may complicate agreement on which food

---

[101] As regards the use of additives, see D.L. Leive, *supra* n. 16, pp. 498-499.

[102] Codex Stan 108-1981, Rev. 1-1997, Art. 2.1(b) and (e) [emphasis added].

[103] Report of the 5th Session of the Codex Committee on Natural Mineral Waters, Thun, Switzerland, 3-5 October 1996, ALINORM 97/20, paras. 18-20. In other regions sources are often situated in natural parcs which did not allow the construction of plants directly next to the source. Presentation during the 2003 Centre of Studies at the Academy for International Law by mr. van Hoogstraten, former Vice-Chairman of the Codex Alimentarius Commission. See, for other examples on discussions on the scope of commodity standards, such as the standard on fruit cocktail and the standard on corned beef, D.L. Leive, *supra* n. 16, pp. 396-398.

[104] C. Lewis and A. Randell, *supra* n. 75, pp. 288-289.

[105] Ibid.

[106] F.C. Lu, 'The Joint FAO/WHO Food Standards Programme and the Codex Alimentarius', 24 *WHO Chronicle* (1970), p. 205.

safety approach to follow. For instance, when developing the food hygiene portions of several dairy product standards, including cheese, the issue arose as to whether the application of the Codex general principles for food hygiene combined with the HACCP system would be sufficient, or whether mandatory pasteurisation would be required.[107] In the discussion on the MRLs for hormones in beef, too, non-scientific considerations played a major role.[108] Concerns raised by the European Community, among others, which were not included in the scientific report of the Joint Expert Committee on Food Additives (JECFA), included concerns about the different aspects of the use of hormones, about the practical possibilities of enforcing and controlling MRLs, and the concerns of European consumers as regards the safety of meat treated with growth hormones.[109] Agreement was not established and the adoption of the MRLs took place by vote.[110] The controversial discussion on the non-scientific considerations was brought to the attention of the Codex Committee on General Principles in order for it to develop a more structural approach to deal with non-scientific considerations.[111]

As a result, the Codex Alimentarius Commission has adopted the 'Statements of Principles concerning the Role of Science in the Codex Decision-Making Process and the Extent to which other Factors are taken into Account' (hereinafter referred to as 'Statements of Principles') in order to clarify the role that non-scientific concerns are to play in the elaboration of Codex measures.[112] The 'Statements of Principles' clarify that 'legitimate concerns' taken into account by the Codex Alimentarius Commission should not affect the scientific basis of risk analysis, and should include 'other legitimate factors' that are relevant for health protection and fair trade practices.[113] However, the terminology 'other legitimate factors' is not defined in the 'Statements of Principles' and its scope remains unclear.[114] It can be concluded from the 2nd Consideration of the 'Statements of Principles' that 'other legitimate factors' include factors which are relevant to the protection of the health of consum-

---

[107] H.M. Wehr, *supra* n. 24, p. 533. Codex Members of the European Communities were against the adoption of mandatory pasteurization, while the United States was in favour. This clearly reflects the differences in risk perception as regards unpasteurized cheese or other milk products.

[108] See, for detailed discussion on the debate over hormones in beef, D. Jukes, 'The role of science in international food standards', 11 *Food Control* (2000), pp. 183-185. See also, on the adoption of the MRLs and the preceding discussions, D. McCrea, 'A View from Consumers', in N. Rees and D. Watson (eds.), *International Standards for Food Safety* (Gaithersburg, Maryland, Aspen Publishers 2000), pp. 155-162.

[109] D. Jukes, *supra* n. 108, p. 183.

[110] See Report of the 21st Session of the Joint FAO/WHO Codex Alimentarius Commission, Rome, 3-8 July 1995, ALINORM 95/37, para. 45

[111] D. Jukes, *supra* n. 108, p. 186.

[112] Procedural Manual, 16th edn., 2006, pp. 164-165. See, for discussion on the preparation and the adoption of the Statements, D. Jukes, *supra* n. 108, pp. 185-186.

[113] See second and third bullet of the Criteria for the Consideration of the Other Factors Referred to in the Second Statement of Principle, Procedural Manual, 16th edn., 2006, p. 164.

[114] D. Jukes, *supra* n. 108, p. 189, S. Poli, *supra* n. 99, p. 623. H.M. Wehr, *supra* n. 24, p. 532. F. Veggeland and S.O. Borgen, *Changing the Codex: The Role of International Institutions,* Working Paper 2002-16 (Norwegian Agricultural Economics Institute 2002), p. 18.

ers or to the ensuring of fair practices in the food trade.[115] The Codex Secretariat has prepared a document listing the factors that have been used in Codex measures. It represents a rather limited scope of factors, which include:

1. economical sustainability;
2. lack of appropriate methods of analysis;
3. technological achievability; and
4. safety factors, such as the age of the consumer, and dietary habits.[116]

The status of this document is, however, uncertain. During the discussions that took place prior to the adoption of the Codex measures, other factors were invoked, such as environmental concerns, consumer concerns, and animal welfare concerns.[117] However, the acceptability of these factors is even less certain. Nevertheless, the 'Statements of Principles' are clear about the fact that only the factors which can be accepted on a world-wide scale are to be taken into consideration.[118] The factors that are not relevant or are not acceptable world-wide, are for the Codex Members to regulate. For these cases, the Codex Alimentarius Commission has adopted a principle which states that:

> 'when the situation arises that members of Codex agree on the necessary level of protection of public health but hold differing views about other considerations, members may abstain from acceptance of the relevant standard without necessarily preventing the decision by Codex.'[119]

Consequently, contrary to the approach taken with regard to situations in which there is a lack of sufficient information, measures *are* adopted. For all 'other legitimate factors' which result from national circumstances that are not covered by Codex measures (as only world-wide accepted factors are to be included), explicit reference is made to the discretion of Codex Members to deviate from these measures through the Codex acceptance procedure.[120]

4. THE CODEX STANDARD-SETTING PROCEDURE AND THE ACCEPTANCE PROCEDURE

One of the powers delegated to the Codex Alimentarius Commission is the initiation and guidance of the preparation of draft standards, followed by the finalisation

---

[115] Consideration 2 of the Statements of Principle. See also, D. Jukes, *supra* n. 108, p. 189.
[116] D. Jukes, *supra* n. 108, pp. 189-190.
[117] Ibid., pp. 190-192. S. Poli, *supra* n. 99, p. 624.
[118] Criterion 5 of the Criteria for the Consideration of the Other Factors Referred to in the Second Statement of Principle, Procedural Manual, 16th edn., 2006, p. 165.
[119] Consideration 4 of the Statements of Principles [emphasis added].
[120] The Codex acceptance procedure is explained below in Section 4.3 of this chapter.

of standards and their publication.[121] As emphasised in Section 3.1.1.2 of Chapter I, adopted standards do not need the approval of the parent organisations of the Codex Alimentarius Commission. They are directly published. Hence, the Codex Alimentarius Commission can execute its normative powers independently from its parent organisations. Furthermore, this power is strengthened by the fact that the Codex Alimentarius Commission can establish and amend its rule-making procedure without the approval of its parent organisations.[122] This is the result of a proposal of the Codex Alimentarius Commission for an amendment to its Rules of Procedures, which enabled it to establish and revise its standard-setting procedures at its 2nd Session in 1964, which was approved by the Directors-General of the FAO and the WHO.[123] At the same session, it established its 10-step procedure, which comprised both the standard-setting procedure as well as the acceptance procedure. These procedures together constituted the Codex rule-making procedure.[124] Since its establishment, the Codex step procedure has been amended several times.[125] One of the major changes made to the 'Procedures' was the separation of the standard-setting procedure and the publication and acceptance procedure, as a result of the revision of the step-procedure at the 14th Session of the Codex Alimentarius Commission in 1981. The main reason was that no Codex standard had received

---

[121] Art. 1(c) and (d) of the Statutes, Procedural Manual, 16th edn., 2006, p. 3. See also, Section 3.1.1.2 of Chapter I.

[122] See also, Section 3.1.1.2 of Chapter I. The term 'rule-making' procedure is used to cover both the standard-setting procedure that results in the adoption of the standards and the procedure of submitting the standards for consent to Codex Members (the acceptance procedure).

[123] Report of the 2nd Session of the Codex Alimentarius Commission, ALINORM 64/30, para. 6 and Appendix A, para. 17: 'The Working Party considered that it was desirable that a new Rule should be added to the Rules of Procedure to make it clear that the Commission could, subject to the provisions of the Rules of Procedure themselves, lay down the steps to be taken by the Commission, its subsidiary bodies and other bodies assisting in its work in the elaboration and final adoption of both world-wide standards and standards for regions and groups of countries. The Working Party thought that this was necessary because of the inevitable length and complexity of any procedure for the elaboration of standards and the need to have it clearly understood by all members. A permissive or enabling clause... would give the Commission power to amend the procedure in the light of experience at any Session so that the amendments came in force immediately...' See also, J.P. Dobbert, *supra* n. 10, p. 681.

[124] Report of the 2nd Session of the Codex Alimentarius Commission, ALINORM 64/30, Appendix A. See J.P. Dobbert, *supra* n. 10, p. 681.

[125] The most important amendments of the procedure were in 1966, 1969 (to broaden the scope of the step-procedure to other documents prepared by the Codex Alimentarius Commission, such as codes of practice), 1981 (the acceptance procedure was detached from the step-procedure and the Criteria for the Establishment of Work Priorities were introduced to be taken into account during Step 1), 1993 (the alignment of the different Codex procedures through the adoption of a uniform elaboration procedure and the introduction of the responsibility of the Executive Committee to advance standards beyond Step 5), and 2004 (to include the responsibility of the Executive Committee to review the standard-setting progress critically), 2005 (the acceptance procedure was abolished). See, for discussions on older versions of the Codex step procedure, C.H. Alexandrowicz, *The Law-Making Functions of the Specialised Agencies of the United Nations* (Sydney, Angus and Robertson 1973), pp. 77-84. R. van Havere, 'Codex Alimentarius', in R. Kruithof, *Levensmiddelenrecht* (Brussel, Ced-Samson 1979), pp. 31-33. D.L. Leive, *supra* n. 16, pp. 435-459. J.P. Dobbert, *supra* n. 10, p. 680.

sufficient acceptance to be adopted under this procedure.[126] The 10 step-procedure became an 8-step procedure and the publication and acceptance procedures were regulated by separate provisions.[127] In 2005, the acceptance procedure was abolished.[128]

4.1 A uniform 8-step standard-setting procedure

The standard-setting procedure, also referred to as the Codex Elaboration Procedure, applies to the elaboration and adoption of all types of documents, including standards, guidelines and codes of practice, as well as the adoption and modification of documents for internal working procedures, with the exception of the amendment of the Rules of Procedure. In the past, different elaboration procedures existed, according to the type of standards, such as regional standards, MRLs for pesticides and veterinary drugs, world-wide standards, Codex Advisory Specifications for the Identity and Purity of Food Additives, Milk and Milk products Standards, and International Individual Cheese Standards.[129] These procedures were aligned in 1993, and are now contained in one uniform procedure.[130] As briefly indicated in Section 3.1.1.2 of Chapter I, the establishment of the Codex standard-setting procedure has been a means of strengthening the position of the Codex Alimentarius Commission as the co-ordinator of international food standards activities. Instead of taking the international standards developed by other international organisations into account by directly including them in the publication of the Codex Alimentarius, it requires them to be submitted as draft standards under the Codex standard-setting procedure, in order for them to become part of the Codex Alimentarius. The Codex Alimentarius Commission has defined the standard-setting procedure in such a way that these proposed standards enter the procedure at Step 3, requiring the adoption of the Commission as final decision-maker.[131]

The uniform standards-setting procedure of the Codex Alimentarius Commission consists of 8 steps and is illustrated in Figure 3.[132]

---

[126] Report of the 14th Session of the Codex Alimentarius Commission, Geneva 29 June-10 July, 1981, ALINORM 81/39, paras. 159-165.

[127] Part V (Subsequent procedure concerning publication and acceptance of Codex Standards) of the 'Procedures'.

[128] See, for further discussion, Section 4.5 of Chapter IV.

[129] See also, R. van Havere, *supra* n. 125, pp. 31-32. C.H. Alexandrowicz, *supra* n. 125, pp. 82-84. J.P. Dobbert, *supra* n. 10, pp. 699-706.

[130] The establishment of different procedures was mainly chosen to give special recognition to non-Codex bodies in the early stages of the elaboration procedure. See Report of the 10th Session of the Codex Committee on General Principles, Paris, 7-11 September 1992, ALINORM 93/33, paras. 38-39. The recognition of non-Codex bodies is still reflected in the step procedure, such as the International Dairy Federation. See, for comments, Section 2.4 of Chapter V.

[131] Provision 7 of the 'Guidelines on co-operation between the Codex Alimentarius Commission and international intergovernmental organisations in the elaboration of standards and related texts', Procedural Manual, 16th edn., 2006, p. 32.

[132] 'Procedures for the Elaboration of Codex Standards and Related Texts', Procedural Manual, 16th edn., 2006.

Step 1: The procedure starts with the decision to elaborate, revise or amend a standard, and the appointment of the subsidiary body to undertake the work. This decision can be taken by the Codex Alimentarius Commission itself, or by a subsidiary committee, which would, in this case, have to submit this decision for approval by the Commission at the earliest possible opportunity. One of the tasks that falls under Step 1 is the determination of work priorities.[133] Given that this is a long and time-consuming process, particularly when it concerns a food safety standard in need of scientific assessment, it is essential that a well-balanced decision is made with regard to whether the potential Codex standard will, indeed, contribute to the harmonisation process. Furthermore, there needs to be the prospect of completing the elaboration of the standard or related texts within a reasonable period of time. Thus, each proposal for the initiation of new work must be accompanied by a project document prepared by the subsidiary body or the Codex Member submitting the proposal.[134] The proposal to initiate new work is submitted to the critical review of the Executive Committee, which takes the 'Criteria for Establishing Working Priorities' into account before giving its recommendation.[135]

Step 2: Step 1 is followed by the preparation of a proposed draft standard, preferably after having heard either the Joint Meetings of the FAO Panel of Experts on Pesticide Residues in Food and the Environment, and the WHO Panel of Experts on Pesticides Residues (JMPR), or the Joint FAO/WHO Expert Committee on Food Additives (JECFA) or any other expert group or meeting, according to their relevance to the elaboration of the standard in question. In the case of milk, milk products or cheese standards, the relevant Committee recommendations of the International Dairy Federation (IDF) will be requested. Furthermore, during this step, the Codex Secretariat arranges the preparation of the proposed draft standards.[136] Although not formally laid down in the 'Procedures for the Elaboration of Codex Standards and Related Texts' (hereinafter referred to as the 'Procedures'), the preparation of proposed draft standards is often mandated to Codex Members, international governmental organisations or international non-governmental organisations.[137]

Step 3: Next, the proposed draft standard is sent by the Secretariat to the Codex Members and interested international organisations for their comments.

Step 4: During Step 4, the proposed draft standard is discussed in the respective subsidiary body, or in more than one subsidiary body if the substance of the stan-

---

[133] See, for the legal basis of this responsibility, Art. 1(c) of the Statutes, Procedural Manual, 16th edn., 2006, p. 3.

[134] Consideration 1, Part 2 of the Procedures, Procedural Manual, 16th edn., 2006, p. 21.

[135] Consideration 3, Part 2 of the Procedures, Procedural Manual, 16th edn., 2006, pp. 21-22. See, for further discussion, Section 2.2.3 of Chapter V.

[136] Step 2 of the Procedures, Procedural Manual, 16th edn., 2006, p. 23.

[137] See for further discussion on this issue, Section 2.4 of Chapter V.

dard so requires. All comments received under Step 3 are sent by the Chairperson of the subsidiary body in question to Codex Members as well as to observers and interested international organisations two months prior to the opening of the session.[138] The subsidiary bodies may amend the proposed draft standard. Although a voting procedure on amendments is not excluded, it is strongly recommended that the subsidiary body decides on the basis of consensus. In principal, a proposed draft standard may remain at Step 4 for several sessions of the subsidiary body. The responsible subsidiary body decides whether to advance the standard or related text to the next step, in order to submit the draft proposed standard to the Codex Alimentarius Commission for adoption as a *draft* standard.[139]

Step 5: Step 5 starts with the submission of a proposed draft standard by the responsible subsidiary body to the Codex Secretariat, which, in its turn, sends it to the Executive Committee for critical review and to Codex Alimentarius Commission to be adopted as a draft standard. The Codex Alimentarius Commission takes the recommendations of the Executive Committee that result from its critical review, as well as the comments made by the Members and the organisations, into account. It can then take several decisions. It can decide to send the standard or related text back to a previous step in the procedure, or to adopt the standard at Step 5 as a draft standard. Neither the Rules of Procedure, nor the 'Procedure' is clear about how these decisions must be taken. Rule XII of the Rules of Procedures states that 'the Commission shall make every effort to reach agreement on the adoption or amendment of *standards* by consensus'.[140] However, it remains silent on the adoption of 'draft standards'. One may assume that Rule XII applies *mutatis mutandis* to the adoption of draft standards at Step 5. No provision contradicts this assumption. Furthermore, several documents of the Procedural Manual emphasise the importance of consensus.[141]

Step 6: Once the draft standard has been adopted by the Codex Alimentarius Commission, it is then sent back by the Secretariat to the Members and interested international organisations for their comments.

---

[138] 'Guidelines for Codex Committees and Ad Hoc Intergovernmental Task Forces' (hereafter referred to as 'Guidelines for Committees'), Procedural Manual, 16th edn., 2006, p. 51. Although formally the 'Procedures' state that it is the Secretariat of the Codex Alimentarius Commission that is responsible for sending the comments received, in practice it is the secretariat of the hosting country that undertakes this task, as also suggested by the 'Guidelines for Committees'.

[139] See Consideration 3 of the introduction of the 'Procedures for the Elaboration of Codex Standards and Related Texts', Procedural Manual, 16th edn., 2006, p. 19: '… and is then considered … by the subsidiary body concerned which may present the text to the Commission as a "draft standard".' Emphasis added. Or Consideration 5 of the introduction of the 'Procedures': The Commission or the subsidiary body or other body concerned may decide that the draft be returned for further work at any appropriate previous Step in the Procedure.' (Procedural Manual, 16th edn., 2006, p. 20).

[140] 'Rules of Procedure', Procedural Manual, 16th edn., 2006, p. 16. Emphasis added.

[141] See Section 4.2 of this chapter.

Step 7: Step 7 concerns the second reading of the standard in preparation by the subsidiary body or subsidiary bodies. Comments are submitted via the secretariat of the host country, as under Step 4, to Codex Members, observer states and interested international organisations, two months prior to the opening of the session. The draft standard and the submitted comments are discussed in the subsidiary body that is to amend the draft standard in accordance with the outcome of the discussion. As under Step 4, amendments and decisions must preferably be made by consensus. Again, the subsidiary body may send the draft standard back to any previous step of the procedure.

Step 8: Step 8 starts with the submission of the draft standard for a second reading by the Commission, and its submission to the Executive Committee for critical review. Again, the Codex Alimentarius Commission may take several decisions. It may refer the draft standard back to any previous step of the procedure.[142] Alternatively, it may hold the draft standard at Step 8.[143] This means that the draft standard is not sent back to a subsidiary body, and that it will be submitted to the next session of the Codex Alimentarius Commission. Furthermore, it may adopt the draft standard as a Codex standard. When taking a decision, the Codex Alimentarius Commission takes recommendations of the Executive Committee into account, as well as any written proposals received by Members or interested international organisations. Previously, only amendments that were not substantive could be made to a draft standard that is adopted at Step 8.[144] This meant that all substantive amendments of draft standards that had been approved by the Codex Alimentarius Commission had to be referred to a previous step of the procedure.[145] As already briefly mentioned above, Rule XII.2 of the Rules of Procedure emphasises the importance of adopting standards on the basis of consensus. Only when consensus cannot be reached does the Codex Alimentarius Commission proceed to a vote.[146]

There are two possible deviations from this uniform 8-step procedure. First, at Step 1, the Codex Alimentarius Commission may decide to initiate an 'accelerated' procedure if there is urgency to adopt a standard.[147] The Commission can do so on the

---

[142] Consideration 5 of the Introduction or the 'Procedures for the Elaboration of Codex Standards and Related Texts', Procedural Manual, 16th edn., 2006, p. 20.

[143] Ibid.

[144] Consideration 5 of the 'Guide to the Consideration of Standards at Step 8 of the Procedure for the Elaboration of Codex Standards including Consideration of any Statements relation to Economic Impact', Procedural Manual, 14th edn., 2004, p. 27. The 'Guide to the Consideration of Standards at Step 8 of the Procedure for the Elaboration of Codex Standards including Consideration of any Statements relation to Economic Impact' has been deleted at the 2006 Session of the Codex Alimentarius Commission.

[145] Consideration 5 of the 'Guide to the Consideration of Standards at Step 8 of the Procedure for the Elaboration of Codex Standards including Consideration of any Statements relation to Economic Impact'.

[146] Rules of Procedure, Procedural Manual, 16th edn., 2006, p. 16.

[147] Step 1 of Part 4 of the 'Procedures' (Uniform Accelerated Procedure for the Elaboration of Codex Standards and Related Texts), Procedural Manual, 16th edn., 2006, p. 25.

HARMONISATION THROUGH STANDARD-SETTING

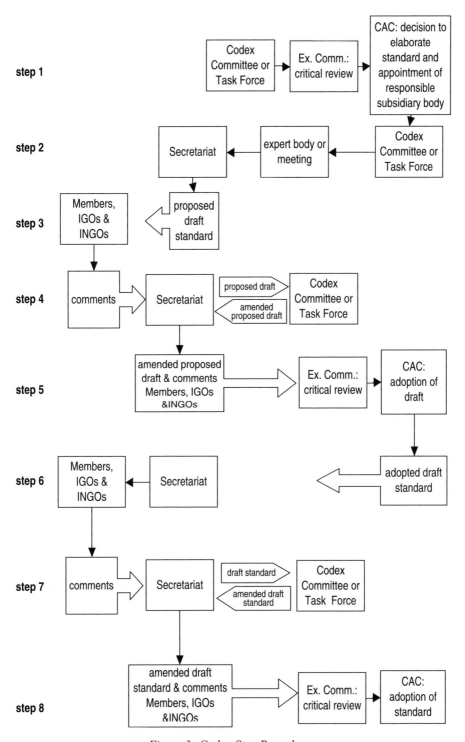

Figure 3. Codex Step Procedure

basis of a two-thirds majority-vote, either on its own initiative or by confirming the initiative of the responsible subsidiary body (the initiative of which has equally to be based on a qualified-majority of two-thirds of the votes cast).[148] The accelerated procedure consists of only one round and the Codex Alimentarius Commission adopts the standard in question at Step 5.

Second, the Commission may decide to omit Steps 6 and 7 and to proceed immediately with the adoption of the standard at Step 8.[149] The omission has to be recommended by the responsible Codex Committee and has to be decided upon by a two-thirds majority-vote.[150]

The accelerated procedure should not be confused with the procedure in which Steps 6 and 7 have been omitted. The difference is that the decision to initiate an accelerated procedure is taken at Step 1, whereas the decision to omit Steps 6 and 7 is made at Step 5. Furthermore, the underlying reasons differ. Whereas it is considerations, such as matters concerning new scientific information, urgent problems related to trade or public health, or the revision or up-dating of existing standards, which trigger the necessity to initiate the accelerated procedure, the decision to omit Steps 6 and 7 often results from the fact that a general consensus has already been achieved at Step 5 and renders further discussion unnecessary.

4.2 The emphasis on consensus-building during the standard-setting procedure

Several provisions indicate the importance of making decisions on the elaboration and adoption of standards and other related texts by consensus. Despite this emphasis, the Procedural Manual lacks a definition of what consensus actually means in the context of the standard-setting procedure. Literally, 'consensus' means 'common feeling' or 'concurrence of feelings'.[151] Often, consensus is defined in a negative way, for example, in terms of 'the absence of any objection expressed by a

---

[148] Ibid.

[149] Consideration 6 of the Introduction to 'Procedures for the Elaboration of Codex Standards and Related Texts', Procedural Manual, 16th edn., 2006, p. 20. The possibility of omitting the second round of the standards-setting procedure was introduced during the 8th session of the Executive Committee, for two reasons: 1). that there might be a general interest to speed up the procedure (mainly applied in the field of food additives or the selection of suitable methods of analysis) and 2). some commodity standards might prove entirely uncontroversial and could be sent out for adoption after only one round of consultations with governments. See Report 4th Session of the Codex Alimentarius Commission, Rome, 7-14 November 1966, ALINORM 66/30, para. 21.

[150] Consideration 6 of the Introduction to the 'Procedures for the Elaboration of Codex Standards and Related Texts', Procedural Manual, 16th edn., 2006, p. 20.

[151] H.G. Schermers and N.M. Blokker, *International Institutional Law*, 4th edn. (The Hague, Martinus Nijhoff Publishers 2003), p. 524. A. D'Amato, 'On Consensus', 8 *Canadian Yearbook of International Law* (1970), p. 107. K. Zemanek, 'Majority Rule and Consensus Technique in Law-Making Diplomacy', in R. St. J. MacDonald and D.M. Johnston (eds.), *The Structure and Process of International Law* (Dordrecht, Martinus Nijhoff Publishers 1983), pp. 871-872. There is, however, no uniform practice in international organisations of what constitutes a 'common feeling' amongst participants.

representative'.[152] An inherent part of decision-making procedures by consensus is the process of active negotiations to achieve concurrence of views.[153] These prior negotiations aim to eliminate controversial points.[154] Within the context of the Codex Alimentarius Commission, this element is also referred to as 'consensus-building'.

As the Codex Alimentarius Commission and its subsidiary bodies are composed of a large number of participants, it can be difficult to achieve consensus. For instance, it took the Codex Committee on Fish and Fish Products 5 years to list all the species that could be called sardines or sardine-type.[155] In the past, decision-making within the Codex Alimentarius Commission was based upon the understanding that all involved parties should refrain from obstructing the adoption of decisions.[156] To provide ample possibilities for the expression of views and in order to look actively for compromises prior to the adoption of decisions, the Codex Alimentarius Commission has opted to structure the standard-setting procedure.[157] Thus, the standard-setting procedure has two important features that stimulate consensus-building: the structure of the two readings of the procedure, and the role of the subsidiary bodies.

As mentioned in the previous section, the procedure consists of two readings: Steps 3-5 and Steps 6-8. During these two rounds, governments and interested observers have at least two occasions upon which to submit their comments, the responsible Codex Committee has two occasions upon which to discuss and amend the draft standard (Steps 4 and 7), and the Commission has two occasions upon which to intervene in the standard-setting procedure (at Step 5 and as final decision-making body at Step 8).[158] Although long and complicated, the structure of the two rounds allows all interested Members and observers to submit their comments and to participate in the standard-setting procedure.[159] On top of this, the

---

[152] H.G. Schermers and N.M. Blokker, *supra* n. 151, p. 524. See, for instance, definition of consensus in the Footnote 1 to Rule IX.1 of the WTO: 'The body concerned shall be deemed to have decided by consensus on a matter submitted for its consideration, if no member, present at the meeting when the decision is taken, formally objects to the proposed decision.'

[153] H.G. Schermers and N.M. Blokker, *supra* n. 151, p. 524. K. Zemanek, *supra* n. 151, p. 875.

[154] K. Zemanek, *supra* n. 151, p. 863. M. Footer, 'The Role of Consensus in GATT/WTO Decision-making', 17 *Northwestern Journal of International Law and Business* (1996-1997), p. 659.

[155] G.E. Spencer, et al., *supra* n. 64, p. 178.

[156] F. Veggeland and S.O. Borgen, 'Negotiating International Food Standards: The World Trade Organization's Impact on the Codex Alimentarius Commission', 18 *Governance: An International Journal of Policy, Administration, and Institutions* (2005), p. 683.

[157] M.E. Bredahl and K.W. Forsythe, 'Harmonizing Phyto-sanitary and Sanitary Regulations', 12 *The World Economy* (1989) pp. 194-195.

[158] See also, F.C. Lu, *supra* n. 106, p. 200.

[159] S. Shubber, 'The Codex Alimentarius Commission under International Law', 21 *International and Comparative Law Quarterly* (1972), p. 641 citing Kermode: As one commentator put it: 'While...[the Codex] procedure looks to be rather a lengthy one, it has been deliberately designed to give governments the fullest opportunity to comment on standards while they are still in draft, and to enable the Commission to satisfy itself that the standards are being prepared along the right lines.' He went on to say: 'Any procedure for the elaboration of international standards which fails to give governments adequate time to reflect on and to consider the standards from all aspects would in the long run be self-

Commission chose to adopt a flexible standard-setting procedure to this end. This is reflected in the absence of a provision that regulates the discontinuation of the procedure if no consensus is achieved at Step 8. In fact, the Codex Commission, the subsidiary committee, or another body can decide to refer standards back to a previous step of the procedure, or to hold it at Step 5 or at Step 8 for further work.[160] Formally speaking, however, the steps in these two rounds do not differ. Codex Members are not precluded from raising new issues during the second round. The lack of a provision which comprises an automatic discontinuation of the elaboration procedure implies that the elaboration procedure may last a long time. The reduction of lengthy procedures is sought in a preventive manner, namely, in the initiation phase of the procedure, as it requires the determination of whether the standard or related text can be finished in reasonable time.

Furthermore, subsidiary bodies play an important role in the standard-setting procedure. Consequently, Codex Committees have an important task in consensus-building. The chairpersons of the Codex Committees should always try to achieve a consensus among the Members before proposing to send the (proposed) draft standard to the Codex Alimentarius Commission level for approval.[161] The reason behind this is to facilitate the work of the Codex Alimentarius Commission and to prevent long discussions about the content of standards from having to take place at Commission level.

Chairpersons play an important role in the promotion of consensus. On the one hand, they should ensure that all issues have been fully discussed and that all written comments of Codex Members not present at the meeting have been taken into account.[162] On the other hand, given the time constraints and the full agenda of most of the Codex Committees, the chairperson also needs to pay attention to the progress of the work in hand. However, the Codex Committees have the means at their disposal to facilitate consensus-building, such as the establishment of (electronic) working groups. These working groups, both electronic working groups as well as physical working groups, allow Codex Members to exchange ideas and come to a provisional agreement that contributes to consensus-building at Committee level.[163] Working groups must be established on an *ad hoc* basis as the Codex Committee may not establish standing sub-committees without the approval of the Codex Alimentarius Commission.[164] Furthermore, neither electronic working groups

---

defeating. The lack of proper procedures has been the rock on which many previous attempts to secure international agreement on food standards have foundered.' See also, C.C. van der Tweel, *supra* n. 23, p. 36.

[160] Consideration 3 of the Introduction of 'Procedures for the Elaboration of Codex Standards and Related Texts', Procedural Manual, 16th edn., 2006, p. 19.

[161] 'Guidelines for Codex Committees and *ad hoc* Intergovernmental Task Forces', Procedural Manual, 16th edn., 2006, p. 58.

[162] 'Ibid., pp. 57-58.

[163] The CCFAC, for instance, established an Electronic Working Group on Quality control of food additives. Although the CCFAC itself was hosted by the Netherlands, the US organised and chaired the electronic working group. The Electronic working group lasted in this case about 4 months.

[164] 'Guidelines for Codex Committees and ad hoc Intergovernmental Task Forces', Procedural Manual, 16th edn., 2006, p. 50-51.

nor physical working groups are allowed to take any decisions with regard to proposals to be submitted to the Codex Alimentarius Commission or other related decisions which are mandated to the Codex Committee in question.[165]

4.3 The publication and acceptance procedure

During the first 18 years of the Codex Alimentarius Commission, the procedure used to consist of ten steps, of which the submission to governments for acceptance of the standard constituted Step 9, followed by Step 10, which was the publication of the standard in question. Since 1981, the acceptance and the publication procedure have no longer been part of the step-procedure and have been regulated by separate provisions.[166] The acceptance procedure was abolished during the 28th Session of the Codex Alimentarius Commission in 2005.[167] The reason for the abolition, as expressed by many Codex delegations, was the fact that the acceptance procedure had never really worked, and was no longer relevant within the framework of the WTO SPS (Agreement on the Application of Sanitary and Phytosanitary Measures) and TBT (Technical Barriers to Trade) Agreements.[168] In spite of the fact that the Codex acceptance procedure no longer exists, it is useful to discuss the provisions that regulated the procedure. The different methods of acceptance, and the resulting obligations for Codex Members, give us an insight into the way that the competences to advance harmonisation were divided between the Codex Alimentarius Commission and the Codex Members (See also, Sections 5 and 6 of this chapter).

Under the acceptance procedure, adopted standards were formally submitted for acceptance by Codex Members.[169] Non-Codex Members could also accept Codex standards, provided that they were members of the FAO or of the WHO.[170] Only standards and MRLs could be submitted for acceptance. Consequently, recommended codes of practice and guidelines were not subject to acceptance.

During the discussions that took place before the establishment of the acceptance procedure, it was recognised that a flexible acceptance procedure would contribute to the achievement of a 'wider measure of agreement at international level'.[171] Consequently, the Codex Alimentarius Commission adopted an acceptance proce-

---

[165] Report of the 13th Session of the Codex Committee on General Principles, Paris, 7-11 September 1998, ALINORM 99/33, para. 27.
[166] See Section 4 of this chapter.
[167] Report of the 28th Session of the Codex Alimentarius Commission, Rome, 4-9 July 2005, ALINORM 05/28/41, para. 36.
[168] Report of the 22nd Session of the Codex Committee on General Principles, Paris, 11-15 April 2005, ALINORM 05/28/33A, paras. 77 and 80.
[169] See Part V of the 'Procedures', Procedural Manual, 14th edn., 2004.
[170] J.P. Dobbert, *supra* n. 10, p. 692. C.H. Alexandrowicz, *supra* n. 125, p. 78, footnote 10.
[171] Report of the 5th Session of the Codex Alimentarius Commission, Rome, 20 February-1 March 1968, ALINORM 68/35, para. 14.

dure that consisted of three ways according to which Members could accept Codex standards.[172]

The first way was the method of 'full acceptance'. The 'General Principles' contained separate provisions for full acceptance of commodity standards, general standards and the individually published MRLs (for pesticides and veterinary drugs) respectively.[173] Full acceptance of commodity standards obliged the Codex Member to ensure that a domestic and imported product which was in conformity would be distributed freely and would not be hindered by administrative or legal provisions related to the food issues covered by the standard. In addition, acceptance obliged the Member to ensure that products which were not in conformity with the standard in question would not bear the name or description laid down in the standard. Similar obligations applied for a Codex Member that had accepted a general standard. They were obliged to ensure that food products complied with the relevant requirements contained in the standards and that any sound food product that conformed with the standard would not be hindered by any legal or administrative provisions. Full acceptance of maximum levels for residues of pesticides or veterinary drugs in food resulted in similar obligations. However, in contrast to full acceptance of commodity standards and general standards, this method of full acceptance allowed Codex Members to apply the obligations to imported food alone.[174]

A second method of acceptance was acceptance with specific deviations.[175] This way of acceptance allowed the Member to accept the Codex standards with specific derogations. However, this method did not apply to the acceptance of Codex Maximum Limits for Residues of Pesticides and Veterinary Drugs in Food. With regard to Commodity Standards, acceptance with specific deviations required Codex Members to ensure the free distribution of all products that complied with the standard, with the exception of the deviations provided. They were under the obligation to indicate the reasons for the deviations, whether products that fully applied with the standard could be distributed freely, and whether they expected to accept the standard fully at a future moment in time. With regard to General Standards, acceptance with specific deviations entailed the commitment to ensure that products conformed to the standard as qualified by the deviations. With regard to these standards, Codex

---

[172] Provisions 4-7 of the 'General Principles of the Codex Alimentarius', Procedural Manual, 14th edn., 2004, pp. 31-35. See also, D.G. Victor, *supra* n. 94, p. 192.

[173] Respectively, Provisions 4, 5 and 6 of the 'General Principles of the Codex Alimentarius', Procedural Manual, 14th edn., 2004, pp. 31-35.

[174] Provision 6.A of the 'General Principles of the Codex Alimentarius', Procedural Manual, 14th edn., 2004, pp. 34-35: 'A Codex maximum limit for residues of pesticides or veterinary drugs in food may be accepted by a country in accordance with its established legal and administrative procedures in respect of the distribution within its territorial jurisdiction of (a) home-produced and imported food or (b) imported food only, ...'

[175] This method had replaced the method of acceptance with minor deviations for the simple reason that it would be complicated for the Codex Alimentarius Commission to determine whether deviations were indeed 'minor'. The replacement by 'specific' deviations avoided the need for this discussion. On former methods of acceptance, see, for instance, C.H. Alexandrowicz, *supra* n. 125, pp. 78-80, J.P. Dobbert, *supra* n. 10, pp. 690-698, R. van Havere, *supra* n. 125, pp. 32-33.

Members were also under obligation to indicate their reasons and whether they expected to give the standard full acceptance in the future.

A third way of acceptance was by a declaration of free distribution, which meant that food products that complied with the Codex standard could freely circulate in the territory of the country.

Acceptance of General Standards and Commodity Standards in any of these three ways triggered the obligation to apply the relevant provisions of the Codex standard in a uniform and impartial way to both imported and domestic products.

A Codex Member was under no obligation to accept the standards in any of these ways. No statement of a delegate at any earlier step could lead to the conclusion that the Codex Member had accepted the standard in question.[176] In addition, a Codex Member that had accepted a standard could withdraw or amend its acceptance at any time.[177] They had to notify the Secretariat on the status and use of the relevant standard by the Codex Member within its regulatory system.[178] The Secretariat was to examine deviations from Codex measures as notified by Codex Members and report to the Codex Alimentarius Commission on the possible need for amendments or revisions of the relevant Codex standards. Details on the notifications received from Codex Members were to be part of the publication of the Codex Alimentarius.

As briefly indicated above, until 1981, the publication phase constituted the 10th Step of the elaboration procedure. This fell under the responsibility of the Codex Alimentarius Commission. The publication of standards was dependent upon the number of Members which had accepted the standard in question (which constituted the 9th Step of the procedure). When deciding on whether a sufficient number of Members had accepted the standard, the Commission took account of the capacities of these Members as producers or consumers of the commodity in question.[179] This gave the Commission a great deal of leeway to determine whether it was appropriate or not to publish a given standard.[180] However, since 1981, the publication of standards has taken place independently of whether or not Codex standards have been accepted.[181]

No separate decision of the Codex Alimentarius Commission is required for the publication of standards and other decisions.[182] The publication of Codex measures

---

[176] Report 4th Session of the Codex Alimentarius Commission, Rome, 7-14 November 1966, ALINORM 66/30, para. 14.

[177] Provision 7 of the 'General Principles of the Codex Alimentarius', Procedural Manual, 14th edn., 2004, p. 35.

[178] Part 5 of the 'Procedures for the Elaboration of Codex Standards and Related Texts', Procedural Manual, 14th edn., 2004, p. 25.

[179] F.C. Lu, *supra* n. 106, p. 202.

[180] S. Shubber, *op. cit supra* n. 159, p. 644.

[181] W.H Vermeulen, 'De Codex Alimentarius Commissie en de Europese Commissie ter bestrijding van mond- en klauwzeer, twee volwassen dochterinstellingen van de Wereldvoedselorganisatie', 10 *Sociaal Economische Wetgeving* (1983), p. 598.

[182] In July 2005, as a result of the abolition of the acceptance procedure, Art. 1(d) of the Statutes of the Codex Alimentarius Commission (recognised as the mandate of the Codex Alimentarius Commis-

is merely triggered by their final adoption under the standard-setting procedure. According to the 'Procedures for the Elaboration of Codex standards and Related Texts', adopted Codex standards and related texts are published and distributed to all members and associate members of the FAO and/or the WHO and to all interested international organisations.[183] As a consequence of the abolition of the acceptance procedure, the publications no longer contain the notifications received by Codex Members.[184]

5. THE CODEX ALIMENTARIUS: LEVELS OF HARMONISATION

As stated above in Section 4.3, the obligation to apply the provisions contained in the Codex standards did not result from the adoption of the standards itself. Consequently, the actual decision to proceed to harmonisation was not collectively taken by the Codex Members through the final adoption of measures. This decision was taken by individual Codex Members through their explicit consent, which was given in the context of the acceptance procedure and through implementation at national level. As a result, the harmonisation process was an 'indirect' harmonisation process.[185]

Hence, the aim of the Codex Alimentarius is to assist in the harmonisation of food requirements, as formulated by Provision 1 of the 'General Principles'.[186]

In 1968, it was recognised by the Codex Alimentarius Commission that full acceptance of a Codex standard reflecting a highly harmonised content was not possible for all Codex Members, precisely because of their different circumstances.[187] A solution which would allow for other, less constraining, acceptance methods would be sought. This solution would:

---

sion, see Section 3.1.1.2 of Chapter I) has been amended in the following way (Report of the 28th Session of the Codex Alimentarius Commission, Rome 4-9 July 2005, ALINORM 05/28/41, Appendix III): 'finalizing standards elaborated under (c) above, and (after acceptance by governments), publishing them in a Codex Alimentarius …'.

[183] Part 5 of the 'Procedures for the Elaboration of Codex Standards and Related Texts', Procedural Manual, 14th edn., 2004, p. 25 in conjunction with Report of the 28th Session of the Codex Alimentarius Commission, Rome 4-9 July 2005, ALINORM 05/28/41, Appendix IV (the amendments made to the Procedural Manual resulting from the abolition of the acceptance procedure).

[184] See Report of the 28th Session of the Codex Alimentarius Commission, Rome 4-9 July 2005, ALINORM 05/28/41, Appendix IV (the amendments made to the Procedural Manual resulting from the abolition of the acceptance procedure).

[185] R. van Havere, *supra* n. 125, p. 23: 'Teneinde dit doel op wereldvlak te bereiken, werd niet gekozen voor de directe harmonisatie van wetgeving, zoals bijvoorbeeld in de EEG het geval is, maar voor de indirecte standardisatie. Hierbij worden voor de internationale handel normen opgesteld, die éénmaal aanvaard door regeringen, op een indirecte wijze tot een zekere harmonisatie moeten leiden.'

[186] Procedural Manual, 16th edn., 2006, p. 30.

[187] Report of the 5th Session of the Codex Alimentarius Commission, Rome, 20 February-1 March 1968, ALINORM 68/35, para. 14.

'afford governments the means of accepting standards in ways applicable to their particular circumstances and which would still go a considerable way to the achievement of the objectives of the Codex Alimentarius.'[188]

In other words, the Codex Alimentarius Commission opted to lay down possibilities for deviation from Codex standards in the acceptance procedure, rather than include them in the Codex standards themselves. This decision still has consequences for harmonisation levels as reflected in the Codex Alimentarius, despite the fact that the acceptance procedure has been abolished. The level of harmonisation, as aimed for by the Codex Alimentarius Commission, cannot be fully appreciated, nor can the impact of the entry into force of the WTO agreements on its harmonisation process,[189] without examining the harmonisation levels as reflected by the acceptance procedure.

Hence, given the understanding that Codex standards have to be read in conjunction with the provisions on the acceptance procedure, this section examines the harmonisation levels in both the Codex measures and the acceptance procedure. It is based upon the situation prior to July 2005, when the acceptance procedure was still in force.

5.1 Codex standards and other Codex measures

Levels of harmonisation are reflected through the formulation used throughout Codex measures. The relevant questions, for instance, include: Are the provisions intended to be mandatory of nature? Do the provisions consist of different options from which the Codex Members can choose when implementing them at national level?

Examination of Codex measures illustrates the following: Codex standards often contain the term 'shall' in order to emphasise the intention to entail obligations. For instance, most of the Codex Commodity Standards include the requirement that products covered by its scope are in compliance with the MRLs for applicable contaminants, pesticides veterinary drugs or additives as established by the Codex Alimentarius Commission. Alternatively, with regard to a limited permitted use of additives, Provision 4 of the Codex Standard for Cheese, for example, states that 'only those food additives listed below may be used and only within the limits specified'.[190] In general standards, the obligatory nature of the provisions (once fully accepted by Codex Members) is also to be found in numerous places. For example, Provision 4 of the Codex General Standard for the Labelling of Pre-packaged Foods is entitled 'mandatory labelling of pre-packaged foods' and most of its sub-provisions contain the term 'shall'.[191] This commitment expressed in the language is, however, limited to Codex standards. In line with the acceptance proce-

---

[188] Report of the 5th Session of the Codex Alimentarius Commission, Rome, 20 February-1 March 1968, ALINORM 68/35, para. 14.
[189] To be discussed in Chapter IV.
[190] Codex General Standard for Cheese, Codex Stan A-6-1978, Rev. 1-1999, Amended in 2001.
[191] See Codex Stan 1-1985 (Rev. 1-1991) which is contained in Annex III of this book.

dure, recommended codes of practice and guidelines clearly remain advisory, as is reflected in the use of the terms 'should' or 'may' throughout the texts.[192]

Several Codex standards explicitly include the intention of the Codex Alimentarius Commission that the Codex provisions are to prevail over national food requirements. For instance, Sections 4.1.1.1 to 4.1.1.4 of the General Standard for the Labelling of Pre-packaged Foods indicate the importance of names, as established in the Codex commodities standards, and exempts the application of national regulations in cases where such a name has been established.[193] A similar provision is included in the General Standard for the Labelling of Food Additives when sold as such, and in the General Standard for the Labelling of and Claims for Pre-packaged Foods for Special Dietary Uses.[194]

Sometimes Codex standards contain a provision that leave discretion to Codex Members. An example is the above-mentioned provision of Section 4.1.1.1 of General Standard for the Labelling of Pre-packaged Foods. It implies that, in situations where several names are established by Codex commodity standards, Codex Members have the possibility of choosing from these several names. It states that:

'where a name or names have been established for a food in a Codex standard, at least one of these names shall be used.'[195]

The Codex Standard for Canned Sardines and Sardine-type Products contains a provision establishing several names from which the Codex Member may choose. Another example of an 'optional clause' is provision of Section 7.1.1 of the Codex standard for butter that states: 'Butter may be labelled to indicate whether it is

---

[192] As explained in Section 2.3, codes of practice are addressed to producers in order to assist them in ensuring the production and distribution of wholesome foods. Guidelines provide assistance to the governments of Codex Members for the establishment of enforcement mechanisms, such as methods of analysis and sampling.

[193] Codex Stan 1-1985 (rev. 1-1991):
'4.1.1.1 Where a name or names have been established for a food in a Codex standard, at least one of these names shall be used.
4.1.1.2 In other cases, the name prescribed by national legislation shall be used.
4.1.1.3 In the absence of any such name, either a common or usual name existing by common usage as an appropriate descriptive term which was not misleading or confusing to the consumer shall be used.
4.1.1.4 A 'coined', 'fanciful', 'brand' name, or 'trade mark' may be used provided in accompanies one of the names provided in sub-sections 4.1.1.1 to 4.1.1.3'. See Annex III of this book.

[194] Codex Stan 107-1981, for instance, para. 4.1(a): 'The name of each food additive present shall be given. The name shall be specific and not generic and shall indicate the true nature of the food additive. Where a name has been established for a food additive in a Codex list of additives, that name shall be used. In other cases the common or usual name shall be listed or, where none exists, an appropriate descriptive name shall be used.' Provision 5.2.4 of the General Standard for the Labelling of and Claims for Pre-packaged Foods for Special Dietary Uses (Codex Stan 146-1985): 'Claims as to the suitability of a food defined in Section 2.1 for use in the prevention, alleviation, treatment or cure of a disease, disorder or particular physiological condition are prohibited unless they are: (a) in accordance with the provisions of Codex standards or guidelines for foods for special dietary uses, and follow the principles set forth in such standards or guidelines; or (b) in the absence of an applicable Codex standard or guideline, permitted under the laws of the country in which the food is distributed.'

[195] Section 4.1.1.1 of Codex Stan 1-1985 (rev. 1-1991).

salted or unsalted according to national legislation.'[196] Likewise, some Codex standards include the explicit possibility of deviating from these norms, mainly by providing options from which manufacturers may choose. For instance, Provision of Section 2.1.2 of the Codex Standard for Canned Sardines and Sardine-type Products states that: 'Head and gills shall be completely removed; scales and/or tail may be removed. The fish may be eviscerated. If eviscerated, it shall be practically free from visceral parts other than roe, milt or kidney...'[197] This provision leaves the manufacturers with a restricted choice on how to prepare fish. However, in 1985, the Commission adopted the following conclusion with regard to these optional clauses in Codex standards:

> 'Codex standards, being of a mandatory nature, should not include optional clauses providing for agreement between buyer and seller in regard to quality factors of an aesthetic nature, like styles, types of packs, *etc.*, as this would not provide consumer protection and would not ensure fair practice in the food trade, especially when dealing with products where such criteria are important.'[198]

As may be concluded from the above, the term throughout Codex standards express the mandatory nature of the provisions and the fact that 'optional clauses' are only rarely used. Consequently, in cases where standards become legally-binding, there is not much room for flexible application of the standards. However, this is different for other Codex measures, as Codes of Practice and Guidelines are clearly advisory in their language.

5.2 The acceptance procedure

The obligations under the method of full acceptance of Commodity Standards and General Standards were result-oriented. Codex Members, having fully accepted a standard, had to ensure that all foods, home-produced and imported alike, distributed within their territory, comply with the Codex standard. At the same time, no legal or administrative provisions relating to consumer health or to other food-standard matters that hinder the free distribution of food products which were in conformity with the accepted Codex standards could be adopted. In other words, although full acceptance would, in most cases, result in the need to adapt national regulations, the Codex Member in question was free to choose how to implement the obligations.

During the discussion prior to the establishment of the acceptance procedure, it was considered necessary to include a type of 'safeguard clause'.[199] For instance,

---

[196] Codex Standard for Butter, Codex Stan A-1-1971, Rev. 1-1999.
[197] Codex Standard for Canned Sardines and Sardine-type Products, Codex Stan 94-1981, Rev. 1-1995.
[198] Report of the 16th Session of the Joint FAO/WHO Codex Alimentarius Commission, Geneva, 1-12 July 1985, ALINORM 85/47, para. 107.
[199] Report of the 5th Session of the Codex Alimentarius Commission, Rome, 20 February-1 March 1968, ALINORM 68/35, para. 16.

in cases where there is a need for protection against the introduction of disease affecting livestock or humans, Codex Members should have the possibility of being allowed to limit the distribution of food products that comply with the Codex standards. However, the resulting 'safeguard clause' was only applicable if the considerations of human, plant or animal health had not been specifically dealt with by the Codex standard.[200]

Full acceptance of the maximum levels for residues of pesticides or veterinary drugs in food resulted in similar obligations. However, this method of full acceptance allowed the Codex Member to apply the obligations to imported food alone.[201] The reason behind this so-called partial harmonisation was to allow Codex Members to enforce good agricultural practices that would result in lower levels of pesticide residues than those allowed for in the Codex MRL.[202] Considering the lack of the possibility of accepting specific deviations for the MRLs, this option was thought to provide a solution. Acceptance of MRLs for imported food alone would result in two lines of distribution: one relating to home-produced foods that would have to comply with the lower level of residues as permitted by the national regulations of the Codex Member, and one relating to imported foods that have to comply with the Codex MRLs.

Acceptance with specific deviations for both Commodity Standards and General Standards entails the same obligations as under the method of full acceptance, with the exception of such deviations that have been specified by the Codex Member. Codex Members are, however, free to determine for themselves the requirements of the standards to which they wish to make specific deviations.

The declaration of free distribution entails the guarantee, on the part of a Codex Member, that food products that comply with the Codex standard may be freely distributed within the territory of the country. The difference between a declaration of free distribution and the acceptance of a standard, with or without specific deviations, is that, in the former case, the Codex Member is not required to enforce the Codex standard, in accordance with the obligations under the methods of acceptance.[203] Consequently, this method of acceptance does not require the Codex Member to adapt national requirements.[204] This option leaves the manufacturers (of im-

---

[200] Art. 4.A.(i)(c) of the General Principles of the Codex Alimentarius. Art. 5.A.(i) of the General Principles: 'It also means that distribution of any sound products conforming with the standard will not be hindered by any legal or administrative provisions in the country concerned, which relate to the health of the consumer or to food standard matters and which are covered by the requirements of the general standard.' [emphasis added]

[201] Art. 6.A. of the General Principles of the Codex Alimentarius: 'A Codex maximum limit for residues of pesticides or veterinary drugs in food may be accepted by a country in accordance with its established legal and administrative procedures in respect of the distribution within its territorial jurisdiction of (a) home-produced and imported food or (b) imported food only, ...'

[202] Report of the 4th Session of the Codex Committee on General Principles, ALINORM 74/36, Paris, 4-8 March 1974, paras. 37-47.

[203] Consideration 7 of the 'Guidelines for the Acceptance Procedure for Codex Standards'.

[204] D.G. Victor, *supra* n. 94, p. 192.

ported or domestic products) the choice between the Codex standards and the national legislation in place.

6. Legal status of the Codex Alimentarius under the acceptance procedure

The Procedural Manual says little about the legal status of the measures contained in the Codex Alimentarius. Provision 2 of the 'General Principles' makes this distinction between standards and provisions of an advisory nature. It states that:

> 'It also [in addition to standards] includes provisions of an *advisory* nature in the form of codes of practice, guidelines and other recommended measures.'[205]

This brings to the fore the question of the legal status of the Codex standards, which was the result of the acceptance procedure. As briefly mentioned in Section 4.3 of this chapter, only standards and MRLs were submitted for acceptance. It has been acknowledged that the acceptance procedure has been abolished and the legal status of standards no longer results from the acceptance procedure. However, it is necessary to discuss the acceptance procedure to understand fully the changes in the legal status of Codex measures resulting from the entry into force of the *SPS Agreement* and the *TBT Agreement* and the consequences thereof for the Codex Alimentarius. This change of legal status will be further discussed in Sections 3 and 4 of Chapter IV.

What were the implications of the acceptance procedure for the legal status of the Codex standards? In an opinion submitted to the Codex Alimentarius Commission, the legal counsels of the FAO and the WHO stated that:

> 'standards by their very nature are recommendations addressed to governments and, in effect, they are binding on those governments which have formally accepted them.'[206]

As the acceptance procedure only covered standards and MRLs, only these measures were intended to become legally-binding upon acceptance by Codex Members.

In order to structure the relationship of trust and commitment between countries, reciprocity of obligations resulting from the binding nature of legal provisions is essential.[207] In most of the sources of international law, one can find the element of

---

[205] General principles of the Codex Alimentarius, Procedural Manual, 16th edn., 2006, p. 30. [Italics and additions are inserted].

[206] Report of the 6th Session of the Codex Alimentarius Commission, Geneva, 4-14 March 1969, ALINORM 69/67, Appendix III, Opinion of the legal counsels of FAO and WHO on Codes of Practice in relation to the Codex Alimentarius. See also, C.H. Alexandrowicz, *supra* n. 125, p. 78.

[207] See, on the concept of reciprocity, B. Simma, 'Reciprocity', in: R. Bernhardt (ed.), *Encyclopedia of Public International Law*, Instalment 1, (1981), p. 400-404. The concept of reciprocity in this

reciprocity, which enables countries to commit themselves to certain international obligations, knowing that they can rely on the fact that other countries, having accepted these requirements, have the same obligations from which they may derive certain rights. For instance, the legally-binding status of treaties results from the entry into force of the treaty in question, which depends on a certain number of countries having consented to the requirements of the treaty, or even to the reservations in the treaties made by one party to the treaty. Such reservations influence the validity of the provision between the party making the reservation and the other parties of the treaty to the extent of the reservation.[208] The acceptance procedure neglected this element of reciprocity.[209] There was no provision in the Procedural Manual of the Codex Alimentarius Commission that allowed for the operation of the principle of reciprocity with regard to obligations resulting from acceptance, nor do the specified deviations from acceptance result in reciprocal reservations for other Codex Members. If one Codex Member accepts a codex standard, the resulting obligations are of a unilateral nature. It cannot invoke these same obligations with regard to any other Codex Member unless this Member has accepted the same Codex standard under the same conditions.[210] The possibility of accepting standards with specific deviations complicates the relationship of commitment and trust between Codex Members even further, as does the possibility of withdrawing unilaterally from the acceptance of the standards.[211]

The lack of reciprocity in the relationship between Codex Members requires a higher level of trust. The acceptance procedure never really functioned.[212] The level of acceptance, in particular, the acceptance of Codex MRLs by the United States

---

context refers to the 'political' concept, as basis for most international agreements in order to function. It rests upon the expectations of countries that when committing themselves to certain agreements, they can expect that these same commitments apply to the other countries that have ratified the agreement (See also, Chapter V).

[208] Art. 21 of the Vienna Convention on the law of treaties, 1155 UNTS, I-18232 (1980), 8 ILM (1969).

[209] As remarked by the German delegate during the 6th Session of the Codex Alimentarius Commission in 1969, see Report of the 6th Session of the Codex Alimentarius Commission, Geneva, 4-14 March 1969, ALINORM 69/67, para. 13.

[210] Historical records, however, illustrate that the element of reciprocity was to a certain extent meant to form a conditions for the obligations arising from the publication of Codex standards. As explained above in Section 4.3 of this chapter, during the first 10-15 years of the Joint FAO/WHO Food Standards Programme, the publication phase constituted the 10th step within the elaboration procedure. The Commission had the power to determine whether it was appropriate or not to publish a given standard. Publication depended, for instance, on the number of member countries having accepted the standard in question (which constituted the 9th step of the procedure), taking into account the capacities of these Members as being producer or consumer of the commodity in question. Published standards were in this context seen as standards, which had acquired a certain recognition, a certain status. There were, however, no explicit legal consequences accorded to the publication of Codex standards, such as an 'entry into force' of the published standards.

[211] See, on the comparison of the legal status of Codex standards with the status of treaties, C.H. Alexandrowicz, *supra* n. 125, p. 80. See also, H.G. Schermers and N.M. Blokker, *supra* n. 151, p. 773.

[212] D.G. Victor, *supra* n. 94, pp. 204-205.

and the Members of the European Community always remained low.[213] Despite the fact that various Codex Members have incorporated provisions of Codex measures into their national legal and administrative provisions, the lack of success of the acceptance procedure illustrates that they have been reluctant to convert these actions into international commitments *vis-à-vis* other Members.

Consequently, the majority of the Codex standards and MRLs never became legally-binding. Only a few standards and MRLs became legally-binding upon some Codex Members.

7. CONCLUSIONS

The Codex Alimentarius Commission was set up to assist its Members in the harmonisation of food legislation through the establishment of the Codex Alimentarius, a collection of uniformly-defined standards. The approach of developing the Codex Alimentarius was initially commodity-based, meaning that standards were developed per commodity or per commodity group. However, over the years, the Codex Alimentarius Commission has moved towards a more horizontal approach, and has emphasised the preference for the inclusion of essential and general aspects which cover more than one commodity group. Furthermore, the initial important role for regional standards has been reduced by the clear preference for establishing worldwide standards. Besides standards, other measures have also been adopted, such as recommended codes of practice and guidelines. Over the years, the Codex Alimentarius has developed into a complex system of food requirements of which individual Codex measures form an integral part.

The substance of Codex measures reflects the large range of the issues covered, which are often laid down in a very detailed manner. Furthermore, as regards Codex standards, it can be stated that the language used throughout the texts is clearly mandatory in nature. 'Optional clauses' and 'safeguard clauses' that provide Codex Members with some discretion are not common practice. Overall, examination of the Codex measures illustrates the favouring of a Codex Alimentarius Commission that reflects a high level of integration of different circumstances between Codex Members, and also expresses the aim of a high level of harmonisation.

However, reaching agreement on provisions that need to respond to a large variety of national circumstances is not an easy task. The Codex Alimentarius Commission has applied several instruments to facilitate agreement. It has adopted a standard-setting procedure that is composed of two readings, both at a Codex Committee level as well as at a Codex Alimentarius Commission level. Emphasis is put on consensus-building and voting is very much discouraged. The importance of consensus is emphasised by the flexible character of the standard-setting procedure: a standard or other related text in progress of elaboration can be sent back to previous

---

[213] See Consultant's report, Review of the Working Procedures of the Joint FAO/WHO Meeting on Pesticide Residues (JMPR), 2002, p. 6. D.G. Victor, *supra* n. 94, pp. 204-208.

stages of the procedure. Furthermore, to overcome an *impasse* in situations where the diversity of national circumstances renders agreement highly improbable, the Codex Alimentarius Commission has made several statements. Examples include the 'Statement to Respond to Situations of Insufficient or Incomplete Scientific Data' or the 'Statement of Principles concerning the Role of Science in the Codex Decision-Making Process and the Extent to which Other Factors are Taken into Account' to address controversy on 'other legitimate factors'.

However, probably the most important aspect that has facilitated agreement on detailed food provisions was the establishment of a voluntary and flexible acceptance procedure. The acceptance procedure was abolished during the session of the Codex Alimentarius Commission in 2005 because it had not really been used by its Members and was considered no longer relevant within the framework of the *SPS Agreement* and the *TBT Agreement*. However, the presence of the acceptance procedure had long been the driving force to develop international measures which reflected a high level of harmonisation. Obligation to apply Codex measures did not result from their adoption as a result of the standard-setting procedure. The actual application of the harmonisation process depended on the willingness of Codex Members to implement Codex standards, codes of practice and guidelines into their regulatory framework. Codex Members had to give their consent through the acceptance of each individual Codex standards in order to be bound by them. Furthermore, Codex Members were permitted to accept with specific deviations which, in a way, meant that they could determine their own optional clauses and safeguard clauses. As a result, the Codex Alimentarius was, indeed, only meant to assist in the harmonisation of food requirements, as expressed in Principle 1 of the General Principles of the Codex Alimentarius. Although it has been abolished, the consequences of the long-existed acceptance procedure are still present: the Codex standards in place today, still reflect a high level of harmonisation.

Chapter III
THE CODEX AND THE EC

1. INTRODUCTION

This chapter examines the status accorded to Codex measures by EC institutions in the process of the EC harmonisation of food requirements over the years.

Before the establishment of the European Economic Communities, European countries were searching for a way to reduce the trade barriers that resulted from the diversity of food requirements through the adoption of common standards. Between 1954 and 1958, European countries, on the initiative of the Austrian government, made efforts to create a European Food Code (the *Codex Alimentarius Europaeus*).[1] The Council of the *Codex Alimentarius Europaeus* sought support from other organisations in order to develop an international food code, in particular, the Food and Agriculture Organisation of the UN (FAO). During the first FAO Regional Conference for Europe, it was expressed that:

> 'The desirability of international agreement on minimum food standards and related questions (including labelling requirements, methods of analysis, *etc.*) ... as an important means of protecting the consumer's health, of ensuring quality and of reducing trade barriers, *particularly in the rapidly integrating market of Europe.*'[2]

It was upon the basis of this initiative of European countries that the Codex Alimentarius Commission and the Joint FAO/WHO Food Standards Programme were established.

With the establishment of the European Economic Communities (EEC) in 1958, the means to respond to the need for the harmonisation of food requirements were also adopted at regional level. The Community institutions had always had the normative competence to adopt harmonisation measures. This meant there was an overlap with regard to this mandate between the Codex Alimentarius Commission and the EC institutions. One might expect the EC institutions to be capable of adopting harmonisation measures that respond more specifically to the concerns of the EC member states than the Codex Alimentarius Commission could, especially when regional Codex measures started to diminish. However, Regulation 178/2002, adopted in 2002, on the general principles and requirements of food law, contains a requirement for the EC institutions to take international standards into account.[3]

---

[1] Understanding Codex Alimentarius. Food and Agriculture Organisation of the United Nations. World Health Organization, 2000.

[2] Ibid. [emphasis added].

[3] Art. 5(3) of Regulation 178/2002 laying down the general principles and requirements of food law, establishing the European Food Safety Authority and laying down procedures in matters of food safety [2002] L 31/1.

What is the reason for this requirement and what are the legal consequences (Section 6)? In order to respond to this question, this chapter will analyse to what extent Codex measures played a part in the EC harmonisation process of food requirements prior to 2002. It will start with a comparison of the harmonisation process of the EC over the years with that conducted by the Codex Alimentarius Commission (Section 2). In the following sections, the use of the Codex acceptance procedure by the EC institutions to promote the application of Codex standards by EC member states (Section 3), the use of Codex measures by the EC institutions when preparing and adopting secondary food legislation (Section 4), and the use of Codex measures by the European Court of Justice (Section 5) will be analysed. This chapter concludes with an analysis of the relationship between the Codex Alimentarius Commission and the EC Commission, and responds to the question of whether this relationship has undergone changes that can be linked with recent developments (Section 7).

2. Overview of the harmonisation process of the European Community and the Codex Alimentarius Commission compared

As the influence and the position of Codex measures on EC harmonisation of food requirements is closely connected with the progress of EC harmonisation through secondary food legislation itself, this section will give a short overview of the key phases of the EC harmonisation process and their characteristics, and compare them with the developments that have been taken place in the context of the Codex Alimentarius Commission. The development of EC harmonisation of food legislation can, for these purposes, be divided into four phases. These phases are each characterised by the important institutional changes that have been particularly important to the development of EC food legislation. They do not necessarily correspond to the institutional changes that are generally considered as milestones in the development of the EU.

The first phase encompasses the period prior to the 1987 Single European Act. The second phase covers the period from the new approach, as introduced in 1985, and ends with the BSE crises of 1996. The third phase covers the period 1997-2002 and starts with the impact of the BSE crises on the characteristics of Community food regulation. The last phase addresses recent developments which have taken place from 2002 onwards, the year in which Regulation 2002/178, or also referred to as the General Food Law, was adopted.

2.1 **Prior to 1987**

The initial phase of the harmonisation process of EC food law is characterised by a problematic and slow progress of adopting EC secondary legislation on foodstuffs. Prior to 1987, secondary EC legislation on foodstuffs was adopted on the basis of Article 100 EEC Treaty. At that time, this provision required unanimity of votes of the member states. Consequently, the harmonisation process depended largely on the willingness of the member states themselves. Apart from the first EC secondary legislation on food colourings, the Council of the European Communities struggled to arrive at decisions by unanimity and was not very successful in adopting secondary legislation for foodstuffs.[4] The fact that the harmonisation process was based on a vertical approach and, as a result, required agreement on what were, at times, very detailed prescriptions for foodstuffs, did not contribute to the matter. For instance, the proposal for a directive on cocoa and chocolate took almost 10 years to be adopted.[5] From the 43 vertical directives, which were intended to be adopted by 1971 (as stated in a Council Resolution in 1969),[6] only 14 were actually adopted by 1985.[7] The Council's struggle to adopt secondary legislation was not only characteristic for the area of foodstuffs, but existed in a wider range of EEC areas, even in these areas that did not require unanimity of vote.[8]

Furthermore, the European Commission lacked the power to promote the adoption of Community food legislation. It was not only the difficulty of the Council to adopt the Commission's initiatives which undermined its capacity, but also the Commission's lack of both expertise and institutional resources, which rendered it unable to implement its ambitious programme adequately. As stated by Diana Welch:

'The Commission has been trying too much, too quickly, with inadequate resources. Thus, the ambitious harmonization programmes have, largely due to a sterile and inefficient procedure failed... The number of Commission staff allotted to food law harmonization is inadequate if they are to prepare new legislation as well as managing existing directives. The tasks of these Commission staff have developed more rapidly than the structure and working of their department.'[9]

Likewise, for scientific expertise, the Commission depended largely on the member states. Although the Commission had already investigated the possibilities of setting up a scientific committee in 1964,[10] the establishment of the first scientific committee, the Scientific Committee for Food, had to wait until 1974.[11] Until then,

---

[4] See also, on the problems and criticism of actions taken during this period, D. Welch, 'From "Euro Beer" to "Newcastle Brown", A Review of European Community Action to Dismantle Divergent 'Food' Laws', 22 *Journal of Common Market Studies* (1983) pp. 52-59.

[5] Council Directive 73/241/EEC, 1973 OJ L 228/23.

[6] See Council Resolution of 28 May 1969, OJ C 76/5.

[7] Communication from the Commission to the Council and the European Parliament on the completion of the internal market: Community legislation on foodstuffs, COM(85)603 final 3, see also, E. Vos, *Institutional Frameworks of Community Health and Safety Regulation: Committees, Agencies and Private Bodies* (Oxford, Hart Publishing 1999), p. 133.

[8] This was mainly due to the 'empty chair crisis' and the Luxembourg Compromise of 1966; see, for a detailed overview of the history of European Integration, D. Dinan, *Ever Closer Union. An Introduction to European Integration* (NewYork, Palgrave 1994), pp. 46-49.

[9] D. Welch, *supra* n. 4, pp. 56-57. See also, E. Vos, *supra* n. 7.

[10] Europees Parlement, Schriftelijke vragen en antwoorden, Schriftelijke vraag No. 3 van de heer Bergmann van 26 maart 1964, Publicatieblad 1217/64, 22.5.64.

[11] Commission Decision relating to the institution of a Scientific Committee on Food, OJ L 136/1, 20.5.1974. The Scientific Committee for Pesticides was established by a Commission Decision of 21 April 1978.

it consulted experts on an *ad hoc* basis and depended on the Standing Committee on Foodstuffs, which was created in 1969 by the Council to enable the Commission to consult experts.[12] Even when the Scientific Committee for Food was established in 1974,[13] the structure of the Scientific Committee still reflected the dependence on member states for expertise.[14]

In contrast, during this first phase, the Codex Alimentarius Commission had less problems adopting standards and other measures, as its measures were not binding *per se*. It had adopted about 180 standards, 900 MRLs for pesticides and almost 40 codes of practice by 1981.[15] The standards which were initially prepared by the Committee of government experts on the Code of Principles concerning Milk and Milk Products, and which were submitted to the Codex Alimentarius Commission for adoption, were an important contribution to these measures.[16] Furthermore, from the beginning, the Codex Alimentarius Commission relied for scientific expertise on the (JECFA) and the (JMPR) both already operational before the Codex Alimentarius Commission was jointly established under the auspices of the FAO and the WHO.

2.2 1987-1997

At EC level, a major change was the adoption of the Single European Act and the introduction of Article 95 EC (former Article 100a) as an additional legal basis for the adoption of EC harmonisation legislation. Article 100a allowed community legislation to be adopted by majority-voting of the Council, instead of the required unanimity-voting under Article 94 EC (former Article 100). In addition, Article 95 allowed for the adoption of regulations in addition to directives (Article 94 only allowed for the adoption of directives) and the requirement for the adoption of secondary legislation was less strict ('measures ... which have as their object' the establishment of the internal market, instead of the required 'direct effect' on the Common Market of Article 94).[17] With the introduction of a 'new approach', the Commission aimed to restrict harmonisation legislation exclusively to the provi-

---

[12] Council Decision 69/414/EEC setting up a Standing Committee for Foodstuffs [1969] OJ L 291/9.

[13] See on the early work of the SCF, R. Haigh, 'The activities of the Scientific Committee for Food of the Commission of the European Communities', in C.L. Galli, et al. (eds.), *Chemical Toxicology of Food* (Amsterdam, Elsevier 1978), pp. 81-97.

[14] E. Vos, *supra* n. 7.

[15] W.H Vermeulen, 'De Codex Alimentarius Commissie en de Europese Commissie ter bestrijding van mond- en klauwzeer, twee volwassen dochterinstellingen van de Wereldvoedselorganisatie', 10 *Sociaal Economische Wetgeving* (1983), p. 599.

[16] Since 1958, the Committee of government experts on the Code of Principles concerning Milk and Milk Products has been working on the adoption of food standards in the area of milk products. It had adopted 16 individual standards on different types of cheese. See J.P. Dobbert, 'Le Codex Alimentarius vers une nouvelle méthode de réglementation internationale', 15 *Annuaire Français de Droit International* (1969), pp. 701-706. See also, Section 2 of Chapter I on the historical background of the Joint FAO/WHO Food Standards Programme.

[17] D. Vignes, 'The Harmonisation of National Legislation and the EEC', 15 *European Law Review* (1990), p. 360.

sions that were necessary to achieve common and essential objectives, such as public health and consumer protection.[18] On the basis of the 'new approach', several 'horizontal' directives were adopted, which covered the requirements for a wide range of food products.[19] Furthermore, at institutional level, resources were created to allow the Commission to fulfil its function more appropriately. In the view of the increasing workload, the Commission extended the number of members of the Scientific Committee for Food from 15 to 18 in 1986.[20] Likewise, the structure of the Scientific Committees was adjusted.[21] Instead of establishing an independent agency,[22] the Commission preferred to establish a small supporting staff, complemented with a mechanism for having the preparatory work done by the existing institutions of the member states.[23]

As discussed in Section 2.1 of Chapter II, the Codex Alimentarius Commission, too, decided to shift to a 'horizontal approach' in order to progress more rapidly in the completion of the Codex Alimentarius. However, the acceptance of the adopted standards and MRLs by Codex Members still lagged behind. By 1993, only 12% of Codex standards had been accepted.[24] References to the Codex Alimentarius Commission in the framework of the GATT were welcomed as means of promoting the use of Codex standards.[25] The Uruguay Round of negotiations led to references to outside international standards and meant an increased status of Codex standards under both the adopted *SPS Agreement* and the *TBT Agreement*.[26]

2.3 1997-2002

Between 1997 and 2002, both the Codex Alimentarius Commission and the EC institutions faced distrust and decreased legitimacy. With regard to the EC institu-

---

[18] A. Swinbank, 'The EEC's policies and its food', 17 *Food Policy* (1992), p. 60.

[19] E. Vos, *supra* n. 7, p. 134.

[20] 86/241/EEC: Commission Decision amending Decision 74/234/EEC with respect to the number of members of the Scientific Committee for Food, OJ L 163/40, 19.6.1986.

[21] Proposal for a Council Directive on assistance to the Commission and cooperation by the member states in the scientific examination of questions to food, COM(91)16 def. –SYN 332, 3.4.1991.

[22] The structure of the scientific committee was based on the principle of subsidiarity. See COM(91)16 def. –SYN 332, pp. 8-9, see also, E. Vos, *supra* n. 7. The option to establish an independent Community institution responsible for scientific evaluations is not given effect until the establishment of the European Food Safety Authority (EFSA) in 2002.

[23] ESC Opinion on the proposal for a Council Directive on assistance to the Commission and cooperation by the member states in the scientific examination of questions relating to food, OJ C 14/6, 20.1.1992.

[24] S. Suppan, *Governance in the Codex Alimentarius Commission* (Consumers International 2005), p. 22.

[25] Report of the 18th Session of the Joint FAO/WHO Codex Alimentarius Commission, Geneva, 3-12 July 1989, ALINORM 89/40, paras. 78 and 83.

[26] See, for further discussion, Chapter IV. Such references were already included in the former *TBT Agreement* (also referred to as the TBT Code) resulting from the Tokyo Round. However, the TBT Code did not have the status that the *TBT Agreement* and the *SPS Agreement* have now, being an integral part of the WTO Agreement, binding on all WTO Members and subject to the WTO dispute settlement mechanism.

tions, the BSE (Bovine Spongiform Encephalopathy disease) crises resulted in a lack of confidence in the EU institutions on the part of the public.[27] In order to respond to this lack of confidence which resulted from the food crises, emphasis was placed on the precedence of public health protection over economic considerations. Although the protection of consumers' health was already incorporated in Articles 129 and 129(a) (now Article 153), as inserted by the Maastricht Treaty, the amendments of these Articles and of Article 95, as laid down in the Amsterdam Treaty, resulted in a reinforcement of the Community powers to regulate, through the recognition that consumer protection forms an integrated part of EC policies.[28] The BSE crises also gave rise to the need to re-organise the framework of scientific committees, and the division between risk assessment and risk management. They illustrated that the outcome of scientific assessment can still be largely influenced by political constraints.[29] As a consequence of the reform, the Scientific Committees were transferred from different Directorates General (DG), under which each of them had been functioning, to one DG (DG XXIV, which is responsible for consumer interests, which is now termed DG Health and Consumer Protection). The reason for this shift was not only to co-ordinate the work between the committees better,[30] but also to ensure the application of the so-called principles of independence, excellence and transparency of the expertise, in order to avoid the influential dominance of industry interests at the cost of consumer interests.[31] These principles, designed to promote the objectivity of scientific basis of EC food legislation, are also reflected in Commission Decision 97/579/EC and Commission Decision 97/404/EC, which restructure the framework of scientific committees.[32]

---

[27] EP resolution, OJ 1997, C222/53: 'Extremely concerned at the loss of confidence which the preliminary report has already caused amongst the general public, consumers and farmers within the EU, particularly in the wake of the BSE crisis.' See, for detailed discussion on the BSE crises and their aftermath, K. Vincent, 'Mad Cows and Eurocrats-Community Responses to the BSE Crisis', 10 *European Law Journal* (2004), pp. 499-517.

[28] E. Vos, *supra* n. 7, p. 36.

[29] E. Vos, *supra* n. 7. See Report on the alleged contraventions or maladministration in the implementation of Community law in relation to BSE, without prejudice to the jurisdiction of the Community and national courts, A4-0020/97/A, PE 220.544/fin/A, 7 February 1997. Criticism has been expressed with regard to the fact that the relevant Scientific Committee (the Scientific Veterinary Committee) was chaired by a UK national and its composition existed out of a substantial number of British scientists. The minutes were made by a former official of the British Ministry of Agriculture, and appeared to contain omissions and discrepancies. The Report of the Inquiry Committee calls the correspondence between the Chairman and the reporter of the Scientific Veterinary Committee 'a typical example of correspondence between two officials (one from the Commission, one from a national government), rather than an instance of co-operation between an independent scientist and a Community institution.' Most importantly, the inclusion of minority opinions was highly discouraged and action was undertaken to prevent that further scientific discussion on the issue would take place.

[30] Commission Green Paper on the General Principles of Food Law in the European Union, COM (97) 176 final.

[31] Ph. James, et al., A European Food and Public Health Authority: the future of scientific advice in the EU (Brussels, EU Commission, 1999).

[32] Commission Decision 97/579/EC of 23 July 1997 setting up Scientific Committees in the field of consumer health and food safety, OJ 1997 L 237/18. Commission Decision 97/404/EC of 10 June 1997 setting up a Scientific Steering Committee, OJ 1997 L 169/85.

Besides the BSE crises, the WTO Panel report in *EC-Hormones* also had an impact on the authority of the Community institutions. The Panel's decision confronted them with the limits of their discretionary powers to adopt food legislation as imposed by an external source of law.[33]

As mentioned above, due to the adoption and entry into force of the WTO Agreement, the measures adopted by the Codex Alimentarius Commission obtained increased status. However, this increased status raised questions of legitimacy with regard to the decision-making powers of the Codex Alimentarius Commission. As will be explained in Chapter V, the increased status of Codex measures politicised the standard-setting procedure and complicated the reaching of consensus. Furthermore, one of the related expert bodies (the Joint Meeting for Pesticides Residues) encountered a serious distrust with regard to the independence of the scientific experts.[34]

2.4 Post-2002

The 'post-2002' phase is characterised by the adoption of various measures to respond to the lack of trust and the decreased legitimacy. Within the European Union, Regulation 178/2002 was adopted in 2002, to regulate the objectives of the general food law of the European Union.[35] The Regulation provides for the establishment of an independent European Food Safety Authority that breaks with the old way of the EC institutions, in which it depended on the co-operation of the member states for its scientific expertise.[36] Furthermore, the Regulation sets out the general principles for food law, which are meant to serve as an overarching framework for all EC food legislation and are meant to link all different regulations, directives and decisions together as a system. With the adoption of Regulation 178/2002, the EC institutions clearly aim to increase the confidence of consumers in its food law.[37]

To respond to the questions of legitimacy with regard to the Codex Alimentarius Commission, its parent organisations, the FAO (Food and Agriculture Organisation) and the WHO (World Health Organisation) launched an evaluation undertaken by a group of independent experts in 2002.[38] The publication of the evaluation report resulted in the adoption of multiple important amendments of the Codex Procedural Manual. Furthermore, a Trust Fund was established by the FAO and the WHO to stimulate the participation of developing Codex Members. Likewise, both the

---

[33] See, for further discussion, Section 6 of this chapter.

[34] See Section 2.3.2 of Chapter V.

[35] European Parliament and Council Regulation 178/2002/EC laying down the general principles and requirements of food law, establishing the European Food Safety Authority and laying down procedures in matters of food safety [2002] L 31/1.

[36] Preamble 33 of Regulation 178/2002/EC. See also, Section 6 of this chapter. See, on the EFSA, also, E. Vos, 'Risicobeheersing door de EU: het nieuwe beleid op het gebied van voedselveiligheid', in G. van Calster and E. Vos (eds.), *Risico and voorzorg in de rechtsmaatschappij* (Antwerpen, Intersentia 2005).

[37] See, for instance, Preamble 22, 23 and 35 of the Regulation.

[38] See also, Section 1 of Chapter V.

FAO and the WHO adopted internal rules to ensure the independence of the scientific experts who participate in the relevant expert bodies.[39]

3. THE PROMOTION OF THE ACCEPTANCE OF CODEX STANDARDS AND MRLS

As discussed in Chapter II, the way to take on the commitment to use the adopted Codex standards and MRLs prior to 2005, as promoted by the Codex Alimentarius Commission, was through acceptance.[40] During the early phase of the EC harmonisation process, the Codex acceptance procedure was an important means in the hands of the European Commission. Codex standards could be used as a harmonisation instrument in the absence of EC harmonisation legislation. Consequently, the Codex Alimentarius Commission found a co-operator in the European Commission in order to promote acceptance of Codex standards by EC member states.

The *de jure* recognition of Codex standards as a harmonisation instrument in EEC food law was not, however, successful. This can be illustrated by the fact that neither the EEC Council nor the member states dedicated great importance to the acceptance procedure of the adopted Codex standards. Acceptance of Codex standards by the Council of Ministers turned out to be a slow and complicated issue.[41] Despite the fact that, in 1977, the EEC accepted four Codex standards,[42] the value of the standards was still largely reduced by the many reservations made. For instance, the acceptance of some standards on sugar by the EC included reservations regarding the definition of white sugar and brown sugar, the criteria for the naming 'sound and fair marketing quality', and regarding the requirements for labelling.[43] In 1990, the European Commission attempted to proceed towards a more structural way of acceptance, as is reflected in its proposal for a Regulation related to the acceptance of Codex pesticides MRLs and MRLs for veterinary drugs.[44] The con-

---

[39] See, for a more detailed discussion on the expert bodies, Section 3.5 of Chapter I and Section 2.3.2 of Chapter V.

[40] See Sections 4.3 and 5 of Chapter II.

[41] Communication de la Communauté Economique Européenne concernant l'acceptation des normes Codex y compris les limites maximales pour les residues de pesticides. SEC (81) 1007, document de travail des services de la Commission, 19 juin 1981. During the 14th Session of the Codex Alimentarius Commission, this communication was distributed to the Members of the CAC. See Report of the 14th Session of the Joint FAO/WHO Codex Alimentarius Commission, 1981, para. 22.

[42] Proposal for a Council Regulation (EEC) on the acceptance by the European Economic Community of standards or maximum limits for pesticide residues or maximum limits of veterinary medical drugs residues drawn up under the Joint FAO/WHO Food Standards Programme (Codex Alimentarius), COM(90)216 final, 21 May 1990.

[43] W.H. Vermeulen, *supra* n. 15, p. 599. See also, Report of the 13th Session of the Joint FAO/WHO Codex Alimentarius Commission, 1979, para. 20.

[44] Proposal for a Council Regulation (EEC) on the acceptance by the European Economic Community of standards or maximum limits for pesticide residues or maximum limits of veterinary medical drugs residues drawn up under the Joint FAO/WHO Food Standards Programme (Codex Alimentarius), COM(90)216 final, 21 May 1990.

sequence of the adoption of this Regulation would have been that MRLs consistent with the EC secondary legislation could be subjected to an acceptance procedure, after which they would become legally-binding on the EC member states. However, the Regulation was not adopted and the historical record of EC legislation does not show any additional attempts to adopt this proposal.

4. THE USE OF CODEX ALIMENTARIUS IN SECONDARY EC FOOD LEGISLATION PRIOR TO 2002

As mentioned above in Section 3, the number of accepted Codex standards by the EC member states through the Codex acceptance procedure has been insignificant. However, the failure to integrate Codex standards as *binding measures* into EC law does not preclude the fact that the Community institutions have not used the Codex Alimentarius in other ways. Reference to the work of the Codex Alimentarius Commission and the relevant Joint FAO/WHO expert bodies can be found in documentation that indicates that the work has been taken into account during the legislative process, and in secondary EC legislation.

4.1 The use of Codex Alimentarius in the preparation of EC food law

During the 1970s and 1980s in particular, Codex measures and the work of the related expert bodies had an impact on the content of EC legislation.[45] During the 13th Session of the Codex Alimentarius Commission in 1979, the representative of the European Community confirmed this impact by stating that 'Codex work was very closely followed by the EEC, and that Codex standards had a significant influence on the content of the Community standards'.[46] A clear example of this influence is Directive 79/112/EEC, which was largely based upon the General Standard for the labelling of pre-packaged foods (CAC/RS 1-1969).[47] The impact of this

---

[45] Ch. Lister, *Regulation of Food Products by the European Community*, Current EC Legal Developments Series (London, Butterworth 1992), B. Vroom-Cramer, *Productinformatie over levensmiddelen. Etiketteringsvraagstukken naar Europees en Nederlands recht* (Amsterdam, Koninklijke Vermande 1998), p. 136. R. van der Heide, 'The Codex Alimentarius on Food Labelling', *European Food Law Review* (1991), p. 293. R. McKay, 'The Codex Food Labelling Committee - maintaining International Standards relevant to changing consumer demands', *European Food Law Review* (1992), p. 71, R. van Havere, 'Codex Alimentarius', in R. Kruithof, *Levensmiddelenrecht* (Brussel, Ced-Samson 1979), pp. 17-49.

[46] Report of the 13th Session of the Codex Alimentarius Commission, Rome 3-14 December 1979, ALINORM 79/38, para. 20. See also, Communication de la Communauté Economique Européenne concernant l'acceptation des normes Codex y compris les limites maximales pour les residues de pesticides. SEC (81) 1007, document de travail des services de la Commission, 19 juin 1981, p. 4: 'Au surplus, la législation communautaire est généralement très proche des normes Codex qui lui servent en fait de modèle. L'acceptation formelle de ces normes n'aurait dès lors guère d'effet supplémentaire sur les échanges avec la Communauté.'

[47] Report of the 15th Session of the Joint FAO/WHO Codex Alimentarius Commission, Rome 4-15 July 1983, ALINORM 83/43, para. 68: 'He [representative of the EEC] indicated that the Codex Gen-

Codex standard on the Directive 79/112/EEC was also ascribed to the fact that the same officers had been involved in the decision-making procedure of both the Codex and the EC in this matter.[48]

Likewise, to assemble scientific conclusions to be used as a basis for EC food standards, the European Commission relied on the work of the related expert bodies of the Codex standards: the JECFA and the JMPR. Particularly in the early years, when the Commission did not have its 'own' scientific committees, EC food legislation was mainly based upon scientific evaluations conducted at national level or scientific evaluations conducted by the JECFA or JMPR. The discussion prior to the adoption of the first EC Directive on foodstuffs indicated this link, when the Commission was questioned as to whether it had taken into account the report of the JECFA relating to food colourings.[49] From its answer, it can be concluded that the Commission had been advised by national experts, two of whom had also been involved in the adoption of the relevant JECFA report.[50]

Even after the establishment of the Scientific Committee for Food (SCF) in 1974 and the Scientific Committee for Pesticides in 1978, the JECFA and JMPR reports were often used as a basis for scientific discussions in both committees. The Reports of the SCF illustrate that JECFA's ADIs were taken into account during its risk assessments. Especially at the beginning of the existence of the SCF, the ADIs recommended by the JECFA were often adopted by the SCF without modification, and were submitted to the Commission as advice.[51] Although not formally established, co-operation with the expert bodies of the Joint FAO/WHO Food Standards programme was taking place, and the scientific opinions of the JECFA and the JMPR were important for the adoption of EC secondary legislation on foodstuffs.[52] The experts consulted, often formed part of a network of scientists who were also involved in international scientific discussions.[53]

---

eral Standard for the Labelling of Prepackaged Foods had been used as a model for the EEC Directive on Food Labelling'. See also, B. Vroom-Cramer, *supra* n. 45, p. 136. R van der Heide, *supra* n. 45, p. 293. R. McKay, *supra* n. 45, p. 71. See, more generally, on Council Directive 79/112/EEC, R. Streinz, 'Diverging risk assessment and labelling', 2 *European Food Law Review* (1994), pp. 155-179.

[48] R. van Havere, *supra* n. 45, p. 44.

[49] Written question No. 101 of Mr. De Block to the Commission of the European Economic Community, OJ 20.11.1962, p. 2716.

[50] Ibid., p. 2717.

[51] See Reports of the Scientific Committee on Food, 4th and 5th Series, 1977 and 1978.

[52] Opinion of the Economic and Social Committee, 92/C 14/03, OJ C 14/6. See also, Ch. Lister, *supra* n. 45, p. 17. R. van Havere, *supra* n. 45, p. 44: 'Wat de impact van de Codex Alimentarius op de EEG-harmonisatie betreft, kunnen we besluiten dat, hoewel de Codex Alimentarius op de EEG-harmonisatie nog geen volledige vat heeft, het toch verheugend is te kunnen vaststellen dat: er voor sommige produkten reeds een acceptatie als Gemeenschap bestaat; dat er met de Codex-werkzaamheden binnen de EEG rekening wordt gehouden; dat, wat pesticidenresidu's en additieven betreft, de toxicologische evaluatie van respectievelijk JMPR en JECFA, de FAO/WHO-deskundigen op dat vlak, van een bijzondere betekenis zijn voor wetenschappelijke en administratieve kringen van de EEG.' See also, J. Abraham and E. Millstone, 'Food additive controls. Some international comparisons', 14 *Food Policy* (1989), p. 46.

[53] P. Gray, 'Food law and the internal market, 15 *Food Policy* (1990), p. 115: Previously the opinions of SCF 'were largely based on these summary reports previously carried out at national level.

However, with the completion of the internal market in view, and a stronger position of the Commission in the legislative procedure as initiator, the influence of the Codex Alimentarius on the content of EC legislation seems to have diminished. In its 1986 Communication, the Commission explicitly excluded the Codex Alimentarius from the programme for the completion of the internal market as expressed by the Commission, as, according to the latter, the Codex did not have a direct connection with the concept of 'internal market'.[54] This was an important statement, as it formally abandoned the initial idea of the important role that the Codex Alimentarius was to play in the EC harmonisation process, which had implicitly been promoted by the Commission for several decades, in order to complement the EC harmonisation instruments. Consequently, the increased competence of the Community institutions to adopt EC harmonisation measures and a stronger institutional framework diminished the function of the Codex Alimentarius. The acceleration of the harmonisation process in the European Community also meant an increasing number of issues which appeared on the EC agenda before they had been dealt with in the Codex Alimentarius Commission. For instance, the use of growth hormones is an example of EC regulation that preceded Codex standard-setting. The first EC Directive on this issue was adopted in 1981,[55] while the responsible Codex Committee of Veterinary Drugs was only established in 1985, and held its first meeting in 1986, which was mainly dedicated to the discussion on the principles, procedures and organisation of its future work.[56] Commenting on the influence of another Codex Committee (Codex Committee on Food Labelling) and its work on labelling, van der Heide stated that:

> 'one gets the impression that the lead which the committee [Codex Committee on Food Labelling] had in the development of labelling formats has been diminished gradually.'[57]

For instance, later changes made by the EEC to Directive 79/112/EEC, which was largely based upon the Codex Standard for the Labelling of Pre-packaged Foods as mentioned above, were brought to the attention of the Codex Committee on Food

---

In addition, scientists from these institutes provided a logistic food science support for all government activities such as general policy decisions or international discussions, both bilateral and multilateral.' See, for a similar opinion, R. Haigh, *supra* n. 13, p. 83.

[54] Communication from the Commission to the Council and the European Parliament on the completion of the internal market: Community legislation on foodstuffs, COM(85)603 final 3, p. 3 footnote 4: 'De international activiteiten, in het bijzonder die met betrekking tot de Codex Alimentarius, zijn niet vermeld aangezien zij geen rechtstreeks verband houden met het begrip 'interne markt'.

[55] Council Directive 81/602/EEC of 31 July 1981 concerning the prohibition of certain substances having a hormonal action and of any substances having a thyrostatic action, OJ L 222/32, 07.08.1981.

[56] Answer given by Mr. Andriessen on the written question by Mrs. Beate Weber related to the fraudulent use of hormones in livestock breeding – work on the Codex Alimentarius, OJ C 292/1, 2.11.1987. In 1986 the US already requested that these EC directives would be withdrawn. The existence of the EC directives was largely responsible for the slow progress and controversial discussion on MRLs of hormones.

[57] R. van der Heide, *supra* n. 45, p. 297.

Labelling and was taken into consideration during the revision of Codex General Standard on Food Labelling.[58]

Similar conclusions can be drawn from other documents. For instance, the majority of the secondary legislation on foodstuffs adopted after 1987 did not contain any indication that the Codex Alimentarius had been taken into account. In some cases where a reference had been inserted, the legislative history of the secondary legislation in question indicated that this seemed to be the direct result of insistence by the Council or the Economic and Social Committee.[59] For instance, the explicit reference to the FAO/WHO Food Standards Programme contained in EC Council Regulation 2377/90 which laid down a Community procedure for the establishment of maximum residue limits of veterinary medicinal products in foodstuffs of animal origin is mainly due to the insistence of the Economic and Social Committee.[60]

Furthermore, the expertise and independence of the Scientific Committee for Food has grown over the years. As the SCF responds in an increasingly more specific manner to the concerns of the EC member states, the SCF has become more of a 'competitor' of the JECFA, instead of relying on the work of the latter.[61] Whereas, in 1975,[62] 42 out of the 52 colouring matters discussed in the SCF were based upon the data included in the monographs of the JECFA, more recent reports indicate that the discussion and evaluation has become increasingly based upon independent scientific studies and data collected from the EC member states.[63]

4.2 Reference to the Codex Alimentarius in secondary EC food legislation

Most of the references to the Codex Alimentarius laid down in EC Directives are related to the promotion of uniformity in the member states' provisions on compli-

---

[58] Report of the 15th Session of the Joint FAO/WHO Codex Alimentarius Commission, 1983, para. 68, where the representative of the EEC indicates: 'Since then [the adoption of the EEC Directive on Food Labelling], the EEC had made some changes in the rules of food labelling and had brought these changes to the attention of the Codex Committee on Food Labelling, which was now revising the General Standard for the Labelling of Pre-Packaged Foods. The revised version of the Codex General Standard, as it was now emerging, was very similar to the EEC Directive and this would be a big step in facilitating international trade in food.'

[59] For instance, Commission Directive 90/612/EEC amending Council Directive 78/663/EEC laying down specific purity criteria for emulsifiers, stabilizers, thickeners and gelling agents for use in foodstuffs (consideration 3, OJ L 336/58, 24.11.1990) and Commission Directive 92/4/EEC amending Council Directive 78/663/EEC (consideration 3, OJ L 55/96, 29.2.1992) illustrate that the Codex Alimentarius standard had been taken into consideration with respect to the authorisation of new techniques.

[60] OJ L 224/1, 18.08.1990. Opinion Economic and Social Committee, OJ C 201/89. See, more specifically, on this regulation, R. Ancuceanu, 'Maximum Residue Limits of Veterinary Medical Products and Their Regulation in European Community Law', 9 *European Law Journal* (2003), pp. 215-240.

[61] This development is equally reflected in the opinion of Advocate-General Gullman in Nitrates cases (*Commission* v. *Italian Republic (nitrates)*, Case C-35/89, [1992] ECR I-4545) as discussed below in Section 5.1.2 of this chapter.

[62] Reports of the Scientific Committee for Food, 1st Series (1975).

[63] For instance, Reports of the Scientific Committee for Food, 22nd Series (1989).

ance procedures or verification procedures that give effect to EC law. These provisions are predominantly of a voluntary nature, and rely on Codex codes of practice or general principles which relate to food hygiene practices, or compliance assessment methods as established by the Codex Alimentarius Commission.[64] For instance, the preamble and Article 5(2) of the Directive on the hygiene of foodstuffs indicate that member states should encourage the use of the Codex Code of Practice on the General Principles of Food Hygiene as a basis for voluntary guides to be applied by food businesses in order to comply with the EC requirements on food hygiene.[65] Commission Decision 95/149/EC which fixed the total volatile basic nitrogen (TVB-N) limit values for certain categories of fishery products, and which specified the analysis methods to be used, left the option for the member states to use the conformity assessment procedure as established by the Codex Committee on Fish and Fishery Products.[66] Furthermore, the Directive on honey requires member states, whenever possible, to use 'internationally recognised validated methods such as those approved by Codex Alimentarius to verify compliance with the provisions of this Directive'.[67] However, this requirement is only valid until the EC Commission itself has adopted methods to permit the verification of compliance of honey with the provisions of EC law.[68]

A more stringent provision is reflected in the Directive on the approximation of the laws of the member states concerning foods and food ingredients treated with ionising radiation. This directive regulates that irradiation facilities may only receive approval for operation if the facility meets the requirements of the Codex Code of Practice for the operation of the irradiation facilities used for the treatment of foods and the requirements on the keeping of records for each batch of foodstuffs treated as established by the Commission.[69]

The only specifications to use the requirements contained in Codex standards are laid down in several EC directives concerning infant food used for food aid or for export to other countries, and state that member states only export products or provide products for food aid which are in compliance with the relevant Codex standard.[70]

---

[64] See, on early EEC policy with regard to food inspection and hygiene, L.J. Schuddeboom, 'Food inspection and EEC policy', 2 *Food Policy* (1977), pp. 17-26.

[65] Art. 5.2 of the Council Directive 93/43/EEC, OJ L 175/1, 19.7.1993.

[66] Art. 2(3) Commission Decision 95/149/EC, OJ L 97/84, 29.04.1995.

[67] Art. 4 of the Council Directive 2001/110/EC of 20 December 2001 relating to honey, OJ L 10/47, 12.1.2002.

[68] Ibid.

[69] Art. 7.2, Art. 8 and Art. 12 of Directive 1999/2/EC of the European Parliament and of the Council, OJ L 66/16, 13.3.1999.

[70] Council Directive 92/52/EEC on infant formulae and follow-on formulae intended for export to third countries, OJ L 179/129, 1.7.1992, which leaves member states the choice to comply with either the EC rules or the Codex standards, Commission Regulation EEC No. 944/92 on the supply of food for infants in Moscow and St Petersburg as urgent aid under Council Regulation EEC No. 330/92, OJ L 101/44, 15.4.1992, Commission Regulation (EEC) No. 1799/91 laying down detailed rules for the supply of infant food as urgent aid for the people of the Soviet Union provided for by Council Regulation (EEC) No. 598/91, OJ L 161/5, 26.6.1991 and Commission Regulation (EEC) No. 3421/91 laying down detailed rules for the application of Council Regulation (EEC) No. 597/91 as regards the supply of infant milk and whole-milk powder to Romania, OJ L 324/19, 26.11.1991.

5. THE CODEX ALIMENTARIUS AND THE EUROPEAN COURT OF JUSTICE

Throughout the jurisprudence of the European Court of Justice reference to the work of the Joint FAO/WHO Food Standards Programme (both to the opinions of the Joint FAO/WHO expert bodies and to the Codex measures themselves) can be found. Codex measures or the scientific opinions of the Joint FAO/WHO expert bodies are used by the parties to the dispute, by Advocates-General and by the Court itself. The references can be found in a wide range of food issues. There are numerous cases that deal with food safety issues, such as food additives, contaminants, microbiological risks, food hygiene and conformity assessment procedures. Even in cases that relate to trade descriptions, reference to Codex measures is to be found, such as the use of the trade descriptions of Edam, Emmenthaler, 'deep-frozen yoghurt' and 'naturally pure jam'. It is clear from the decisions of the Court that it accords a function to Codex measures and their scientific basis. This function differs according to the context of its decisions:

- decisions relating to the conformity of national measures with Articles 28 and 30 EC;
- decisions related to conformity of national measures with secondary EC legislation, and
- decisions related to the judicial review of Community measures.

The following section aims to define the role as accorded by the Court to Codex measures and their scientific basis.

5.1 The European Court of Justice and its reference to the Codex Alimentarius in the context of Article 28 and 30 EC

Parallel to the Community's activities to approximate national food legislation, the active role of the ECJ and its jurisprudence on the elimination of 'quantitative restrictions or measures having equivalent effect' in the context of Articles 28 and 30 EC constituted another way of working towards the completion of the common market.[71] Article 28 precludes measures which have an equivalent effect to quantitative restrictions on trade, be they discriminatory or non-discriminatory. Article 30 EC contains exceptions to the prohibition laid down in Article 28 EC. According to Article 30 EC, measures aimed to achieve precise objectives, such as the protection of public health, are accepted as long as they do not constitute arbitrary discrimination or a disguised restriction to trade. Furthermore, measures which need to be 'justified', or, in other words, they need to fulfil several conditions which, taken together, are often also referred to as the principle of proportionality.[72] Cases relat-

---

[71] A. Swinbank, *supra* n. 18, p. 60.

[72] J. Wiers, *Trade and Environment in the EC and the WTO. A Legal Analysis* (Groningen, Europa Law Publishing 2002), p. 72. See, more generally, on the principle of proportionality, G. de Burca, 'The Principle of Proportionality and its Application in EC Law', 13 *Yearbook of European Law* (1993), pp. 105-150.

ing to national food measures have played an important role in the development of the jurisprudence in the context of Articles 28 and 30 EC.[73]

It is in this context that the work of the Joint FAO/WHO Food Standards Programme came to play an explicit role. A binding effect of Codex standards in EC law, however, was rejected by the European Court of Justice in *Deserbais*.[74] In this case, France invoked the Codex standard on Edam (and the relevant provisions of the Stresa Convention) to justify its requirement of a minimum level of fat in order for products to bear the name 'Edam' and wished to prohibit the distribution of food products that did not comply with the international provisions. In other words, it invoked the international obligation that would result from the acceptance of the Codex standard on Edam. The Netherlands, as third party to the dispute, made the observation that the Codex Alimentarius consisted of internationally-agreed quality-standards and deserved to be protected under Community law.[75] Advocate-General Slynn, however, indicated that, given the fact that some member states had not accepted the Codex standard, the standard could not be considered as a Community rule overriding Article 28.[76] The Court built upon the opinion of Advocate-General Slynn and stated that:

> 'It must be observed that the rules of the *Codex Alimentarius* on the composition of certain foodstuffs are, in fact, intended to provide *guidance* for defining the characteristics of those foodstuffs. However, the mere fact that a product does not wholly conform with the standard laid down does not mean that the marketing of it can be prohibited.'[77]

While rejecting the direct effect of Codex Alimentarius in EC law, the Court thus recognised the Codex Alimentarius as important tool for guidance. In what manner has the Codex Alimentarius provided guidance to the European Court of Justice in interpreting and applying Articles 28 and 30 of the EC Treaty? The jurisprudence, examined in this sub-section, results from cases brought to the Court under Article 234 EC (preliminary questions) or Article 226 EC (infringement procedures against member states). This means that it is not only the interpretation of the Court, but also its application to the facts at hand that contribute to define the role of Codex Alimentarius in the context of Articles 28 and 30 EC.

---

[73] Numerous important decisions can be named, such as *Procureur du Roi* v. *Benoit and Gustave Dassonville*, Case 8/74, [1974] ECR 631. Rewe-Zentral AGV Federal Monopoly Administration for Spirits (Cassis de Dijon), Case 120/78, [1979] ECR 649. Commission v. Germany (Reinheitsgebot) Case 178/84 [1987], ECR 1227. See also, A. Swinbank, *supra* n. 18, p. 60.

[74] *Ministère public* v. *Gérard Deserbais*, Case 286/86 [1988] ECR 4907.

[75] Ibid. p. 4919: 'The Netherlands Government has argued that the Stresa Convention, along with the *Codex Alimentarius*, represents an internationally agreed quality standard which deserves to be protected under Community law. Since a number of Member States, particularly the Federal Republic of Germany, have not accepted the standard thus proposed, I find it difficult to say that they are rules of Community law which override or qualify the rule in Article 30.'

[76] Ibid.

[77] *Ministère public* v. *Gérard Deserbais*, Case 286/86 [1988] ECR 4907, para. 15.

5.1.1 The explicit role of the work of the Joint FAO/WHO Food Standard Programme in the jurisprudence on Article 30 EC

The first case in which reference to the Codex Alimentarius can be found was *Eyssen* (1981).[78] In this case, the ECJ was faced with preliminary questions from the Dutch Court as to whether the general prohibition on additives, including the prohibition on nisin (an antibiotic) was incompatible with Article 30 EC, in its entirety, or, at least, with regard to the prohibition of adding nisin to processed cheese. The Court used the studies that had been conducted by the JECFA as guidance to conclude that the consumption of the antibiotic nisin presents or may present a risk to human health. Based on the studies conducted by the JECFA, the Court also considered that the assessments of these risks still contained difficulties and uncertainties due to the variability of circumstances.[79] It stated that:

> '... In the particular case of the addition of nisin to products intended for human consumption, such as processed cheese, it is indeed accepted that the increasingly widespread use of that substance...has revealed the need, both at national level in certain countries and at international level, to study the problem of the risk which the consumption of products containing the substance presents, or may present, to human health and has led certain international organizations, such as the Food and Agriculture Organization of the United Nations and the World Health Organization, to undertake research into the critical threshold for the intake of that additive. Although those studies have not as yet enabled absolutely certain conclusions to be drawn regarding the maximum quantity of nisin which a person may consume daily without serious risk to his health, this is essentially due to the fact that the assessment of the risk connected with the consumption of the additive depends upon several factors of a variable nature...'[80]

According to the Court, the difficulties and uncertainties inherent in such a risk assessment may justify the lack of harmonisation.[81] This lack of harmonisation and the presence of uncertainties regarding the proper maximum level of nisin to be prescribed led the Court to decide that the Dutch prohibition did not constitute a

---

[78] *Officier van Justitie* v. *Koninklijke Kaasfabriek Eyssen BV* (preliminary ruling requested by the Gerechtshof, Amsterdam), Case 53/80 [1981] ECR 409.

[79] See also, on Eyssen and the finding on 'difficulties and uncertainties', Ch. Joerges, *Integrating Scientific Expertise into Regulatory Decision-Making. Scientific Expertise in Social Regulation and the European Court of Justice: Legal Frameworks for Denationalized Governance Structures*, EUI Working Paper RSC 96/10, (San Domenico, European University Institute 1996), p. 13.

[80] *Officier van Justitie* v *Koninklijke Kaasfabriek Eyssen BV* (preliminary ruling requested by the Gerechtshof, Amsterdam), Case 53/80 [1981] ECR 409, consideration 13.

[81] Ibid., consideration 14: 'The difficulties and uncertainties inherent in such an assessment may explain the lack of uniformity in the national laws of the Member States regarding the use of this preservative and at the same time justify the limited scope which the prohibition of the use of the additive in a given product, such as processed cheese, has in certain Member States, which prohibit its use in products intended for sale on the domestic market while permitting it in products intended for export to other Member States where the requirements for the protection of human health are assessed differently according to dietary habits of their own population.'

'means of arbitrary discrimination or a disguised restriction on trade between member states' in the sense of Article 30 EC.[82] The preliminary conclusions of the JECFA thus served as justification for the prohibition of nisin in processed cheese, which, as a result, was found to be in conformity with Article 30 EC.

In the following cases,[83] *Motte* and *Muller*, the Court decided that the member states had to take the results of international scientific research into account when deciding on the authorisation of the marketing of foodstuffs to which additives had been inserted,[84] and that it was up to them to justify the necessity of the refusal of such an authorisation.[85] The fact that the competent national authorities have to show the necessity of their measures 'with due regard to the results of international scientific research' leads to the suggestion that 'international scientific research' is actually to be taken as a reference point in the context of the principle of proportionality.[86] This is, at least, what Advocate-General Darmon clearly had in mind when he stated in his opinion in *Muller* that:

> 'The research undertaken by the international organizations and by the Scientific Committee for Food could constitute a *solid foundation* for any application for authorization to market additive E 475 which the Member State *could not reject except* by showing that, having regard to French eating habits, the additive meets no real need and, even if the daily acceptable intake is not exceeded, constitutes a danger to health.'[87]

In *Reinheitsgebot*, the Court confirmed this line and referred explicitly to the 'Codex Alimentarius Committee of the Food and Agriculture Organisation of the United Nations (FAO) and the World Health Organisation'.[88] Consequently, it left no doubt that the Codex is understood to be 'international scientific research' as an integral part of the following formulation of the criteria according to which member states need to justify the necessity of their prohibitions:

> 'the use of a specific additive which is authorized in another Member State *must* be authorized in the case of a product imported from that Member State where, in view,

---

[82] Ibid., consideration 16. Advocate-General Warner accords importance to the fact that equally the Codex Alimentarius Commission had not succeed to adopt standards yet. He stated: 'this is a matter over which, plainly, pending the enactment of a 'harmonizing' directive, or at all events the adoption of a *generally accepted international standards*, Member States retain a measure of discretion ...'

[83] In *Sandoz*, relevant Codex standards on vitamin A and D were invoked in the dispute. However, neither the Advocate-General, nor the Court referred to these standards in this case. See more generally on Sandoz, K.J. Mortelmans, 'SANDOZ-arrest. De Warenwet tussen DE-regulering en EG-regulering', 33 *Ars Aequi* (1984), pp. 100-109.

[84] *Criminal proceedings* v. *Léon Motte*, Case 247/84 [1985] ECR 3887, paras. 24-25.

[85] *Ministère public* v. *Claude Muller and Others*, Case 304/84 [1986] ECR 1511, para. 26.

[86] Whereas the Court did not require the Netherlands to justify the necessity of its measures, in following cases, member states were required to justify the necessity of their measures in the light of the principle of proportionality. See more generally on the principle of proportionality, G. de Burca, *supra* n. 72, pp. 105-150. J. Wiers, *supra* n. 72, pp. 72-83 and 102-116.

[87] *Ministère public* v. *Claude Muller and Others*, Case 304/84 [1986] ECR 1511, p. 1518.

[88] See also, on the Reinheitsgebot 'formula', Ch. Joerges, *supra* n. 79, pp. 11-12.

on the one hand, of the findings of international scientific research, and in particular of the work of the Community's Scientific Committee for Food, *the Codex Alimentarius Committee of the Food and Agriculture Organization of the United Nations (FAO) and the World Health Organization*, and, on the other hand, of the eating habits prevailing in the importing Member State, the additive in question does not present a risk to public health and meets a real need, especially a technical one.'[89]

Furthermore, the Court held, in this same case, that, when assessing the technological need of an additive, 'account must also be taken of the findings of international scientific research and in particular the work of the Community's Scientific Committee for Food, the *Codex Alimentarius Committee* of the FAO and the World Health Organization.'[90] This finding was confirmed by the Court in *Debus*.[91] However, in a later case, *Commission v. Denmark*, the Court specified that the lack of a technological need alone does not suffice to ensure conformity with Articles 28 and 30 EC, even though the national measure is in line with the Codex Alimentarius.[92] It may play a role in a risk assessment, but does not justify a total ban by itself if it is not based upon a case-by-case risk assessment that identifies the risks for human health for each of the substances in question.[93]

When it comes to the risks that the consumption of additives may represent, several cases demonstrate that international scientific research upon which Codex standards are based have indeed been used by the Court as a 'solid foundation' for defining the necessity of trade-restrictive measures which aim to protect human health. To be precise, the results of international scientific research upon which Codex standards are based have been used as a reference point to examine the justification of member states which deviate from the scientific basis of the Codex standards and serve to define whether member states have provided sufficient proof.

---

[89] *Commission v. Germany* (Reinheitsgebot) Case 178/84 [1987], ECR 1227, p. 1274, para. 44 [emphasis added]. Recited in Criminal Proceedings against Michel Debus, Joined Cases C-13/91 and C-113/91 [1992] ECR I-3617 para. 17.

[90] *Commission v. Germany* (Reinheitsgebot) Case 178/84 [1987], ECR 1227, p. 1276, para. 52. Advocate-General Slynn in his opinion in *Reinheitsgebot* explicitly refers to the General Principles for the Use of Food Additives adopted by the Codex Alimentarius Commission for justifying its interpretation of the concept of 'technologically need': 'Preferences differ as to colour and flavour of beers; if additives are needed to achieve the desired taste and colour or amount of foam they are technologically necessary for this purpose. *That view is consistent with the "General Principles for the Use of Food Additives", adopted by the ninth session of the Codex Alimentarius Commission…It thus seems to me* that, for example, substances needed to assist the malting or the brewing processes or to affect the quality of the beer, emulsifiers, foam stabilizers, flavouring agents and colourants are all capable of being technologically necessary for the making of particular beers even if they are not needed or used in the making of German beer.' (*Commission v. Germany* (Reinheitsgebot) Case 178/84 [1987], ECR 1227, p. 1254 [emphasis added]).

[91] *Criminal Proceedings v. Michel Debus*, Joined Cases C-13/91 and C-113/91 [1992] ECR I-3617, para. 29.

[92] *Commission v. Denmark*, Case C-192/01 [2003] ECR I-09693, paras. 54-56. In this case, Denmark invoked several criteria contained in the Codex General Principles for the Addition of Essential Nutrients to Foods as requirements for authorisation of the addition of the ingredients. See also, *Ministère public v. John Greenham and Léonard Abel*, Case C-95/01 [2004] ECR I-01333.

[93] *Commission v. Denmark*, Case C-192/01 [2003] ECR I-09693, paras. 54-56.

In *Reinheitsgebot*, as a result of an infringement procedure made against Germany by the Commission, the Court was requested to examine the necessity of, *inter alia*, the German prohibition of the use of specific additives in beer. In its argumentation, the Commission relied on the ADIs as established by the JECFA and adopted by the SCF. It was, first and foremost, Advocate-General Slynn, who explicitly used ADIs accepted by the Commission and internationally (Codex ADIs) as reference point to examine the justification for the German prohibition. He stated that:

> 'there is no real evidence that the additives in themselves have been shown to interact adversely or that they are subject to suspicion based on concrete evidence. Nor is there any convincing evidence to show that the quantities of each additive likely to be ingested through imported beer is such that, taken with additives in other foods, a real risk to health is created by the acceptable daily intake (ADI) of each additive being exceeded. In my opinion, the Federal Republic *dismisses too readily and without producing satisfactory reasons the system of ADIs accepted by the Commission and internationally* in respect to some if not all additives.'[94]

Advocate-General Slynn used the ADIs as established by the JECFA, and which had served as the basis for the Codex MRLs, as the reference point for a scientific basis for a measure that was necessary to protect human health. In his opinion, Germany had not provided proof that indicated any risks that were not covered by the international ADIs. Although the Court did not refer specifically to the reasoning of the Advocate-General, it is clear that it agrees with the fact that Germany had not provided sufficient and specific evidence to justify its strict rules. It stated that '… mere reference to the potential risks of the ingestion of additives in general and to the fact that beer is a foodstuff consumed in large quantities does not suffice to justify the imposition of stricter rules in the case of beer'.[95]

In *Debus*, where the Codex MRLs were invoked to challenge the necessity of the measures at issue, the Court more clearly used the scientific basis of the Codex MRL as reference point:

> 'The need for such a prohibition for the purposes of health protection *has not been demonstrated. Quite to the contrary*, it appears from the Commission's uncontested statements that the absorption of sulphur dioxide as a result of consumption of beer containing 36.8 mg/l of that additive does not involve a serious risk that the limits to the maximum daily dose of sulphur dioxide authorized by the FAO and the WHO will be exceeded.'[96]

---

[94] *Commission v. Germany* (Reinheitsgebot) Case 178/84 [1987], ECR 1227, p. 1258.

[95] Ibid., para. 49.

[96] *Criminal proceedings v. Michel Debus*, Joined Cases C-13/91 and C-13/91 [1992], ECR I-3617, para. 24 [emphasis added]. The fact that the case *Debus* appeared before the Court by means of preliminary question did not withhold the Court from *applying* the principle of proportionality (See also the opinion of Advocate-General Darmon, in: *Van der Veldt*, para. 26). Advocate-General van Gerven in Debus was more careful with regard to this point and explicitly saw it as the responsibility of the

All the cases cited above were related to the issue of food additives. However, with regard to other issues of food safety, the Court also used scientific research conducted by the Joint FAO/WHO expert bodies as a reference point for evaluating the evidence brought before it.

For instance, in *Commission v. Italy (fish containing nematode larvae)*, the Codex Alimentarius was invoked by the Commission to argue that the measures were inconsistent with Article 28 of the EC Treaty.[97] The Court used the Codex Alimentarius as a reference point to determine the necessity of the national trade-restrictive measures. In this case, the Commission relied on a Codex standard to argue that the Italian measures which prohibited the consignments of fish from Norway in order to prevent risk to human health resulting from nematode larvae, were not justifiable under Article 30. The Court concluded that Italy had not adduced any evidence to refute the arguments of the Commission that international scientific research indicated that dead or devitalised nematode larvae posed no risk to human health.[98] Amongst other considerations, this led the Court to conclude that the Italian measures were not in conformity with Article 28 EC.[99]

In another case, *Commission v. Italy (health checks on imports of curds)*, the invocation of the Codex Alimentarius was also successful. This time, the Codex Alimentarius was invoked by Italy to justify national measures restricting free circulation of curds.[100] This case resulted from an infringement procedure against Italy with regard to its health checks on imports of curds. The justification for the frequency and long duration of the checks of imports into Italy was based upon the discovery of a very high level of *E coli*, which was also, according to the Codex standards, unacceptable.[101] The fact that the consignments did not satisfy the Codex standards on hygiene was sufficient for the Court to assume that the checks conducted by Italy were aimed to ensure the protection of public health. According to the Court, the Commission had not been able to refute these allegations.[102] The reliance of Italy on the Codex standards was, therefore, sufficient to justify the necessity of its measures.[103]

With regard to the trade descriptions of food products to ensure fair practices in the food trade, the Court also used the Codex Alimentarius as guidance. For instance, in *Smanor*, the Court referred to a characteristic feature of 'yoghurt' as defined by the relevant Codex standard (the presence in abundant quantities of live lactic bacteria) as a criterion to determine whether deep-frozen yoghurt was sub-

---

national court to assess whether the national regulation had met the requirements of the principle of proportionality. To that end, he only enumerated some aspects, which should be taken into account by the national courts.

[97] *Commission v. Italy* (fish containing nematode larvae) Case C-228/91, ECR [1993] I-02701, para. 7.
[98] Ibid., para. 28.
[99] Ibid., paras. 42-43.
[100] *Commission v. Italy* (health checks on imports of curds) Case 35/84, ECR [1986] 545, para. 4.
[101] Ibid., para. 4.
[102] Ibid., para. 11.
[103] Ibid., para. 13.

stantially different from yoghurt.[104] Only when deep-frozen yoghurt was substantially different from yoghurt, and when appropriate labelling was insufficient to ensure that consumers were properly informed, could a member state reserve the right to use the name 'yoghurt' only for fresh yoghurt.

In a later case, *Guimont*, the Court faced a preliminary question as to whether the prohibition of a cheese without rind under the designation 'Emmenthal' was in conformity with Article 28 EC.[105] The fact that the relevant Codex standard allowed the designation 'Emmenthaler' also for cheese without rind restricted the member state from prohibiting the designation for these products.

5.1.2   *Developments reducing the role of the* Codex Alimentarius *under Article 30*

The function of the work of the Joint FAO/WHO Food Standards Programme as a reference point for examining the justification of national measures in the context of Article 30 EC is, however, restricted by two elements. First, the increasing research undertaken by the Community's scientific committees has the tendency to reduce the relevance of the findings of the Joint FAO/WHO expert bodies as reference point. The fact that the role of the international scientific research of the joint FAO/WHO expert bodies may be limited by the increase in the scientific evaluations undertaken by the European scientific committees has been implied by Advocate-General Gulmann in his opinion in the Nitrate cases. When referring to the conclusion of the Court in previous cases that the question of health risks must be judged on the basis of international scientific research, in particular the conclusions drawn by the Community's Scientific Committee for Food and the 'Codex Alimentarius Committee', he stated that:

'I consider that, in the cases before the Court, it is right to attach *most* weight to the Report of Nitrate and Nitrite drawn up by the Scientific Committee for Food.'[106]

---

[104] *Proceedings for compulsory reconstruction* v. *Smanor SA* (Smanor), Case 298/87 [1988], ECR 4512, paras. 22 and 23: 'It must be stated that it is clear from both the Codex Alimentarius drawn up by the Food and Agriculture Organization (FAO) and the regulations of several Member States, referred to by the Commission, that the characteristic feature of the product marketed as "yoghurt" *is the presence in abundant quantities of live lactic bacteria*.

In those circumstances, the prohibition by national rules of the use of the name "yoghurt" for the sale of deep-frozen products appears to be disproportionate in relation to the objective of consumer protection, when the characteristics of the deep-frozen products are not substantially different, *particularly as regards the quantity of bacteria*, from fresh products ...'. See, also on Smanor, Ch. Lister, 'The naming of foods: the European Community's rules for non-branc food product names', 18 *European Law Review* (1993), pp. 179-201.

[105] *Criminal Proceedings* v. *Jean-Pierre Guimont*, Case C-448/98 [2000] ECR I-10663.

[106] *Commission* v. *Italian Republic* (nitrates), Case C-35/89, [1992] ECR I-4545, para. 19. The reasons for his choice are threefold: 1. the articles and monographs on which the scientific report of the SCF is based are more recent, 2. the SCF has sought information concerning the use of nitrates and nitrites from both invited experts and the member states, 3. the report of the SCF is an expression of a unanimous opinion of the experts.

Although, from the findings of the Court in its joint decision, it is not clear what weight it accorded to the work of the SCF or the JECFA, it no longer referred to the 'Codex Alimentarius Committee of FAO and the WHO' when it stated that:

> 'the existence of a risk arising from the use of an additive must be assessed in the light of international scientific research, in particular the work of the *Scientific Committee for Food*.'[107]

In a later case, *Hahn*, the Court did, indeed, only refer to the work of the relevant EC expert body, the Scientific Committee on Veterinary Measures Relating to Public Health, and no trace can be found in its findings that it accorded any weight to the work of the relevant joint FAO/WHO expert body, to which Advocate-General Geelhoed referred several times in his opinion.[108]

A second element is the simple fact that the scope of Articles 28 and 30 EC themselves are restricted. As the application of the Codex Alimentarius and the related work – as reference point for the Court under Article 30 EC – is inherent to the scope of this article, the increasing adoption of EC secondary food legislation which diminished the role of Articles 28 and 30 EC in food issues, also diminished the relevance of the use of the Codex as a solid foundation from which to examine the justifications for national measures. The rise of EC secondary legislation and their influence on the operation of Article 30 EC is already visible in some of the cases mentioned above, such as, *Smanor* and *Guimont*. Despite the fact that, in both cases, an EC directive applied, the Court applied Article 28 and 30 EC, as the directives in question did not ensure full harmonisation. In both cases, the examination under Article 30 was undertaken in the light of the existing EC directives and clearly illustrated the function of the Codex standards as a means of guidance in interpreting the directive and filling the gaps that had not yet been covered by the secondary EC food legislation. Consequently, the role of the Codex Alimentarius under Article 30 is limited by the presence of an EC directive and determined by the terms and the coverage of that directive.

5.2 The European Court of Justice and its use of the Codex Alimentarius in the context of secondary food legislation

Several cases indicate that even when secondary legislation has been adopted in such a way that Articles 28 and 30 EC no longer apply, reference to the work of the Joint FAO/WHO Food Standards Programme can still be found in the Court's decisions. It has a function as a means of interpretation in situations where Community legislation is not clear (*ERU Portuguesa*), or not specific enough (*Darbo*), or where Community legislation introduces concepts and provisions that come from the Codex Alimentarius itself (*Monsanto*).

---

[107] *Commission* v. *Italian Republic* (nitrates), Case C-35/89, [1992] ECR I-4545, para. 13.
[108] *Criminal proceedings against* v. *Walter Hahn*, Case C-121/00, ECR [2002] I-09193, paras. 41, 50 and 51.

In *ERU Portuguesa*, Advocate-General Fennelly used the Codex (as invoked by the Commission), as a means of interpretation to determine the level of phosphorus in cheese in order to define the meaning of 'casein' and choose the right classification category for the product in question.[109] The Combined Nomenclature as contained in Annex I to Council Regulation 2658/87 on the tariff and statistical nomenclature and on the Common Customs Tariff did not provide sufficient information to define whether Iceland skimmed milk cheese should be classified as casein or as cheese for processing. Advocate-General Fennelly classified the Codex Alimentarius as a 'source of information as to the objective characteristics and properties of casein'.[110] In *Sachsenmilch*, a similar case, it was the Court itself that used the relevant Codex standard (as invoked by the *Oberfinanzdirektion*) as a means of interpretation. In this case, the Court was faced with a similar question relating to the same EC Regulation 2658/87.[111] The Court referred to the general Codex standard on cheese in order to obtain guidance in defining 'ripened cheese' in order to determine whether mozzarella stored for two weeks could be classified as such.[112]

In *Darbo*, the Court used the Codex MRLs for cadmium and lead in fruit to determine whether the labelling of jam as 'naturally pure jam', while residues of pesticides and lead were traced in the jam, were to be considered as misleading in the sense of Article 2(1) of Directive 79/112 EEC.[113]

In *Monsanto*, the European Court of Justice, on its own initiative, used the Codex Alimentarius as a means of interpretation to define the concept of 'substantial equivalence' in Regulation 258/97,[114] as 'the concept of substantial equivalence should be placed in the context of the work carried out by the international scientific institutions where it was elaborated …'.[115]

5.3 The Codex Alimentarius, the European Court of Justice and the legality of EC measures adopted by the Community institutions

Reference to the work of the Joint FAO/WHO Food Standards Programme is also found in the decisions of the European Court of Justice under the procedure of Articles 232 EC on the legality of the decisions taken by the Community institutions.

---

[109] *Fábrica de Queijo Eru Portuguesa Ld^a v. Tribunal Técnico Aduaneiro de Segunda Instância*, Case C-42/99, ECR [2000] I-07691, para. 33.

[110] Ibid. See also, S. Poli, 'The European Community and the Adoption of International Food Standards within the Codex Alimentarius Commission', 10 *European Law Journal* (2004), p. 617.

[111] *Sachsenmilch AG v. Oberfinanzdirektion Nürnberg*, Case C-196/05, [2006] ECR I-5161.

[112] Ibid., para. 29.

[113] *Verein gegen Unwesen in Handel und Gewerbe Köln eV v. Adolf Darbo AG*, Case C-465/98, ECR[2000] I-02297, para. 31.

[114] *Monsanto Agricoltura Italia SpA and Others v. Presidenza del Consiglio dei Ministri and Others*, Case C-236/01, ECR [2003] I-08105, para. 79. See also, S. Poli, *supra* n. 110, p. 617.

[115] *Monsanto Agricoltura Italia SpA and Others v. Presidenza del Consiglio dei Ministri and Others*, Case C-236/01, ECR [2003] I-08105, para. 75.

In *CEVA Santé animal SA* v. *Commission*, reference is made to the work of the JECFA.[116] The issue raised by the applicants in the dispute was that the Committee for Veterinary Medicinal Products (CVMP) was the sole competent body for formulating scientific opinions on the safety of a product, which thereby implied that the Commission had given too much weight to the work of international scientific bodies (including the JECFA) compared with the reports of the CVMP.[117] The Court decided that, in situations where a matter was scientifically and politically complex and sensitive, the Commission could ask the CVMP for a re-evaluation (even though it had given a positive opinion with regard to the authorisation of the substances).[118] However, the Court found that the inaction of the Commission after receiving a complete confirmation of its first opinion from the CVMP, constituted a clear and serious breach of the principle of sound administration.[119] This conclusion of the Court clearly entails a subordinate position of the work of the JECFA, in this case *vis-à-vis* the CVMP. However, it also illustrates that the function of the reports of the Joint FAO/WHO expert bodies can consist of confirming the scientific complexity of the case at hand, which may justify requesting a re-evaluation from the Scientific committee in question.[120]

---

[116] *CEVA Santé animale SA* v. *Commission*, Joined Cases T-344/00 and T-345/00 [2003] ECR II 00229, paras. 25, 29.

[117] Ibid., para. 60. In *CEVA Santé animale SA* v. *Commission*, the decision at stake was the decision of the Commission to postpone a positive decision on the inclusion of progestone in the list of authorised veterinary drugs, despite the positive conclusion of the Committee for Veterinary Medicinal Products (CVMP), the responsible scientific committee advising the Commission. In fact, it requested the CVMP to provide for a further opinion on the issue. When evaluating the risks relating to the veterinary drug progesterone, the Commission used the work of other scientific committees, such as the International Agency for Research on Cancer (IARC) and the JECFA as reason for this decision. The applicants held that by not adopting the relevant MRLs under a rapid procedure upon the reception of the scientific opinion of the CVMP, the Commission had manifestly failed to act (paras. 61-62).

[118] *CEVA Santé animale SA* v. *Commission*, Joined Cases T-344/00 and T-345/00 [2003] ECR II 00229, para. 99.

[119] Ibid., paras. 101-103.

[120] This latter role can equally be found in another case, *Pfizer*, in which reference was not made to the work of a Joint FAO/WHO expert body, but to a closely related body, a WHO expert meeting (*Pfizer Animal Health SA* v. *Council*, Case T-13/99 [2002] ECR II 03305, para. 36). In *Pfizer*, the issue was raised that the Community institutions had erred by basing their risk assessment on conclusions and recommendations of international research institutions. (In this case, a report from a WHO Expert Conference), Community and national bodies other than the opinion of the Scientific Committee for Animal Nutrition (SCAN), the relevant competent European scientific committee. The Court found that there was nothing that would preclude the Community to take into account international studies, in fact it was considered to be in line with the approach to ensure that most recent results of international research is taken into account (*Pfizer Animal Health SA* v. *Council*, Case T-13/99 [2002] ECR II 03305, para. 309). This decision rested upon three reasons. First, the reports of amongst others the WHO Conference was the result of a large group of scientists consulted and thus to be considered as 'best available data available at international level' (para. 307). Second, the reports dealt specifically with risks associated with the use of antibiotics (para. 306). Third, the Commission did not replace the reports of the SCAN by the conclusions of the reports of international research in its considerations preceding the establishment of the directive in question. Its decision to deviate from the SCAN opinion was not founded on the conclusions of the international reports, but on the aspects of the SCAN opinion itself (para. 305). See, on the Pfizer case, E. Vos, 'Antibiotics, the Precautionary Principle and the Court of First Instance', 11 *Maastricht Journal of European and Comparative Law* (2004) pp. 187-200.

In *Affish*,[121] another way of invoking Codex measures in order to contest the legality of Community decisions was found: through the invocation of the international obligations resulting from the WTO agreements, in particular, through the *SPS Agreement* or the *TBT Agreement*. If the Court were to apply the WTO obligations directly in order to examine the legality of EC measures, the function of the Codex Alimentarius would also increase. The opinion of Advocate-General Cosmas in *Affish* implied this additional way of invoking Codex standards. Although he rejected the direct effect of the *SPS Agreement*, he completed his examination in the event that the Court would rule the contrary. In his examination, he took account of the fact that the Directive was based upon and in conformity with the principles and findings of the Codex Alimentarius Commission as required by Article 3.1 of the *SPS Agreement*.[122]

However, in former cases, the Court rejected the WTO rules as a basis for examining the legality of EC measures.[123] Despite the fact that WTO law is binding on the European Community, and, as such, has become an integral part of the EC legal order,[124] this does not imply that it can be directly invoked before the Court. In fact, in *Portugal* v. *Council*, the ECJ explicitly rejected the direct effect of the provisions of the WTO agreements, and, therefore, they do not form part of the rules by which the Court can review the legality of acts adopted by the Community institutions.[125] The direct effect of Codex standards to examine the non-conformity of EC measures will, most probably, be rejected. This conclusion is based upon the decision of the Court and the Advocate-General in *Portugal* v. *Commission*, in which Portugal directly invoked the guidelines of the International Office of Epizootics (OIE), claiming that the provisions of EC law were not consistent with these guidelines.[126] The interesting aspect of this claim was that it implied the direct effect of the OIE guidelines, which are, given their status under international law and the reference to them in the *SPS Agreement* and the *TBT Agreement*, comparable to the Codex Alimentarius. Both the Court[127] and Advocate-General Mischo[128] rejected the binding effect of the OIE Guidelines.

---

[121] This case resulted from a preliminary question under the procedure of Art. 234 EC.

[122] *Affish BV* v. *Rijksdienst voor de keuring van Vee en Vlees*, Case-183/95, [1997] ECR I-4315, paras. 134 and 136. However, he does not examine whether the EC Directive is actually based and conform with Codex principles but relies on the argument of the Commission.

[123] See, on the ECJ's jurisdiction in relation to WTO law, J.H.J. Bourgeois, 'The European Court of Justice and the WTO: Problems and Challenges', in J.H.H. Weiler, *The EU, the WTO and the NAFTA* (New York, Oxford University Press 2000), pp. 71-123.

[124] Haegeman, Case 181/73 [1974] ECR 449, Kupferberg, Case 104/81 [1982] ECR 3641, Demirel, Case 12/86 [1987] ECR 3719. On the basis of Art. 300(7) TEC.

[125] *Portugal* v. *Council*, Case C-149/96 [1999] ECR I-8395. Also the direct effect of decisions of the WTO 'Dispute Settlement Body' is rejected by the Court (Biret International SA v. Council, Case T-174/00, [2002] ECR II-00017, para. 67). This consideration is in line with an earlier finding of the ECJ in *Atlanta* v. *European Community* [1999] ECR I-6983, in which the invocation of the DSB decision on the inconsistency of the common organisation of the market in the area of bananas with the 1994 GATT provisions was raised as a matter of appeal to a decision of the Court of First Instance concerning the lack of direct effect of the 1994 GATT provisions.

[126] *Portugal* v. *Commission*, Case C-356/99 [2001] ECR I-05645, para. 14.

[127] Ibid., para. 34.

[128] Ibid., para. 42 and 43.

The Court's refusal to accord binding effect to international obligations in order to examine the legality of EC measures is unlikely to change in the near future. However, this does not preclude the Court from using Codex standards as 'guidance' in future cases when conducting judicial review of EC measures. An important incentive to do so can be found in the adoption of EC Regulation 178/2002, in particular, its Article 5(3), which will be discussed in the following section.

6. 2002 AND BEYOND: INCREASING RESORT TO CODEX STANDARDS?

The previous sections indicated the decreased use of the Codex Alimentarius at all levels of Community decision-making. This decreased use of the Codex Alimentarius would most probably have persisted if the EC had not been faced with the consequences of the adoption of the WTO agreements as a result of the *EC-Hormones* case. During the negotiations that preceded the adoption of the WTO agreements, attention had already been given to the possible consequences that the adoption would have on the EC food policy.[129] However, it was the publication of the preliminary report of the WTO Panel in *EC-Hormones* that triggered the awareness of the increased status of the Codex Alimentarius under the WTO agreements and the resulting consequences for the EC. Striking, in this respect, is the resolution adopted by the European Parliament as a result of the preliminary Panel report in *EC-Hormones*, which stated that it:

> 'Challenges the *use of the Codex alimentarius standards as the sole reference criteria* for WTO arbitration, particularly in view of the fact that they are already scientifically outmoded and to the undemocratic and obscure procedural rules of the Codex; notes that the scientific arguments which were decisive in determining the panel's decision are largely based on evidence supplied by the industry itself.'[130]

With the consequences of the entry into force of the WTO agreements, the Community institutions face a divergence between the diminishing use of Codex measures and the responsibilities which originate from the international obligations to use Codex measures. Prior to 2002, the EC system did not provide for structural legal instruments to correct this disparity. However, several recent developments have aimed to address this concern and to augment the use of Codex measures by both the Community institutions and EC member states.

---

[129] See questions by Llewellyn Smith No. 1644/92, 1645/92, 1646/92, 1649/92 and 1650/92, C 333/2, 8.12.93.

[130] EP resolution, OJ 1997, C222/53 [emphasis added].

6.1 General principles of food law and the reference to the Codex Alimentarius

First and foremost, the adoption of Regulation 178/2002 responds partly to this concern.[131] It is meant to serve as a common basis of concepts, principles and procedures for measures taken in the member states and at Community level.[132] It states in Article 5(3), which regulates the objectives of the general food law of the European Union, that:

> 'Where international standards exist or their completion is imminent, they shall be taken into consideration in the development or adaptation of food law, except where such standards or relevant parts would be an ineffective or inappropriate means for the fulfilment of the legitimate objectives of food law or where there is a scientific justification, or where they would result in a different level of protection from the one determined as appropriate in the Community.'[133]

This provision is important for two main reasons. First, as mentioned above in Section 4.2, several EC regulations, directives and decisions explicitly refer to Codex standards and other related texts, and demonstrate that they have been taken into account during the adoption procedure. However, this does not constitute a coherent approach, and the use or reference to Codex measures rests upon the willingness of the Community institutions during the preparation and adoption of each separate EC measure. Article 5(3) of the Regulation, however, signifies an important breakthrough with this rather *ad hoc* approach of taking Codex measures into account. According to Article 5(3), international standards must be taken into consideration in the development or adaptation of food law. Article 3 of the Regulation defines 'food law' as: 'the laws, regulations and administrative provisions governing food in general, and food safety in particular, *whether at Community or national level.*'[134] Consequently, the obligation laid down in Article 5(3) applies at both national as well as Community level.

Second, as appears from the discussion in Section 5.3, the European Court of Justice declined to review the legality of Community institutions' actions on the basis of international standards or of the international obligations resulting from the

---

[131] European Parliament and Council Regulation 178/2002/EC laying down the general principles and requirements of food law, establishing the European Food Safety Authority and laying down procedures in matters of food safety [2002] L 31/1. The Regulation provides in the establishment of an independent European Food Safety Authority that breaks with the old way of the EC institutions depending on the co-operation with the member states for its scientific expertise. Furthermore, the Regulation sets out the general principles for food law, which are meant to serve as an overarching framework for all EC food legislation and to link all different regulations, directives and decisions together as a system.

[132] Consideration of the Preamble.

[133] European Parliament and Council Regulation 178/2002/EC laying down the general principles and requirements of food law, establishing the European Food Safety Authority and laying down procedures in matters of food safety [2002] L 31/1.

[134] [emphasis added].

WTO agreements. It is clear from the terms that Article 5(3) reflects an 'implementation' of the requirements of Article 2.4 of the *TBT Agreement*, and Article 3.1 in conjunction with Article 3.3 of the *SPS Agreement*.[135] As part of EC legislation, it constitutes a provision which enables the actions of Community institutions and national food legislation to be challenged before the European Court of Justice. However, this 'implementation' contains significant shortcomings, as the term of Article 5(3) differs from the terms of the relevant provisions of the *SPS Agreement* and the *TBT Agreement*.

A first difference is related to the scope of the Codex measures covered by the provisions. Article 5(3) only refers to 'international standards', whereas both the *SPS Agreement* and the *TBT Agreement* include guidelines and recommendations in the requirements.[136]

A more crucial difference is the one which relates to the nature of the obligation as contained in the provisions. Both Article 3.1 of the *SPS Agreement* and Article 2.4 of the *TBT Agreement* require that international standards *shall be used as a basis* for national regulations.[137] However, Article 5(3) of the EC Regulation requires that international standards 'shall be taken into consideration'. The difference between the obligations is that the obligation under the *SPS Agreement* and the *TBT Agreement* contains an obligation based on a substantive approach, whereas Article 5(3) of the Regulation is clearly restricted to an obligation based upon a procedural approach. The following illustrates this. The term 'taken into consideration' laid down in Article 5(3) implies that Codex standards are just one of the elements to be considered during the legislative process. It does not require that the resulting food law substantively embodies any element of the Codex standard. In contrast, the obligation to use international standards as a basis as required by the *SPS Agreement* and the *TBT Agreement* suggests the requirement to incorporate some elements of the Codex standard into the food measure to be adopted.[138] The obligation is clearly substantive in nature and goes beyond an obligation to take Codex standards merely into account during the legislative process. As a result, the

---

[135] See, for a detailed discussion on the obligation laid down in Art. 3.1 of the *SPS Agreement* and Art. 2.4 of the *TBT Agreement*, Chapter IV.

[136] The *TBT Agreement* makes a distinction between international standards to which its Art. 2.4 refers and international guidelines and recommendations as referred to in Art. 5.4 of the Agreement. See also, Section 3.1 of Chapter IV.

[137] Art. 3.1 of the *SPS Agreement*: 'To harmonize sanitary or phytosanitary measures on as wide a basis as possible, members *shall base* their sanitary measures or phytosanitary measures on international standards, guidelines or recommendations....' [emphasis added]. Art. 2.4 of the *TBT Agreement*: 'Where technical regulation are required and relevant international standards exist or their completion is imminent, Members *shall use them*, or the relevant parts of them, *as a basis* for their technical regulations...' [emphasis added]. See also, Section 2.2.1 of Chapter IV.

[138] Report of the Appellate Body, *EC-Measures Concerning Meat and Meat Products (Hormones)*, WT/DS26/AB/R, WT/DS48/AB/R (1998), para. 163. The Appellate Body rejected the obligation to incorporate all elements of the Codex standards as would be required under an obligation *to conform* measures *with* international standards. A similar finding has been concluded as regards the *TBT Agreement*, Report of the Appellate Body, *European Communities – Trade Description of Sardines*, WT/DS231/AB/R, 26 September 2002, para. 248. See also, Section 2.2.1 of Chapter IV.

obligation laid down in Article 5(3) of the Regulation is not as strict as the obligations contained in the *SPS Agreement* and the *TBT Agreement* and may very probably result in a Community or national decision in which Codex provisions are outweighed by other elements prevailing in the EC legal order.[139] This means that, even though Article 5(3) lays down an obligation which can be invoked before the European Court to examine the legality of the Community institutions' decisions and the decisions of member states, it actually provides the Court with few incentives to overrule the actions on the basis of not having considered Codex Alimentarius.[140]

6.2 EC food law and the consideration of Codex standards and other related texts

Since the adoption of Regulation 178/2002, an increasing number of EC regulations and directives has taken Codex measures into account, which has led to different results. These results vary from a reference to the work of the Codex Alimentarius Commission and the relevant expert bodies and the fact that this has been taken into consideration, to an actual adjustment of EC legislation as a consequence or the inclusion of specific justifications to deviate from Codex work.

Several Commission directives merely indicate the importance of the specifications and analytical techniques for additives as drafted by the JECFA and set out in the Codex Alimentarius.[141] Likewise, Consideration 6 of the Common Position

---

[139] See also, for a similar conclusion, the Opinion of the Economic and Social Committee, OJ 2001/ C 155/08, 29.5.2001, para. 3.4: 'Article 5 ("General Objectives"), paragraph 3, while generally in line with the basic principles of the SPS Agreement, is loosely phrased enabling the EU to opt out of international obligations seemingly without having to provide detailed justification.'

[140] In this context, reference can be made to the Court's decision in CEVA Santé Animale (*CEVA Santé animale SA* v. *Commission*, Joined Cases T-344/00 and T-345/00 [2003] ECR II 00229), where the case concerned Council Regulation 2377/90 which refers in the third recital of preamble explicitly to the Codex Alimentarius: 'whereas in order to protect public health, maximum residue limits must be established in accordance with generally recognised principles of safety assessment, taking into account any other scientific assessment of the safety of the substances concerned which may have been undertaken by international organisations, in particular the Codex Alimentarius or, where such substances are used for other purposes, by other scientific committees established within the Community.' This explicit reference did, however, not withhold the Court to give overriding force to the reports of the European Scientific Committee in question. Also the opinion of Advocate-General Cosmas in *Affish* indicates the hesitant position to test the decisions of the Community institutions on conformity with Codex Alimentarius. Even though he paid attention to the fact that the EC measures were in conformity with the provisions of the relevant Codex standard, he did so on the basis of the EC claim and did not actually verify whether the claim was correct. The adoption of Art. 5(3) of the Regulation 2002/178 most likely will not change this approach for the reasons as mentioned in this section.

[141] Commission Directive 2002/82/EC amending Directive 96/77/EC laying down specific purity criteria on food additives other than colours and sweeteners, OJ L 292/1, 28.10.2002, Commission Directive 2001/50/EC amending Directive 95/45/EC laying down specific purity criteria concerning colours for use in foodstuffs, OJ L 190/14, 12.7.2001, Commission Directive 2001/52/EC amending Directive 95/31/EC laying down specific criteria of purity concerning sweeteners for use in foodstuffs, OJ L 190/18, 12.7.2001.

with a view to adopting a Regulation on nutrition and health claims made on foods makes reference to the adopted Codex measures in the field of nutrition and health claims and mentions that the definitions and conditions set out in these measures have been taken into consideration.[142] Other EC legislation or proposals thereof also mention that Codex standards or other related texts have been taken into consideration.[143]

Some EC regulations and directives illustrate that the consideration of Codex measures has actually led to an adjustment of EC provisions so that they are in accordance with Codex requirements. For example, the Directive 2002/63/EC establishing Community methods of sampling for the official control of pesticides residues in and on products of plant and animal origin considers:

> 'Methods of sampling for the determination of pesticides residues for compliance with maximum residue levels (MRLs) were developed and agreed by the Codex Alimentarius Commission. The Community supported and endorsed the recommended methods. It is *appropriate to replace* the existing sampling provisions with those developed and agreed by the Codex Alimentarius Commission.'[144]

Likewise, Commission Regulation 1989/2002 amending Regulation 2568/91 on the characteristics of olive oil and olive-pomace oil and on the relevant methods of analysis, indicate that certain limits concerning the characteristics of olive oil and olive-pomace oil as laid down in the old regulation should be adjusted in order to be in line with the international standards as laid down by the Codex Alimentarius and the International Olive Oil Council.[145] Similarly, Commission Directive 2003/40/EC establishing the list, concentration limits and labelling requirements for the constituents of natural mineral waters adjusted several maximum limits of contaminants in natural mineral waters in order to bring them into line with the maximum limits as adopted by the Codex Alimentarius Commission and included in the Co-

---

[142] Common position (EC) 3/2006 of 8 December 2005 adopted by the Council, with a view to adopting a Regulation of the European Parliament and of the Council on nutrition and health claims made on foods, OJ C 80E/43, 4.4.2006, consideration 6: 'At the international level the Codex Alimentarius Commission has adopted General Guidelines on claims in 1991 and Guidelines for the use of nutrition claims in 1997. An amendment to the latter has been adopted by the Codex Alimentarius Commission in 2004. That amendment concerns the inclusion of health claims in the 1997 Guidelines. Due consideration is given to the definitions and conditions set in the Codex guidelines.'

[143] Regulation 852/2004 of the European Parliament and the Council on the hygiene of foodstuffs, OJ L 139/1, 30.4.2004. Commission Directive 2004/2/EC amending Council Directives 86/362/EEC, 86/363/EEC and 90/642/EEC as regards maximum residue levels of fenamipos, OJ L 14/15, 15.01.2004. Commission Directive 2003/113/EC amending the Annexes to Council Directives 86/362/EEC, 86/363/EEC and 90/642/EEC as regards the fixing of maximum levels for certain pesticide residues in and on cereals, foodstuffs of animal origin and certain products of plant origin, including fruit and vegetables, OJ L 324/24, 11.12.2003.

[144] Consideration 4 of the preamble, Commission Directive 2002/63/EC repealing Directive 79/700/EC. OJ L 187/30, 16.7.2002.

[145] Consideration 2 of the preamble, OJ L 295/57, 13.11.2003.

dex standard for natural mineral waters.[146] The reasons for the adjustments were that the Codex maximum limits afforded sufficient protection for public health.[147]

Although Commission Regulation 1181/2003 only briefly refers to the fact that the Codex Stan 94 is to be taken into account, it is clear that the reason for the amendment of Council Regulation 2136/89 laying down common marketing standards for preserved sardines is directly linked to the WTO decision in *EC-Sardines*.[148] It adjusts the 'old' regulation in order to establish Codex Stan 94 as its basis.[149] For instance, the title of the regulation is changed to include sardine-type products and the definition of 'preserved sardine-type products' includes 'Sardinops sagax' (the product at issue in *EC-Sardines*). Furthermore, a new provision was inserted to allow the labelling of Sardinops sagax as 'sardines' in combination with the scientific name of the species.[150]

In addition, some recent proposals for EC secondary legislation explicitly justify deviations from Codex measures. For instance, the proposal for a regulation on the addition of vitamins and minerals and of certain other substances to foods indicates the necessity to go beyond the definition of fortification as given in the 1987 Codex guidelines.[151]

These examples indicate that, in accordance with Article 5(3) of Regulation 178/2002, the Community institutions take Codex measures into consideration during the development and adoption of secondary food legislation. Several regulations indicate that the consideration of Codex standards during the legislative process has actually led to a substantive impact on the resulting regulations. In other cases,

---

[146] Commission Directive 2003/40/EC of 16 May 2003 establishing the list, concentration limits and labelling requirements for the constituents of natural mineral waters and the conditions for using ozone-enriched air for the treatment of natural mineral waters and spring waters, OJ L 126/34, 22.5.2003. Codex Stan 108-1981, Rev. 1-1997, amended in 2002 on natural mineral waters regulates e.g. health-related limits for certain substances. Most of these maximum limits have been included in the EC Commission Direct 2003/40/EC, such as arsenic, cadmium, chromium, copper, nitrate, lead, mercury and nickel.

[147] Consideration 4, 8 and 9 of Commission Directive 2003/40/EC of 16 May 2003 establishing the list, concentration limits and labelling requirements for the constituents of natural mineral waters and the conditions for using ozone-enriched air for the treatment of natural mineral waters and spring waters, OJ L 126/34, 22.5.2003.

[148] See also, S. Poli, *supra* n. 110, p. 616. See Chapter IV on a detailed discussion of the Panel and the Appellate Body decisions in EC-Sardines. Codex Stan 94 regulates the naming, packaging and labelling, quality factors and health protection issues related to sardines and sardine-type products. In Art. 6.1.1., it allows for the labelling of sardine-type products under the name 'sardines' in combination with a specification as regards the area of origin, the species or a common name used in the country in question. The issue in dispute concerned the fact that the former EC regulation did not allow sardine-type products to be labelled under the name 'sardine'.

[149] Commission Regulation 1181/2003 amending Council Regulation 2136/89 laying down common marketing standards for preserved sardines, OJ L 165/17, 3.7.2003. See also, answer by the Commission of 13 November 2002 to a Parliamentary question (E-2724/02) on the term 'sardines', OJ C11E/15, 15.1.2004.

[150] New Art. 7a.

[151] Consideration 9 and 10 of the proposal for a European Parliament and Council Regulation, COM(2003)671 final, 10.11.2003.

deviation from Codex measures is justified. These two actions may not necessarily be required by Article 5(3) of Regulation 178/2002 as discussed above in this Section and may partly be explained as being 'court-driven' in the sense that they anticipate and avoid complications that might have to be faced in the context of the dispute settlement mechanism of the WTO.

6.3 The European Food Safety Authority and the Joint FAO/WHO Food Standards Programme

This 'court-driven' approach was also taken into consideration in the preparation of new structures of scientific committees in order to be able to justify the scientific basis for EC measures in the light of the international obligations that result from the *SPS Agreement* and the *TBT Agreement*. This incentive can be traced back to the 1997 Green Paper on food law where it was mentioned that:

> 'The Community will increasingly be required to provide scientific justification for its measures at international level. The suggestions set out above for prior consultation of scientific committees should also be considered as proposals in connection with changes required of the Community at international level....Since the Community must be able to justify measures which diverge from the relevant international standards, it is important to take account of the international dimension in the Community's scientific assessment work.' [152]

In the light of the necessity for the scientific committees of the Community institutions to assist the latter in justifying its measures at international level, the 1997 Green Paper on food law emphasised the need for the Community scientific committees to establish close collaboration with the international scientific committees, such as Joint Expert Committee on Food Additives (JECFA) and the Joint Meeting on Pesticide Residues (JMPR):

> 'The members of Community scientific committees must be able to co-operate with their other colleagues, particularly within the context of international committees of scientific experts like the Joint Expert Committee on Food Additives (JECFA) and the Joint Meeting on Pesticide Residues (JMPR), which are responsible for preparing the scientific assessments of hazards which form the scientific basis for preparation of Codex standards....'[153]

---

[152] Commission Green Paper on the General Principles of Food Law in the European Union, COM (97)176 final, p. 61.

[153] Ibid. In practice, close collaboration is taken place. See S. Slorach, 'European Community Legislation on Limits for Additives and Contaminants in Food' in N. Rees and D. Watson (eds.), *International Standards for Food Safety* (Gaithersburg, Maryland, Aspen Publishers 2000), p. 54: 'Limits for many substances mentioned above (e.g., ochratoxin A. lead.cadmium, and some pesticides) are being discussed in parallel in the EC and in Codex and often by the same people. Such discussion facilitates coordination of the work in these different fora and should expedite the establishment of limits that can be widely accepted.'

With the adoption of Regulation 178/2002, the independent European Food Safety Authority (EFSA) was established.[154] However, none of the issues as laid down in the Green Paper mentioned above can be found in Regulation 178/2002 and the implementing regulations relating to the EFSA, Commission Regulation 2230/2004 or Commission Regulation 1304/2003.[155]

The networking of organisations to facilitate scientific co-operation, as mentioned in Article 36 of Regulation 178/2002 and elaborated in its implementing rules laid down in Commission Regulation 2230/2004, only concern organisations within the member states and do not include international expert bodies.[156] No explicit mention is made with regard to co-operation with international expert bodies. However, co-operation with international expert bodies is clearly one of the activities envisaged by the EFSA. Its Management for 2006 indicates EFSA's intention to increase collaboration activities, such as working jointly on projects of mutual relevance and benefit, in particular, data collection and the assessment of global scientific issues.[157] In practice, informal structures that have a great influence on the institutional relationships between the EU and the Codex Alimentarius Commission are already in place. For instance, the possibility of inviting external

---

[154] See, on the EFSA, also, E. Vos, *supra* n. 36. K. Vincent, *supra* n. 27, pp. 499-517. S. Krapohl, 'Credible Commitment in Non-Independent Regulatory Agencies: A Comparative Analysis of the European Agencies for Pharmaceuticals and Foodstuffs', 10 *European Law Journal* (2004), pp. 518-538. D. Chalmers, 'Food for Thought: Reconciling European Risks and Traditional Ways of Life', 66 *Modern Law Review* (2003), pp. 532-562. See also, Section 2 of this chapter. See, on EU Agencies in general, E. Vos, 'Reforming the European Commission: What role to play for EU Agencies', 37 *Common Market Law Review* (2000), pp. 1113-1134. R.H. van Ooik en W.T. Eijsbouts, 'De wonderbaarlijke vermenigvuldiging van Europese agentschappen. Verklaring, analyse, perspectief', 16 *Sociaal Economische Wetgeving* (2006), pp. 102-111. E. Chiti, 'The emergence of a Community administration: the case of European Agencies', 37 *Common Market Law Review* (2000), pp. 309-343. E. Chiti, 'Administrative proceedings involving european agencies', 68 *Law and Contemporary Problems* (2004), pp. 219-236.

[155] Commission Regulation (EC) No. 2230/2004 of 23 December 2004 laying down detailed rules for the implementation of European Parliament and Council Regulation (EC) No. 178/2002 with regard to the network of organisations operating in the fields within the European Food Safety Authority's mission, OJ L 379/64, 24.12.2004. Commission Regulation (EC) No. 1304/2003 of 11 July 2003 on the procedure applied by the European Food Safety Authority to requests for scientific opinions referred to it, OJ L 185/6, 24.07.2003.

[156] See proposed list of competent organisations in the framework of Art. 36. The network of scientific bodies in the member states aims to ensure co-operation as already envisaged by the former SCOOP. (The SCOOP referred to the system of scientific assistance and co-operation in the scientific examination of food issues which required member states to co-operate with the Commission through their competent authorities. This system was established in 1993 by the adoption of Council Directive 93/5/EEC on assistance to the Commission and co-operation by the member states in the scientific examination of questions relating to food, OJ L 52/18, 4.3.1993. See, on SCOOP, also, E. Vos, *supra* n. 7, pp. 171-173). The advisory forum, an organ of the EFSA composed of representatives from competent bodies of member states, has the task to promote the networking of the nominated competent organisations of the member states (Art. 27(4)(c) of Regulation 178/2002).

[157] European Food Safety Authority, Management Plan of the European Food Safety Authority for 2006, MB 24.01.2006, adopted by the Management Board at its meeting of 24 January 2006 in Parma, paras. 130-131.

experts, as contained in Article 8 of the EFSA decision concerning the establishment and operations of the scientific committee and panels, can be used to construct an informal relationship.[158] The fact that the former chairperson of the Management Board of the European Food Safety Authority (Dr. Slorach) was elected as chairman of the Codex Alimentarius Commission during its session in 2003,[159] illustrates that informal structures have definitely been set in place.

In contrast with the former procedure for scientific committees under Commission Directive 97/579/EC,[160] the current procedure for scientific committees does not foresee an explicit obligation to take the work undertaken by other international expert bodies into account. However, from its Management Plan for 2006, it becomes clear that the EFSA seeks ways of further utilising the outcomes of the work already undertaken by the JECFA, JMPR and JEMRA without putting the quality of EFSA's scientific opinions and its scientific independence in jeopardy.[161] In the field of pesticides residues, several Commission Directives indicate that the EFSA evaluates whether the adoption of Codex MRLs will pose health risks. They state that 'these [Codex MRLs] have been considered in the setting of the MRLs fixed in this Directive...The MRLs based upon Codex MRLs were evaluated in the light of the risks for consumers, and no risk was established'.[162]

---

[158] Art. 8 of the EFSA decision concerning the establishment and operations of the scientific committee and panels, MB 17.10.2002, available at: <www.efsa.eu.int>.

[159] Report of the 26th Session of the Codex Alimentarius Commission, Rome, 30 June-7 July, 2003, ALINORM 03/41, para. 240.

[160] Art. 2(3)(d) of the Commission Decision 97/579/EC of 23 July 1997 setting up Scientific Committees in the field of consumer health and food safety, OJ 1997 L 237/18: 'evaluate the scientific principles on which Community health standards are based taking into account risk assessment techniques developed by the international organizations concerned.'

[161] European Food Safety Authority, Management Plan of the European Food Safety Authority for 2006, MB 24.01.2006, adopted by the Management Board at its meeting of 24 January 2006 in Parma, para. 131.

[162] Consideration 3 of the Commission Directive 2003/118/EC amending the Annexes to Council Directives 76/895/EEC, 86/362/EEC, 86/363/EEC and 90/642/EEC as regards maximum residue levels for acephate, 2,4-D and parathion-methyl, OJ L 327/25, 16.12.2003. Consideration 7 of the Commission Directive 2003/60/EC amending the Annexes to Council Directives 76/895/EEC, 86/362/EEC, 86/363/EEC and 90/642/EEC as regards the fixing of maximum levels for certain pesticide residues in and on cereals, foodstuffs of animal origin and certain products of plant origin, including fruit and vegetables, OJ L 155/15, 24.06.2003. Initially, the proposal for the Council Regulation on maximum residue levels of pesticides in products of plant and animal origin, the more 'horizontal' regulation in the field of pesticides, required the EFSA to take into account MRLs adopted by the Codex Alimentarius Commission (Draft Art. 25, COM(2003)117 final). The reason indicated was: '... under the WTO rules introduced late in the 1990's, Codex MRLs should be respected. Many Codex MRLs are not acceptable to the Community, particularly those set prior to the late 1990's but where the Commission did not formally object to them at that time. Therefore, the Authority will have to examine on a case-by-case basis whether Codex MRLs ensure the same high level of protection that would be expected from Community MRLs.' (explanatory memorandum to the proposal).

7. The changing inter-institutional relationship between the European Commission and the Codex Alimentarius Commission

Even since the establishment of the European Economic Community, it has been clear that Community institutions will adopt legal instruments to harmonise national food measures themselves. This decision to adopt harmonisation instruments turned the Community institutions into institutions that potentially compete with the Codex Alimentarius Commission with regard to the mandate to adopt harmonisation food measures. Given the urgent need for food standards that are applicable on a European market,[163] and given the fact that the Community institutions were less successful in adopting harmonisation measures, the adopted Codex measures were considered to be a useful supplementary means to further harmonisation within the European Community. As a result, co-operation between the Codex Alimentarius Commission and the Commission of the European Community was established, as both institutions had the same interest in promoting the use of Codex measures by the EC member states. The co-operation activities are clearly reflected in the inter-institutional discussions between the Commission and the Codex Secretariat as well as in a letter from the Director-General of the FAO addressed to the Commission.[164] As appears from the documents, the function of the European Commission consisted of stimulating the harmonisation process of the Codex Alimentarius Commission within the European Community through the promotion of the acceptance of Codex standards, either by EC member states individually, or jointly through the European Community.

However, the function of the European Commission within the Codex Alimentarius Commission was limited to observer-status, and thus it had no power as a member in the standard-setting process. Nevertheless, the Agreement of the member states concerning the 'standstill and information for the Commission' sustained the compatibility of the negotiations conducted by the member states during the Codex elaboration procedure with the EC harmonisation process. The Agreement also applied to the work undertaken by the member states in the context of international organisations such as the FAO and the WHO, in which they participate.[165] As a result, during the negotiations in the context of the Codex elaboration

---

[163] Report of the Joint FAO/WHO Conference on Food Standards, Geneva 1-5 October 1962, ALINORM 62/8, para. 35. '... during the first four years of the Commission's work acceptance by European governments alone would be a necessary and sufficient condition for the publication of a standard in the Codex. This clause was intended to underline the urgent need for food standards applicable to the European market and to provide for the publication of European standards even if agreement on a wider basis should prove impracticable in any given case.'

[164] Proposal for a Council Regulation (EEC) on the acceptance by the European Economic Community of standards or maximum limits for pesticide residues or maximum limits of veterinary medical drugs residues drawn up under the Joint FAO/WHO Food Standards Programme (Codex Alimentarius), COM(90)216 final, 21 May 1990, Annex 2 p. 2.

[165] Agreement of the Representatives of the Governments of the member states meeting in the Council of 28 May 1968 concerning standstill and information for the Commission, OJ C 76/9, 17.6.1969. Art. 2 of the Agreement indicates that the procedure of standstill and information equally applies to

procedure, EC member states were also held to take into consideration the *status quo* of the EC food law (in preparation), and only to undertake new regulatory initiatives that would obstruct the vertical harmonisation programme on foodstuffs, as established by the Council Resolution of 28 May 1969, under very strict circumstances.[166] However, in practice, this 'gentlemen's agreement' did not work.[167] It was found to be inadequate and the member states felt justified to act independently.[168]

With the progress of the completion of the internal market, the need for the Commission to accept Codex standards on behalf of the Community and to play a more important role in the Codex elaboration procedure increased. Consequently, on the initiative of the representative of the EC Commission, the Codex Alimentarius Commission decided to modify its rules concerning the acceptance procedure by allowing for the acceptance of Codex standards and MRLs by the international organisations of regional economic communities.[169] Furthermore, in 1993, the European Commission launched a process to amend the rules of procedure of the Codex Alimentarius Commission in order to allow regional economic integration organisations to become Codex Members.

As mentioned above in Section 6, the *EC-Hormones* decisions resulted in the awareness that the entry into force of the WTO agreements had entailed an increased status of Codex standards. This re-inforced the need for an increased input on the part of the EC, as one voice, within the elaboration procedure of the Codex Alimentarius Commission.[170] It was a call for the EC to act as a 'rule-maker' in Codex standard-setting procedure, with the task of ensuring that the European health interests and the interests of the European food industry were advanced more effectively than before. The Commission was also called upon to initiate a discussion on the legitimacy of the Codex procedure.[171] For this reason, full membership of the

---

activities undertaken in the framework of international organisations, such as the FAO and the WHO with the aim the approximation of Legal and administrative provisions in the area of foodstuffs.

[166] Agreement of the Representatives of the Governments of the member states meeting in the Council of 28 May 1968 concerning standstill and information for the Commission, OJ C 76/9, 17.6.1969.

[167] D. Welch, *supra* n. 4, p. 56.

[168] Ibid.

[169] Communication from the Commission to the Council 'Amendment of the Procedural Manual of the Codex Alimentarius Commission (9th Session of the Codex Committee for General Principles, Paris, 24-28 April 1989) COM(89)120 final, 16.3.1989.

[170] Opinion of the Economic and Social Committee on the Commission Green Paper and the Communication from the Commission, C19/61, 21.1.98, para. 3.6.2: 'With regard to the multilateral dimension and the ever-increasing importance of world trade, it is necessary for the European food industry and European consumer interests to be represented as effectively as possible in hard-fought world trade negotiations and in relevant international bodies such as the WTO and the Codex Alimentarius. European interests need to be pushed more effectively than in the past and this can be done by coordinating and pooling efforts…' and para. 3.6.2.2: 'In this context, the Committee would also call upon the Community and the Member States to actively promote the European viewpoint that in cases of scientific doubt consumer interests take precedence over economic considerations.'

[171] See Resolution on the Commission Green Paper on the general principles of food law in the European Union, A4-0009/98, para. 74: 'Notes that the Codex Alimentarius does not accept consumer

EC to the Codex Alimentarius Commission has been advocated. For instance, in its 1997 Green Paper, the Commission stated that:

> 'The Community must also be capable of playing a full role during the negotiations within Codex Alimentarius and other *fora* which lead to the adoption and acceptance of international standards. Full Community participation is essential in order to ensure that the Community interest is taken into consideration during the preparation of international standards which will be used as a reference point for judging the legitimacy of the Community's own legislation. The current situation where the Community is a full party to the WTO agreements, but is accorded only the status of observer during the elaboration of international standards by Codex, constitutes an unacceptable anomaly, which must be remedied as soon as possible by the accession of the Community to full membership of Codex Alimentarius.'[172]

During the Codex Alimentarius Commission meeting of June 2003, the Codex rules of procedure were amended.[173] Through the adoption of the Council Decision of 17 November 2003 and through its request for accession to the Codex Alimentarius Commission, the EC formally became a Member of the Codex Alimentarius Commission.[174] With the adherence of the European Community as a Member to the Codex Alimentarius Commission in 2003, the EC formally broke with its observer-status as an international organisation.

This new relationship characterises the change that Codex measures have acquired in the EC harmonisation process, as discussed in Section 6 of this chapter. The Commission no longer has a supervisory function to promote the use of Codex measures. As a result of the international obligations contained in the *SPS Agreement* and the *TBT Agreement*, it has acquired the responsibility to use the Codex standards as a basis for Community food legislation, and the responsibility to en-

---

concerns as the basis for setting international standards and calls on the Commission to launch a discussion process on how the Commission's new scientific committees and the Codex Alimentarius scientific committees can best coordinate their future activities to ensure that such concerns are taken into account; *calls on the Commission also to make every effort to amend the Codex Alimentarius decision-making process so as to bring it more into line with the real impact of Codex standards since the WTO was established.*' See also question by Marianne Thyssen, E-2929/96, 8 November 1996: 'Since the start of the World Trade Organization (WTO) the degree of enforceability of certain standards in the framework of the Codex Alimentarius has increased. As the Commission is aware, in the Codex Alimentarius decisions are taken by a simple majority of votes cast. As these standards have an indirect influence on the content of European legislation, it would seem appropriate for the Commission to recommend the Member States to press in the Codex Alimentarius for a change in the decision-making procedure so that it would relate to the actual impact which the Codex standards will have in future. Does the Commission intend to take steps in this direction?'

[172] Commission Green Paper on the General Principles of Food Law in the European Union, COM(97)176 final, p. 62.

[173] Report of the 26th Session of the Codex Alimentarius Commission, ALINORM 03/41, Rome 30 June-7 July 2003, paras. 22-24.

[174] Council Decision on the accession of the European Community to the Codex Alimentarius Commission, OJ L 309/14, 26.11.2003.

sure that Community concerns are taken into account during the Codex standard-setting procedure.[175]

8. CONCLUSIONS

Throughout the history of the EC, Codex measures have played a role in the EC harmonisation process of food measures. This role of Codex measures has varied depending on both internal and external developments relating to the EC food legislation.

Prior to 1987, the harmonisation of national food measures at EC level progressed slowly. In contrast, during this period, the Codex Alimentarius Commission had been able to adopt a large number of Codex measures. During this period, the relationship of the EC Commission *vis-à-vis* the Codex Alimentarius Commission was one of co-operation: the EC Commission assisted in promoting the acceptance of Codex standards by the EC member states. Furthermore, reference to the Codex Alimentarius could frequently be found both in the legislative process of EC food regulation, as well as in the jurisprudence of the European Court of Justice in the context of Articles 28 and 30 EC. Although, in both contexts, Codex measures did not acquire a binding status, their role can be defined as being an important reference point. Both the scientific opinions of the Joint FAO/WHO expert bodies and the Codex measures themselves were taken into account on various occasions during the legislative process and had an impact on the content of several EC food measures. In addition, with regard to the justification of national food measures in the context of Articles 28 and 30 EC, the European Court of Justice required that the international scientific findings of the 'Codex Alimentarius Committee' be taken into account whenever decisions were taken as to whether to authorise a substance or not. The Court itself used the Codex Alimentarius as a reference point to examine the justification for national food measures.

During the next period (1987-1997), the role attributed to the Codex Alimentarius diminished in both contexts. With the completion of the internal market in sight and a newly introduced legislative process in 1987, which favours the adoption of EC secondary food legislation, the influence of the Codex Alimentarius on the content of EC food measures diminished. Furthermore, a stronger institutional framework that facilitated the work of the EC Commission, in particular, a stronger Scientific Committee on Food that responded better to the health concerns of the EC, led to a reduced role of the scientific opinions of the Joint FAO/WHO expert bodies. These two developments also had a reduced impact on the function of Codex measures as established in the Court's jurisprudence on Articles 28 and 30 EC. The simple fact

---

[175] The EFSA should assist the Commission in this new role as is reflected in consideration 39 of Regulation 178/2002: 'The Authority should contribute through the provision of support on scientific matters, to the Community's and Member States' role in the development and establishment of international food safety standards and trade agreements.'

that the increasing adoption of EC food regulation diminished the role of Articles 28 and 30 EC in food issues meant that the function of Codex measures as a reference point also became less relevant. Although references to Codex measures can also be found in more recent jurisprudence on the interpretation and application of secondary food legislation, the function of the Codex was clearly reduced to that of an 'interpretive means' to define more precisely the meaning of the relevant EC food measures. Furthermore, the increasing expertise of the expert bodies at Community level has led to the tendency of the Court to attach less weight to the scientific opinions of the Joint FAO/WHO expert bodies.

In 1997, the same year in which the Community institutions were held accountable as a result of the BSE crises, they were also faced with the restrictions on their responsibilities due to their international obligations under the WTO agreements. The WTO Panel decision in *EC-Hormones* had demonstrated the direct consequences of the entry into force of the WTO agreements for the EC, and the resulting increased status of Codex measures. The EC was confronted with the disparity between the lack of legal instruments within the EC legal order, which would require Codex measures to be taken into account, and the international obligations to use them. The adoption of the WTO agreements also led to an additional means of contesting the legality of Community decisions, namely, through the invocation of the WTO agreements as an international obligation. Although this has been unsuccessful to date, if, in future cases, the Court directly applied the WTO agreements, the status of Codex measures would also increase.

As from 2002 *circa*, legal initiatives have been taken that have led to an increased status and use of the Codex within EC legal order. With the adoption of Regulation 178/2002 and its Article 5(3), a general obligation to take Codex standards into consideration, by both member states as well as the Community institutions, when adopting and preparing food law, has been established. Although it cannot be concluded that Article 5(3) provides Codex measures with a binding force in the EC legal order, it does mean that Codex measures have acquired status as a 'reference point' in the justification of measures. Recently adopted EC secondary food legislation demonstrates that the Community institutions have frequently taken Codex measures or their scientific basis into consideration, which has led to the adjustments of several EC measures or an explicit explanation on the reasons for a deviation from Codex measures. Whether Codex measures as a reference point will be a decisive element in the Court's examination of the justifications of food measures remains, however, to be seen. Furthermore, the position of the EC Commission *vis-à-vis* the Codex Alimentarius Commission has changed considerably, since the EC has become a full Member, and, as a result, has formally broken with its observer status. Within the Codex Alimentarius Commission, the EC Commission has clearly acquired the function of ensuring that the Community's concerns are taken into consideration during the standard-setting procedure.

Chapter IV
THE WTO AGREEMENTS AND THE CODEX ALIMENTARIUS

1. INTRODUCTION

This chapter examines the status of Codex measures under the *SPS Agreement* and the *TBT Agreement* and their legal consequences for the inter-institutional relationship between the WTO and the Codex Alimentarius Commission.

The preamble of both the *SPS Agreement* and the *TBT Agreement* recognise the harmonisation of national measures as an instrument to further the objectives of facilitating trade and minimising the negative impact on international trade.[1] The instrument of harmonisation as a complimentary tool to general prohibitive provisions which aim to eliminate trade barriers was already incorporated in the 'Standards Code' (the former *TBT Agreement* resulting from the Tokyo Round).[2] The reason for the introduction of the instrument of harmonisation was a strong increase of newly created technical regulations and standards by national governments,[3] which occurred during the 1960s and the 1970s, which was not followed by efforts on the part of national governments to agree upon internationally acceptable standards. At the same time, the reduction of trade tariffs as a result of the first five negotiation rounds revealed that non-trade barriers also had an important impedimental effect on the international trade of goods.[4]

The relevance of the Codex measures as a means of harmonisation under the *SPS Agreement* and the *TBT Agreement* results from the references to international standards, guides and recommendations that are contained in both agreements, and has been defined as 'the biggest single step in the history of the globalization of food standards'.[5] The work of the Codex Alimentarius Commission is explicitly mentioned in the *SPS Agreement*.[6] Also in the context of the *TBT Agreement*, which does not explicitly mention the work of the Codex Alimentarius Commission, the Panel in *EC-Sardines* recognised the Codex Alimentarius Commission as an international standardising body.[7] According to scholars, the provisions of both agree-

---

[1] Consideration 5 of the *SPS Agreement* and Consideration 3 of the *TBT Agreement*.
[2] See, for instance, Consideration 3, Art. 2.3 and Art. 9.1 of the Tokyo Round *TBT Agreement*.
[3] J. Groetzinger, 'The New GATT Code and the International Harmonization of Products Standards', 8 *Cornell International Law Journal* (1975), p. 170.
[4] R.E. Sweeney, 'Technical Analysis of the Technical Barriers to Trade Agreement', 12 *Law & Policy in International Business* (1980), p. 187. W. Mattli, 'The politics and economics of international institutional standards setting: an introduction', 8 *Journal of European Public Policy* (2001), p. 329.
[5] J. Braithwaite and P. Drahos, *Global Business Regulation* (Cambridge, Cambridge University Press 2000), p. 403.
[6] Annex A.3(a) of the *SPS Agreement*.
[7] Report of the WTO Panel, *European Communities – Trade Description of Sardines*, WT/DS231/R, 29 May 2002, para. 7.66.

ments entail an increased status of the Codex standards.[8] However, the explanation of the reasons, meaning and consequences that are related to this increased status is often neglected.

To assist in defining the increased status of Codex measures under the *SPS Agreement* and the *TBT Agreement*, this chapter starts with an analysis of the harmonisation provisions in both agreements, their objective, their content and their place within the agreements (Section 2). This section is followed by an analysis of the status accorded to Codex standards as a result of the harmonisation provisions and their reference to international standards. It is primarily based upon the panels' and the Appellate Body's decisions in two cases in which Codex standards were invoked by the complainant: *EC-Hormones*, with regard to the status of Codex standards under the *SPS Agreement*, and *EC-Sardines*, with regard to their status under the *TBT Agreement* (Section 3). This status is compared with the original status of Codex measures under the Codex acceptance procedure and the resulting obligations for countries which had accepted them (Section 4). In a last section (Section 5), the consequences of the status accorded to Codex measures for the inter-institutional relationship between the WTO and the Codex Alimentarius Commission are discussed.

2. Harmonisation in the context of the *SPS Agreement* and the *TBT Agreement*

Harmonisation provisions that relate to food products are contained in two of the annexed agreements to the WTO Agreement: the *SPS Agreement* and the *TBT Agreement*. The instrument of harmonisation is not introduced anew by these agreements, but has its origin in the former *TBT Agreement* that resulted from the Tokyo Round (hereinafter referred to as the TBT Code). The 1960s and the beginning of the 1970s were characterised by a strong increase of newly created technical regulations and standards by national governments.[9] This sudden development at national level was not followed by efforts of national governments to agree upon internationally acceptable standards.[10] For these reasons, the TBT Code introduced a completely new approach to facilitate trade within the GATT system: the promotion of international harmonisation to reduce the different national regulatory approaches.[11] Both

---

[8] R. Muñoz, 'La Communauté entre les mains des normes internationales: les conséquences de la décision Sardines au sein de l'OMC', 4 *Revue du Droit de l'Union Européenne* (2003), p. 463: 'En deuxieme lieu, le role des norms internationals au sein de l'accord OTC a été clairement défini. En effet, le statut des normes internationals se trouve considérablement renforcé par les décisions du panel et de l'Organe d'appel.' M. Echols, *Food Safety and the WTO. The interplay of culture, science and technology* (The Hague, Kluwer Law International 2001), p. 100: 'With the harmonization provision the SPS Agreement elevates the Codex Alimentarius Commission from an obscure role to one of potentially immense importance in international food trade.'

[9] J. Groetzinger, *supra* n. 3, p. 170.

[10] R.W. Middleton, 'The GATT Standards Code', 14 *Journal of World Trade Law* (1980), p. 202.

[11] On the TBT Code and harmonisation, see R.E. Sweeney, *supra* n. 4, p. 187. W. Mattli, *supra* n. 4, p. 329. R.W. Middleton, *supra* n. 10, pp. 206-207. J. Groetzinger, *supra* n. 3, pp. 168-188.

the *SPS Agreement* and the *TBT Agreement* have inherited this approach from the TBT Code, and emphasise the importance of harmonisation through the use of international standards as a component of its policies to eliminate trade barriers.[12] Under both agreements, the instrument of harmonisation functions as a complementary tool to the 'regular disciplines' of eliminating unnecessary non-tariff barriers. This section explores the objectives, functions and content of the harmonisation provisions in the context of the *SPS Agreement* and the *TBT Agreement*.

2.1 The objective and the function of harmonisation in the framework of the *SPS Agreement* and the *TBT Agreement*

The *SPS Agreement* and the *TBT Agreement* are an integral part of the multilateral legal system as established by the WTO Agreement. The ultimate objectives of the WTO Agreement are an increase in the standards of living, the attainment of full employment, the growth of real income and effective demand, and the expansion of the production of, and trade in, goods and services.[13] In aiming to achieve these objectives, the WTO takes into account the optimal use of natural resources in accordance with the objective of sustainable development and the integration of developing countries in the world trading system.[14] The instruments used to achieve these objectives are the substantial reduction of tariffs and other barriers to trade and the elimination of discriminatory treatment in international trade.[15]

Both the *SPS Agreement* and the *TBT Agreement* are included in Annex 1A of the WTO Agreement, entitled 'multilateral agreements on trade in goods'. In line

---

J. Nusbaumer, 'The GATT Standards Code in Operation', 18 *Journal of World Trade Law* (1984), pp. 542-552.

[12] Preamble of the *SPS Agreement*, considerations 4 and 5: '*Desiring* the establishment of a multilateral framework of rules and disciplines to guide the development, adoption and enforcement of sanitary and phytosanitary measures in order to minimize their negative effects on trade. *Recognizing* the important contribution that international standards, guidelines and recommendations can make in this regard.' Preamble of the *TBT Agreement*, consideration 3: '*Recognizing* the important contribution that international standards and conformity assessment systems can make in this regard by improving efficiency of production and facilitating the conduct of international trade.'

[13] Consideration 1 of the Preamble to the Agreement establishing the World Trade Organisation. See also, M. Footer, *An Institutional and Normative Analysis of the World Trade Organization* (Leiden, Martinus Nijhoff Publishers 2006), p. 23, P. Van den Bossche, *The Law and Policy of the World Trade Organization. Text, Cases and Materials* (Cambridge, Cambridge University Press 2005), pp. 86-87.

[14] Consideration 2 of the Preamble to the Agreement establishing the World Trade Organisation. See also, P. van den Bossche, *supra* n. 13, p. 86. See also, M. Footer, *supra* n. 13, p. 23.

[15] Considerations 3 and 4 of the Preamble to the Agreement establishing the World Trade Organisation. M. Footer, *supra* n. 13, p. 23. Although the WTO has been described as linkage machine as its scope of activities links to non-trade concerns, the activities of international co-operation of the WTO concentrate on trade liberalisations and non-trade concerns remains in most cases under the responsibility of WTO Members or other international institutions. See special issue on this topic of the (2002) 96 *American Journal on International Law*, pp. 1-159. When addressing the functional approach of determining the mission of the WTO, Charnovitz states that 'the mission of the WTO is self-evident. It's about trade'. See S. Charnovitz, 'Triangulating the World Trade Organization', 96 *American Journal of International Law* (2002), p. 47.

with the objectives of the WTO, the *SPS Agreement* states in its preamble that it desires 'the establishment of a multilateral framework of rules and disciplines to guide the development, adoption and enforcement of sanitary and phytosanitary measures in order to minimize their negative effects on trade'.[16] Likewise, in its preamble, the *TBT Agreement* refers to the objectives of GATT 1994 and the desire to further these objectives: the substantial reduction of tariffs and other barriers to trade and the elimination of discriminatory treatment in international commerce.[17]

Article 2.1 of the *SPS Agreement* recognises the right of WTO Members to take whatever sanitary and phytosanitary measures they deem to be necessary for the protection of human, animal or plant life or health.[18] This right allows individual WTO Members to adopt regulatory measures in order to address situations that endanger human, animal or plant life or health within their territory. However, these measures may easily impede international trade in food and agricultural products. For this reason, the right to take sanitary and phytosanitary measures is not an unlimited right and is conditioned by strict rules and disciplines as laid down in the provisions of the *SPS Agreement*.[19] These rules and disciplines consist of:

1. the requirement of the scientific justification of the taken sanitary and phytosanitary measures;[20]
2. the requirement that measures are not more trade-restrictive than necessary to achieve sanitary and phytosanitary protection;[21] and
3. the requirement that the measures do not arbitrarily or unjustifiably discriminate between members where similar conditions prevail and do not constitute a disguised restriction to trade.[22]

The *TBT Agreement* also recognises the right of WTO Members to take measures.[23] The following objectives for which technical regulations, standards and conformity assessment procedures may be taken are mentioned: the protection of human, animal or plant, life or health, the protection of the environment, the prevention of deceptive practices, and the protection of national security.[24] The *TBT Agreement*

---

[16] Consideration 4 of the Preamble of the *SPS Agreement*. See, for an extensive analysis of the *SPS Agreement*, D. Prévost and P. Van den Bossche, 'The Agreement on Sanitary and Phytosanitary measures', in P.F.J. Macrory, et al. (eds.), *The World Trade Organization: legal, economic and political analysis,* Vol. I (New York, Springer 2005), pp. 231-370.

[17] Consideration 2 of the Preamble of the *TBT Agreement* in conjunction with consideration 3 of the Preamble of *GATT 1994*.

[18] See also, consideration 1 of the Preamble of the *SPS Agreement*: 're-affirming that *no Member should be prevented* from adopting or enforcing measures necessary to protect human, animal or plant life or health, ...'

[19] C. Button, *The Power to Protect. Trade, Health and Uncertainty in the WTO*, Studies in International Trade Law, Vol. 2 (Oxford, Hart Publishing 2004), p. 43.

[20] Art. 2.2 and 5.1 of the *SPS Agreement*.

[21] Ibid., Art. 2.2 and 5.6.

[22] Ibid., Art. 2.3 and 5.5.

[23] Consideration 6 of the Preamble of the *TBT Agreement*.

[24] Considerations 6 and 7 of the Preamble and Art. 2.2 of the *TBT Agreement*.

also restricts the possibility of taking measures which aim to achieve legitimate objectives and indicates that technical regulations must not be more trade-restrictive than is necessary to fulfil a legitimate objective, and must take the risks that non-fulfilment would create into account.[25] The elements that are to be taken into account when examining the risks of non-fulfilment include: the available scientific and technical information, as well as the related processing technology or the intended end-uses of products. Furthermore, the *TBT Agreement* requires that, if, due to a change of circumstances, a less trade-restrictive technical regulations can address the objectives pursued, then, the existent technical regulation shall no longer be maintained.[26] The *TBT Agreement* contains a similar provision with regard to conformity assessment procedures. Article 5.1.2 specifies that conformity assessment procedures:

> 'shall not be more strict or be applied more strictly than necessary to give the importing member adequate confidence that products conform with the applicable technical regulations or standards, taking account of risks non-conformity would create.'

In seeking to achieve the objectives of reducing negative effects on international trade, harmonisation is an important tool recognised by both the *SPS Agreement* and the *TBT Agreement*. The instrument of harmonisation functions alongside the rules and disciplines which restrict the sovereign right of WTO Members to take measures to achieve the legitimate objectives pursued. The preamble of the *SPS Agreement* recognises the important contribution that international standards, guidelines and recommendations can make with regard to the establishment of a multilateral framework of rules and disciplines to guide the development, adoption and enforcement of sanitary and phytosanitary measures in order to *minimise their negative effects on trade*.[27] The Appellate Body in *EC-Hormones* confirms this:

> '… In generalized terms, the object and purpose of Article 3 is to promote the harmonization of the SPS measures of Members on as wide a basis as possible, while recognizing and safeguarding, at the same time, the right and duty of Members to protect the life and health of their people. *The ultimate goal of the harmonization of SPS measures is to prevent the use of such measures for arbitrary or unjustifiable discrimination between Members or as a disguised restriction on international trade*, without preventing Members from adopting or enforcing measures which are both "necessary to protect" human life or health and "based on scientific principles", and without requiring them to change their appropriate level of protection.'[28]

Likewise, the preamble of the *TBT Agreement* states that:

---

[25] Art. 2.2 of the *TBT Agreement*.
[26] Ibid., Art. 2.3.
[27] Consideration 4 and 5 of the preamble of the *SPS Agreement*.
[28] Report of the Appellate Body, *EC-Measures Concerning Meat and Meat Products (Hormones)*, WT/DS26/AB/R, WT/DS48/AB/R (1998), para. 177. [emphasis added]

'recognizing the important contribution that international standards and conformity assessment systems can make in this regard by improving efficiency of production and *facilitating the conduct of international trade.*'[29]

Regardless of any other objectives that the instrument of harmonisation may have in other *fora*,[30] within the context of the *SPS Agreement* and the *TBT Agreement*, it primarily serves the objective of trade liberalisation.[31] This means that international standards function as a ceiling, and not as a floor.[32] In other words, only regulations that exceed the level of protection at which the international standards aim are subject to the disciplines of the agreements. The agreements do not contain an obligation for WTO Members to implement the minimum level of protection as contained in the international standards. The guarantee that the measures are sufficient to achieve, for example, the protection of human health or fair practices in trade, remains the responsibility of the Members themselves. Conflicts with regard to the proper implementation and enforcement of international standards to this end will not be the subject of WTO dispute settlement.

2.2 The provisions on harmonisation under the *SPS Agreement* and the *TBT Agreement*

The instrument of harmonisation to contribute to the elimination of trade barriers is laid down in various provisions of both the *SPS Agreement* and the *TBT Agreement*.[33] In brief, these provisions regulate the requirement to use international standards as the basis for national regulations, the notification procedure for national regulations which deviate from international standards, and the presumption of consistency with the relevant SPS or TBT provisions if national measures are in conformity with international standards.[34]

2.2.1 *The scope of the obligation to harmonise*

Both the *SPS Agreement* and the *TBT Agreement* contain an obligation to apply international standards, guidelines and recommendations. Article 3.1 *SPS Agreement* states that:

---

[29] Consideration 3 of the preamble of the *TBT Agreement*.

[30] For example, within the context of the Codex Alimentarius harmonisation serves the objective to protect human health and fair trade practices. See section 3 of Chapter II.

[31] As regards the *SPS Agreement*, see also, D. Prévost and P. Van den Bossche, *supra* n. 16, p. 275.

[32] See also, ibid., p. 275.

[33] In the *SPS Agreement*: considerations 5 and 6 of the Preamble, Arts. 3, 5.1, 5.7, 6.1, 12, and Annexes A.2 and 3, B.5, and C.1.
  In the *TBT Agreement*: considerations 3 and 4 of the Preamble, Arts. 2.4, 2.5, 2.6, 2.9, 5.4, 5.5, 5.6, 9, 11.2, 11.6, 12.5 and 12.6, and Annexes 3F, G, H and N.

[34] Or as referred to by Echols, the carrot and stick approach as regards harmonisation (carrot stands for a soft approach of encouragement, stick for an approach based on 'punishment'), see M. Echols, *supra* n. 8, p. 99.

'to harmonize sanitary or phytosanitary measures on as wide a basis as possible, members *shall base* their sanitary measures or phytosanitary measures on international standards, guidelines or recommendations....'

Article 2.4 *TBT Agreement* states that:

'Where technical regulation are required and relevant international standards exist or their completion is imminent, Members *shall use them*, or the relevant parts of them, *as a basis* for their technical regulations...'

With regard to conformity assessment procedures, Article 5.4 of the *TBT Agreement* states that:

'In cases where a positive assurance is required that products conform with technical regulations or standards, and relevant guides or recommendations issued by international standardizing bodies exist or their completion is imminent, Members *shall ensure that central government bodies use them*, or the relevant parts of them, *as a basis* for their conformity assessment procedures....'

With regard to Article 2.4 of the *TBT Agreement*, the Panel in *EC-Sardines* points out 'the use of the word "shall" denotes a requirement that is obligatory in nature and that goes beyond mere encouragement'.[35] The terms 'shall' is also expressed in Article 3.1 of the *SPS Agreement* and Article 5.4 of the *TBT Agreement*. The formulation of Article 3.1 of the *SPS Agreement* is very similar to Article 2.4 of the *TBT Agreement* and nothing indicates that the requirement under Article 3.1 of the *SPS Agreement* would not be obligatory in nature. Article 5.4 of the *TBT Agreement* states that Members must ensure that national governmental bodies apply the international standards. Thus, although they are not obliged to apply the international standards themselves, the terms 'shall ensure' indicates that Members are responsible for the application these standards by their governmental bodies. This responsibility clearly goes beyond the encouragement of the application.[36] In contrast to the *SPS Agreement*, the *TBT Agreement* in its Article 2.4 extends this obligation to the application of draft international standards, which adoption is imminent.[37]

The scope of the obligations to use international standards is, however, reduced. This reduced scope is the consequence of two elements:

1. 'the provisions containing the obligation to apply international standards use the terms 'based on' or 'use as a basis', a formulation which is less strict than

---

[35] Report of the WTO Panel, *European Communities – Trade Description of Sardines*, WT/DS231/R, 29 May 2002, para. 7.110.

[36] The following research will concentrate on the obligation as contained in Art. 3.1 of the *SPS Agreement* and Art. 2.4 of the *TBT Agreement* as it is with regard to these provisions that WTO jurisprudence exists.

[37] See, for more detailed discussion, Section 4.4 of this chapter.

the terms ' conform to' as expressed in Article 3.2 of the SPS Agreement and Article 2.5 of the TBT Agreement,[38] and
2. the WTO Members have the autonomous right to deviate from international standards.'[39]

2.2.1.1 The terms 'based on' or 'use as a basis'

The provisions that regulate the obligation to apply international standards include the terms 'based on' (Article 3.1 of the *SPS Agreement*) or 'use as a basis' (Article 2.4 of the *TBT Agreement*). In the first case where violation of this obligation was invoked (*EC-Hormones*), the Panel held that measures 'based on' international standards equate measures which conform to such standards, and referred to the term used in Article 3.2 of the *SPS Agreement*.[40] However, the Appellate Body, in *EC-Hormones*, rejected the Panel's interpretation.[41] It held that the use of different terms in the different provisions (Article 3.1 versus Article 3.2 of the *SPS Agreement*) is deliberate and that the meaning of the terms is distinct.[42] According to the Appellate Body, a measure that is in conformity with a Codex standard is also based upon that same standard. However, the contrary is not necessarily the case, as a measure based on a Codex standard may incorporate *only some, and not all, of the elements* of the standard in question.[43] The Appellate Body stated that a measure that was in conformity with an international standard would be equivalent to the complete embodiment of that standard in the sanitary measure, and, for practical purposes, would convert it into a municipal standard. However, a measure is based on an international standard if it *stands, is founded, or is built* upon the standard. When addressing this obligation under Article 2.4 of the *TBT Agreement*, the Panel in *EC-Sardines* followed the same line as the Appellate Body in *EC-Hormones*, and interpreted the word 'basis' as meaning the 'principal constituent' or 'fundamental

---

[38] Art. 3.1 of the *SPS Agreement*, Art. 2.4 of the *TBT Agreement*.

[39] Art. 3.3 of the *SPS Agreement* and Art. 2.4 of the *TBT Agreement*. See Report of the Appellate Body, *EC-Measures Concerning Meat and Meat Products (Hormones)*, WT/DS26/AB/R, WT/DS48/AB/R (1998), para. 172.

[40] Report of the WTO Panel, *EC-Measures Concerning Meat and Meat Products (Hormones)*, Complaint by the United States, WT/DS26/R/USA (1997), para. 8.72 Report of the WTO Panel, *EC-Measures Concerning Meat and Meat Products (Hormones)*, Complaint by Canada, WT/DS48/R/CAN (1997), para. 9.75 See, on Art. 3.2 of the *SPS Agreement*, Section 2.2.2 of this chapter. This finding led the Panel to conclude that the EC regulations were not based on international standards (the Codex Maximum Residue Levels (MRLs)) as the regulations resulted in a different level of protection than would be achieved by measures based on the Codex standards and consequently it found that the EC Regulations were not in conformity with Art. 3.1 of the *SPS Agreement* (Report of the WTO Panel, *EC-Measures Concerning Meat and Meat Products (Hormones)*, Complaint by the United States, WT/DS26/R/USA (1997), para. 8.77 Report of the WTO Panel, *EC-Measures Concerning Meat and Meat Products (Hormones)*, Complaint by Canada, WT/DS48/R/CAN (1997), para. 9.80.

[41] Report of the Appellate Body, *EC-Measures Concerning Meat and Meat Products (Hormones)*, WT/DS26/AB/R, WT/DS48/AB/R (1998), paras. 163-166.

[42] Ibid., para. 164.

[43] Ibid., para. 163.

principle'.⁴⁴ In *EC-Sardines*, the Appellate Body arrived at a similar conclusion. It stated that the EC regulation cannot be considered as being based on the *Codex Stan 94* if it is *contradictory* to the Codex standard.⁴⁵

In both cases, the Appellate Body made it clear that the formulation 'based on' allows for some margin of discretion that can be exercised by Members. Consequently, the obligations under Article 3.1 of the *SPS Agreement* and 2.4 of the *TBT Agreement* do not aim to establish uniformity of national regulations.

However, as has been criticised by several scholars, the Appellate Body fails to establish the extent to which a national regulation can deviate from an international standard before it is no longer based on it.⁴⁶ This failure creates a significant amount of legal uncertainty with regard to the scope of the obligation.⁴⁷

The issue has been raised by Horn and Weiler that the Appellate Body in *EC-Sardines* omitted to establish more precisely whether their interpretation of the first part of Article 2.4 of the *TBT Agreement* was based on a procedural approach or a substantive approach.⁴⁸ If the interpretation of the Appellate Body was based on a procedural approach, this would mean that the obligation to use the international standard as a basis would be mainly a procedural obligation for Members. Although they would be required to use the international standard as a basis during the legislative process, the resulting content of the regulation itself would not necessarily have to reflect this. On the other hand, an interpretation of the obligation based on a substantive approach would entail that the substance of the regulation itself (and, thus, the end result) be examined in order to define the extent to which it was in conformity with the international standard.⁴⁹ The difference between the two approaches is obvious. An interpretation based on a procedural approach would grant Members a wider margin of discretion with regard to the outcome of the legislative process. Although the Appellate Body does not explicitly recognise it, it is clear that its interpretation is mainly based on a substantive approach. The authors Horn and Weiler themselves indicate that, by its decision that new international standards apply to pre-existing national measures (a response of the Appellate Body to the argument of the EC that the obligation to use international standards only applies to successively adopted national regulations),⁵⁰ the Appellate Body will inevitably

---

⁴⁴ Report of the WTO Panel, *European Communities – Trade Description of Sardines*, WT/DS231/R, 29 May 2002, para. 7.110.

⁴⁵ Report of the Appellate Body, *European Communities – Trade Description of Sardines*, WT/DS231/AB/R, 26 September 2002, para. 248.

⁴⁶ C. Button, *supra* n. 19, p. 60. A.E. Appleton and V. Heiskanen, 'The Sardines Decision: Fish without Chips?', in A.D. Mitchell (ed.), *Challenges and Prospects for the WTO* (London, Cameron May 2005), p. 190.

⁴⁷ R. Muñoz, *supra* n. 8, p. 476. R. Romi, 'Codex Alimentarius: de l'ambivalence à l'ambiguité', *Revue Juridique de l'Environnement* (2001), pp. 201-213. See, for further discussion, Section 3.3 of this chapter on the status of international standards as binding norms.

⁴⁸ H. Horn and J.H.H. Weiler, 'European Communities – Trade Description of Sardines: Textualism and its Discontent', in H. Horn and P. Mavroidis (eds.), *The WTO Case Law of 2002* (Cambridge, Cambridge University Press, 2005), pp. 256-258.

⁴⁹ Ibid., p. 256.

⁵⁰ The Appellate Body in *EC-Sardines* has decided that the obligation to base technical regulations on international standards equally applies to regulations that have been adopted prior to 1995 (the entry

have to adopt the substantive approach.[51] Even more explicit is the examination of the Appellate Body, in *EC-Sardines*, of whether the EC Regulation is based on the international standard at stake: i.e., of whether Codex Stan 94 is exercised by a textual interpretation of Codex Stan 94 in comparison to the substance of the EC Regulation.[52] As stated above, it concluded that the EC Regulation was contradictory to Codex Stan 94 and it clearly referred to the substance of the regulation.[53] In addition, the Appellate Body, in *EC-Hormones*, followed a substantive approach as it held that the obligation to base a measure on an international standard meant that 'only some, not all, of the elements of the standard are incorporated *into the measure*'.[54]

2.2.1.2 The right to deviate from international standards

The preambles of both the *SPS Agreement* and the *TBT Agreement* mention the important contribution that harmonisation can make to trade facilitation, while, at the same time, recognising the right for WTO Members to determine their own appropriate level of protection. This possibility is reflected in the explicit provision that allows for deviation, Article 3.3 of the *SPS Agreement* and Article 2.4 of the *TBT Agreement*. As a result, WTO Members have the permanent possibility of deviating from the international standards.

A. The right to deviate from international standards under the SPS Agreement
Article 3.3 of the *SPS Agreement* explicitly lays down the right for Members to deviate from international standards. The first part of Article 3.3 of the *SPS Agreement* states that:

> 'Members may introduce or maintain sanitary or phytosanitary measures which result in a higher level of sanitary or phytosanitary protection than would be achieved by measures based on the relevant international standards, guidelines or recommendations, ...'

As the term indicates, the right to deviate from international standards is linked to the autonomous right for Members to determine their appropriate level of protection. This link is also clearly established in Consideration 6 of the Preamble of the *SPS Agreement*:

---

into force of the WTO agreements). Furthermore, the temporal scope of the obligation extends to regulations that existed at the time of the adoption of the international standard at hand. Report of the Appellate Body, *European Communities – Trade Description of Sardines*, WT/DS231/AB/R, 26 September 2002, paras. 107-109, 212 and 216.

[51] H. Horn and J.H.H. Weiler, *supra* n. 48, p. 258.

[52] Report of the Appellate Body, *European Communities – Trade Description of Sardines*, WT/DS231/AB/R, 26 September 2002, paras. 254 and 257.

[53] Ibid., para. 257.

[54] Report of the Appellate Body, *EC-Measures Concerning Meat and Meat Products (Hormones)*, WT/DS26/AB/R, WT/DS48/AB/R (1998), para. 163 [emphasis added].

'Desiring to further the use of harmonized sanitary and phytosanitary measures between Members, on the basis of international standards, guidelines and recommendations...without requiring Members to change their appropriate level of protection of human, animal or plant life or health.'

A point of discussion in the *EC-Hormones* dispute was the relationship between Article 3.1 (the obligation to base measures on international standards) and Article 3.3 of the *SPS Agreement* (the right to deviate from international standards). Based on a textual reading of Article 3.1,[55] the Panel defined the relationship between the two provisions as being a relationship between the general rule (Article 3.1) and the exception (Article 3.3). In other words, it classified the right to deviate as an exception to the obligation to base measures on international standards.[56] The Appellate Body, however, overruled the conclusion that the Panel had drawn. It classified the right to deviate as an autonomous right, and not as an exception to a general obligation as contained in Article 3.1.[57]

This finding of the Appellate Body has led some scholars to conclude that the *SPS Agreement* does not contain an obligation to use international standards.[58] Instead, in their view, the *SPS Agreement* merely encourages the use of international standards.[59] However, this conclusion has neither a basis in the text of the *SPS Agreement* nor a basis in the findings of the Appellate Body. As indicated by the Appellate Body:

'Article 3.1 of the SPS Agreement simply *excludes from its scope of application* the kinds of situations covered by Article 3.3 of that Agreement, that is, where a Member has projected for itself a higher level of sanitary protection than would be achieved by a measure based on an international standard. Article 3.3 recognizes the autonomous right of a Member to establish such higher level of protection, provided that that Mem-

---

[55] Art. 3.1 of the *SPS Agreement*: 'To harmonize sanitary and phytosanitary measures on as wide a basis as possible, Members shall base their sanitary or phytosanitary measures on international standards, guidelines or recommendations, where they exist, *except* as otherwise provided for in this Agreement, and in particular in paragraph 3.' (emphasis added).

[56] Report of the WTO Panel, *EC-Measures Concerning Meat and Meat Products (Hormones)*, Complaint by the United States, WT/DS26/R/USA (1997), para. 8.86 Report of the WTO Panel, *EC-Measures Concerning Meat and Meat Products (Hormones)*, Complaint by Canada, WT/DS48/R/CAN (1997), para. 9.89.

[57] Report of the Appellate Body, *EC-Measures Concerning Meat and Meat Products (Hormones)*, WT/DS26/AB/R, WT/DS48/AB/R (1998), para. 172.

[58] D.G. Victor, 'Risk Management and the World Trading System: Regulating International Trade Distortions Caused by National Sanitary and Phytosanitary Policies', *Incorporating Science, Economics, and Sociology in Developing Sanitary and Phytosanitary Standards in International Trade: Proceedings of a Conference* (National Academies Press, 2000), p. 156. D. Prévost and P. Van den Bossche, *supra* n. 16, p. 266. Even the SPS Committee in their response to the Executive Committee of the Codex Alimentarius Commission emphasised that the *SPS Agreement* does not contain an obligation to apply international standards, see Committee on Sanitary and Phytosanitary Measures, *draft response to the Codex Alimentarius Commission,* G/SPS/W/86/Rev.1, 13 March 1998, para. 5.

[59] C. Button, *supra* n. 19, p. 62.

ber complies with certain requirements in promulgating SPS measures to achieve that level.'[60]

In other words, the finding of the Appellate Body *reduced the scope* of the obligation to base measures on international standards. This does not mean that the *SPS Agreement* does not contain such an obligation. For instance, what if a Member has not 'complied with certain requirements in promulgating SPS measures in order to achieve the appropriate level of protection'? In these cases, it can no longer rely on Article 3.3 and, as a consequence, will have to base its measures on international standards in order to comply with Article 3.[61] As a result, the scope of the applicability of Article 3.1 depends on the scope of the right to deviate as laid down in Article 3.3.

Indeed, the right to deviate from international standards is not an absolute or unconditional right, but is conditioned by several criteria. Article 3.3 of the *SPS Agreement* gives two conditions that restrict the right to establish measures which result in higher levels of protection than is reflected in the international standard. First, the WTO Member must, on the basis of an examination and an evaluation of scientific information, justify that the international standard is not sufficient to meet its appropriate level of protection. Second, the WTO Member has established a higher level of protection that is appropriate and in conformity with Article 5 paragraphs 1 to 8 of the *SPS Agreement*.[62] A footnote under Article 3.3 indicates that:

'For the purposes of paragraph 3 of Article 3, there is a scientific justification if, on the basis of an examination and evaluation of available scientific information in conformity with the relevant provisions of this Agreement, a Member determines that the relevant international standards, guidelines or recommendations are not sufficient to achieve its appropriate level of sanitary or phytosanitary protection.'

The Panel in *EC-Hormones* concluded that both criteria come to mean that the measure at issue needs to be in conformity with the provisions of Article 5. It has stated that:

'We find, therefore, that, whatever the difference might be between the two exceptions, a sanitary measure can only be justified under Article 3.3 if it is consistent with the requirements contained in Article 5. If we were to find that the EC measures in dispute are inconsistent with the requirements imposed by Article 5, these measures cannot be justified under Article 3.3. However, even if we find that the EC measures at

---

[60] Report of the Appellate Body, *EC-Measures Concerning Meat and Meat Products (Hormones)*, WT/DS26/AB/R, WT/DS48/AB/R (1998), para. 104.

[61] As regards a similar finding of the Appellate Body with respect to Art. 2.4 of the *TBT Agreement*, see H. Horn and J.H.H. Weiler, *supra* n. 48, p. 267. See also, Section 2.2.1.2 B of this chapter.

[62] The original reason for providing for two options was a political one. The second option was inserted at a latter stage, but could be interpreted as ignoring the scientific justification of a domestic measure. Therefore, explicit mention was made to the concept of scientific justification in the first option.

issue are consistent with the requirements imposed by Article 5, this will still not be sufficient for these measures to be justified under Article 3.3, since to reach that conclusion, we also need to find that the EC measures in dispute fulfil all provisions of the SPS Agreement other than Article 3 and 5 (*in casu* Article 2).'[63]

The Appellate Body agreed and stated that:

'...We are not unaware that this finding tends to suggest that the distinction made in Article 3.3 between two situations may have very limited effects and may, to that extent, be more apparent than real. Its involved and layered language actually leaves us with no choice.'[64]

This interpretation of the Appellate Body has raised the argument that the obligation under Article 3.1 is just an idle advantage for the Members who use the standards as a basis, rather than being an independent obligation.[65] It is true that Article 5 provides a different set of provisions which would have to be evaluated independently of whether or not an international standard is at stake. Therefore, it has been held that Members are in no worse position if they deviate from international standards than if no relevant international standard exists (and, as a consequence, Article 3 does not apply).[66]

---

[63] Report of the WTO Panel, *EC-Measures Concerning Meat and Meat Products (Hormones)*, Complaint by the United States, WT/DS26/R/USA (1997), para. 8.83 Report of the WTO Panel, *EC-Measures Concerning Meat and Meat Products (Hormones)*, Complaint by Canada, WT/DS48/R/CAN (1997), para. 9.86. [emphasis added]. See also, Report of the Appellate Body, *EC-Measures Concerning Meat and Meat Products (Hormones)*, WT/DS26/AB/R, WT/DS48/AB/R (1998), para. 176.

[64] Report of the Appellate Body, *EC-Measures Concerning Meat and Meat Products (Hormones)*, WT/DS26/AB/R, WT/DS48/AB/R (1998), para. 176. It is to be regretted that the Appellate Body does not examine more closely the distinction between the two criteria. A closer look at the two criteria clearly indicates that such a difference indeed exists and may entail consequences for the examination under Art. 5 of the *SPS Agreement*. The first criterion recognises situations in which a Member country determines its appropriate level of protection and organise the risk assessment techniques and procedure accordingly. The justification will concentrate on the sufficiency of the international risk assessment, including the techniques and methodologies used, to respond to the appropriate level of protection as determined by the Member. For instance, in order to respond to a different level of protection, the risk assessment of a Member may concentrate on different factors, such as dietary habits that have not been taken into account during the international risk assessment procedure preceding the adoption of the codex standard. Or it may equally be that there is new scientific information available, which has not yet been taken into account in the international risk assessment. The second criterion, however, concentrates on the determination of the level of protection as such. A domestic measure deviating from the international standard under this second possibility may very well have the international risk assessment conclusions as a basis. Although both criteria imply the requirement to be in accordance with Art. 5, the consequence of the differentiation between the two criteria could be that the examination under the relevant provisions of Art. 5 of the measure at issue would differ. Whereas with regard to the first option a total examination of Arts. 5.1-5.8 would become necessary, under the second option there will be more emphasis to Arts. 5.4-5.6.

[65] See also, J. Pauwelyn, 'The WTO Agreement on Sanitary and Phytosanitary (SPS) Measures as applied in the first three SPS disputes. EC-Hormones, Australia-Salmon and Japan-Varietals', 2 *Journal of International Economic Law* (1999), p. 656.

[66] C. Button, *supra* n. 19, p. 62.

It is also true that the conclusion of the Appellate Body had the consequence that panels may not apply Article 3.1 for reasons of judicial economy, as non-conformity with Article 5 triggers non-conformity with the *SPS Agreement*.[67] It is, however, another question as to whether the obligation under Article 3.1 has become obsolete. Whether Members are in a no worse position if they deviate from international standards cannot be answered on the basis of the Appellate Body's interpretation of Article 3.3. Thus, it essentially depends on the status that has been accorded to international standards under the examination of the measures' conformity with Article 5 of the *SPS Agreement*.[68]

Furthermore, the use of the right to deviate from international standards *has to be justified*. It has been held by some scholars that Members that establish a measure which deviates from the international standard would not have to justify the deviations.[69] This conclusion is based on the Appellate Body's rejection of the Panel's interpretation of the relationship between Article 3.1 and Article 3.3 of the *SPS Agreement*, which led to a reversal of the burden of proof. The reversal of the burden of proof means that it is no longer the defendant (the Member that imposes a measure which deviates from the international standard) that will have to establish a *prima facie* case with regard to the compliance of its measure with the conditions of Article 3.3 of the *SPS Agreement*. Instead, it is the complainant upon which the burden rests.[70] However, the reversal of the burden of proof is not to be confused with the obligation to justify deviations. Article 7 of the *SPS Agreement* in conjunction with Provision 5 of Annex B obliges WTO Members to notify their measures if they are not substantially the same as the content of the relevant international standard, and if the measure may have potential trade effects. Upon re-

---

[67] For instance, this has been the argument of the Panel in *Australia-Salmon* for not addressing Art. 3. Report of the WTO Panel, *Australia-Measures Affecting the Importation of Salmon*, WT/DS18/R (1998), para. 8.47: 'for these reasons mentioned above, even if we were to start our examination of this dispute under Article 3, we would in any event be referred to and thus still need to address Article 2 and 5...'

[68] See, for the discussion on the status of international standards in the context of Art. 5 of the *SPS Agreement*, Section 3.5 of this chapter.

[69] S. Charnovitz, 'Improving the Agreement on Sanitary and Phytosanitary Standards', in G.P. Sampson and W. Bradnee Chambers (eds.), *Trade, Environment and the Millenium* (Tokyo, United Nations University Press 1999), p. 183.

[70] It is to be doubted that the reversal of the burden of proof results in a heavier burden for the complainant. In *EC-Hormones*, the Appellate Body held that 'after careful consideration of the panel record, we are satisfied that the United States and Canada, although not required to do so by the Panel, did, in fact make this prima facie case that the SPS measures related to the hormones involved here, except MGA, are not based on a risk assessment' (Report of the Appellate Body, *EC-Measures Concerning Meat and Meat Products (Hormones)*, WT/DS26/AB/R, WT/DS48/AB/R (1998), footnote 180.) This prima facie case was established by reference to several studies on the use of hormones and the existence of international standards regulating this use. See D. Roberts, 'Preliminary Assessment of the Effects of the WTO Agreement on Sanitary and Phytosanitary Trade Regulations', 1 *Journal of International Economic Law* (1998), p. 388. As held by Walker, the evidence remained the same, however, the burden of persuasion played an important role, V.R. Walker, 'Keeping the WTO from Becoming the "World Trans-science Organization": Scientific Uncertainty, Science Policy, and Factfinding in the Growth Hormones Dispute', 31 *Cornell International Law Journal* (1998), p. 317.

quest, Members are obliged to provide copies of their regulations, and identify the parts that deviate substantively from the international standards.[71] Under Article 5.8 of the *SPS Agreement*, Members are required to provide other Members with reasons for these deviations. The presence of a strengthened dispute settlement mechanism in the context of the WTO means that it is not for the WTO Members themselves to decide whether a deviation is justified. If the reasons provided as a consequence of a request under Article 5.8 of the *SPS Agreement* are not considered to be satisfactory by the Member who requested the explanation, this can lead to the establishment of a Panel, which will be responsible for determining whether the deviations are justified.

B. The right to deviate under the TBT Agreement

The right to deviate from international standards under the *TBT Agreement* is laid down in the second part of Article 2.4 of the *TBT Agreement*:

> 'Where technical regulations are required and relevant international standards exist or their completion is imminent, members shall use them, or the relevant parts of them, as a basis for their technical regulations *except when such international standards or relevant parts would be an ineffective or inappropriate means for the fulfilment of the legitimate objectives pursued…*'[72]

The 'right to deviate from international standards' under Article 2.4 of the *TBT Agreement* is limited by either one of these two criteria;

- the international standard has to be *ineffective* for the fulfilment of the objectives pursued; or
- the international standard has to be *inappropriate* for the fulfilment of the objectives pursued.[73]

In contrast to the *SPS Agreement*, the term of Article 2.4 and the Preamble of the *TBT Agreement* does not provide a clear indication as to whether this right constitutes an autonomous right and whether it is linked to the right for Members to determine their own appropriate level of protection.[74] Despite this difference, in

---

[71] Under the notification procedure, WTO Members are held to justify regulatory choices. Consequently, in contrast to what has been held by some authors (see, for instance, J. Scott, 'International Trade and Environmental Governance: Relating Rules (and Standards) in the EU and the WTO', 15 *European Journal of International Law* (2004), pp. 325-326) WTO Members do have to justify their deviations from international standards, also before a complaint has been filed by another Member and a prima facie case of inconsistency has been established by the complainant.

[72] [emphasis added]

[73] See, for a more detailed discussion on these two criteria and the consequences for the status of international standards under the *TBT Agreement*, Section 3.4 of this chapter.

[74] See also, G. Marceau and J. Trachtman, 'The Technical Barriers to Trade Agreement, the Sanitary and Phytosanitary Measures Agreement, and the General Agreement on Tariffs and Trade. A Map of the World Trade Organization law of Domestic Regulation of Goods', 36 *Journal of World Trade* (2002), p. 840: 'This requirement [Article 2.4 TBT] appears, however, less complex, and less subtle, than that under the SPS Agreement.'

EC-Sardines, the Appellate Body applied its former findings in *EC-Hormones* to the *TBT Agreement*, and concluded that the relationship between the obligation to base measures on international standards and the right to deviate from them is not a 'general rule-exception' relationship. It stated that:

'... In EC-Hormones, we found that a "general rule-exception" relationship between Articles 3.1 and 3.3 of the SPS Agreement does not exist, with the consequence that the complainant had to establish a case of inconsistency with both Article 3.1 and 3.3. We reached this conclusion as a consequence of our finding there that "Article 3.1 of the SPS Agreement simply excludes from its scope of application the kinds of situations covered by Article 3.3 of that Agreement". Similarly, the circumstances envisaged in the second part of Article 2.4 are excluded from the scope of application of the first part of Article 2.4. Accordingly, as with Article 3.1 and 3.3 of the SPS Agreement, there is no "general rule-exception" relationship between the first and the second parts of Article 2.4...'[75]

This finding has been subject to criticism.[76] Various scholars have questioned the fact that, in contrast to many other issues it examined in *EC-Sardines*, the Appellate Body did not apply a textual interpretation and accord more importance to the term 'except' in Article 2.4.[77] This is, indeed, to be considered regrettable, given the fact that the position and status of the right to deviate under the *TBT Agreement* are distinct from these of the right to deviate under the *SPS Agreement*. As mentioned above, in contrast to the *SPS Agreement*, the right to deviate under the *TBT Agreement* is not linked to the autonomous right of WTO Members to determine their own level of protection. Furthermore, unlike Article 3 of the *SPS Agreement*, Article 2.4 of the *TBT Agreement* does not enumerate three options for WTO Members. In other words, the elements that justified the conclusion of the Appellate Body in *EC-Hormones* that the right to deviate under Article 3 of the *SPS Agreement* is an autonomous right, are not present in the *TBT Agreement*.

A controversial consequence of the application of the conclusion in *EC-Hormones* to the *TBT Agreement* is the reversal of the burden of proof. When reversing the Panel's conclusion on the rule-exception relationship, the Appellate Body holds that it is for the complainant to establish a *prima facie* case and demonstrate that the international standards are indeed effective and appropriate.[78] In the case in hand, this meant that Peru, which was the least informed party with regard to the reasons

---

[75] Report of the Appellate Body, *European Communities – Trade Description of Sardines*, WT/DS231/AB/R, 26 September 2002, para. 275. Art. 2.4 of the *TBT Agreement* states that:

'Where technical regulations are required and relevant international standards exist or their completion is imminent, Members shall use them, or relevant parts of them, as a basis for their technical regulations except when such international standards or relevant parts would be an ineffective or inappropriate means for the fulfilment of the legitimate objectives pursued, for instance because of fundamental climatic or geographical factors or fundamental technological problems.'

[76] A.E. Appleton and V. Heiskanen, *supra* n. 46, pp. 183 and 188.

[77] Ibid., p. 183.

[78] Report of the Appellate Body, *European Communities – Trade Description of Sardines*, WT/DS231/AB/R, 26 September 2002, para. 282.

why the EC had not based its regulation on the Codex standard, had to establish a *prima facie* case. Despite this development, which seems to put the WTO Member that deviates from an international standard in an advantageous position, the report of the Appellate Body in *EC-Sardines* demonstrated that this advantage is more apparent than real.

The Appellate Body report in *EC-Sardines* illustrated that the reversal of the burden to provide evidence that now rests upon the complainant, does not necessarily amount to a more difficult task. In its defence of the allocation of the burden of proof to the least informed party, the Appellate Body referred to the obligation, under Article 2.5 of the *TBT Agreement*, which requires the Member preparing, adopting or applying a technical regulation to explain the justification of this regulation upon request by another Member.[79] It went on to state that it had no reason to doubt the functioning of this 'compulsory mechanism' as established by Article 2.5 of the *TBT Agreement*.[80] Another provision that allows for the collection of information, as mentioned by the Appellate Body, is Article 10.1 of the *TBT Agreement*, which contains the obligation for Members to establish an inquiry point that is able to respond to the inquiries of other Members and to provide the relevant documents.[81] These provisions illustrate that, similar to the *SPS Agreement*, the 'right to deviate' under Article 2.4 of the *TBT Agreement* is not a unconditional right. Consequently, although the burden of proof rests upon the complainant, this does not exempt the WTO Member preparing, adopting or applying a regulation that deviates from an international standard from having to justify the deviation. On the contrary, this conclusion may make future panels more attentive when it comes to the obligations under Article 2.5 of the *TBT Agreement*. Furthermore, the Appellate Body mentioned that the dispute settlement procedure itself provides for other means of obtaining information.[82] As mentioned in preceding Appellate Body decisions, the arguments of a party 'are set out and progressively clarified in the first written submissions, the rebuttal submissions and the first and second panel meetings with the parties',[83] and as 'there is no requirement in the *Dispute Settlement Understanding* (hereinafter referred to as *DSU*) or in GATT practice for arguments on all claims relating to the matter referred to the Dispute Settlement Body (hereinafter referred to as DSB) to be set out in a complaining party's first written submission to the panel'.[84]

---

[79] Ibid., para. 277.
[80] Ibid., para. 278.
[81] Ibid., para. 279.
[82] Ibid., para. 280.
[83] Appellate Body Report, *European Communities – Regime for the Importation, Sale and Distribution of Bananas ('EC – Bananas III')*, WT/DS27/AB/R, adopted 25 September 1997, DSR 1997:II, 591, para. 141. See also, Appellate Body Report, *India – Patent Protection for Pharmaceuticals and Agricultural Chemical Products*, WT/DS50/AB/R, adopted 16 January 1998, DSR 1998:I, 9, para. 88 and Appellate Body Report, *Korea – Definitive Safeguard Measure on Imports of Certain Dairy Products ('Korea – Dairy')*, WT/DS98/AB/R, adopted 12 January 2000, DSR 2000:I, 3, para. 139.
[84] Appellate Body Report, *European Communities – Regime for the Importation, Sale and Distribution of Bananas* ('EC – Bananas III'), WT/DS27/AB/R, adopted 25 September 1997, DSR 1997:II, 591, para. 145.

Based on the assumption of the Appellate Body that Members deviating from international standards have information concerning the ineffectiveness and/or inappropriateness of the international standard in question, and that they act in good faith by providing other Members with this information, meeting the burden of proof by the complainant is relatively easy.[85]

In addition, the Appellate Body's examination of the evidence and legal arguments brought forward by Peru reveals that the evidentiary weight to establish a *prima facie* case is so small that it *de facto* nullifies the consequences of the reversal of the burden of proof.[86] The establishment of the *prima facie* case by Peru was based upon the argument that Codex Stan 94 (the Codex standard in question) requires a denomination for *Sardinops sagax* that is distinct from that of *Sardina pilchardus*, and that Codex Stan 94 has the same objective as the EC regulation.[87] Furthermore, the Appellate Body affirmed the Panel's factual finding that:

> 'it has not been established that consumers in most member states of the European Communities have always associated the common name "sardines" exclusively with *Sardina pilchardus*'.[88]

This reference to the Panel's factual finding which did not require any further evidence from Peru implied the limited evidentiary weight of the burden of proof allocated to the complainant.[89]

Under the *TBT Agreement*, an unjustified deviation from an international standard constitutes a violation of the agreement. Article 2.4 is an independent provision and any inconsistency triggers non-conformity with the *TBT Agreement*.[90] The importance of the obligation to base technical regulations on international standards is even augmented by the fact that unjustified deviating national measures do not have to be discriminatory in nature in order to result in non-conformity.[91]

2.2.2 *Harmonisation as encouragement*

Both the *TBT Agreement* and the *SPS Agreement* contain provisions that encourage WTO Members to make their national measures conform with international standards.[92] Under the *SPS Agreement*, national measures which are in conformity with

---

[85] H. Horn and J.H.H. Weiler, *supra* n. 48, pp. 269-270.
[86] Ibid., p. 273.
[87] Report of the Appellate Body, *European Communities – Trade Description of Sardines*, WT/DS231/AB/R, 26 September 2002, para. 290.
[88] Ibid.
[89] H. Horn and J.H.H. Weiler, *supra* n. 48, p. 273.
[90] J. Scott, *supra* n. 71, p. 328.
[91] H. Horn and J.H.H. Weiler, *supra* n. 48, p. 251.
[92] Art. 3.2 of the *SPS Agreement*: 'Sanitary or phytosanitary measures which conform to international standards, guidelines or recommendations shall be deemed to be necessary to protect human, animal or plant life or health, and presumed to be consistent with the relevant provisions of this Agreement and of GATT 1994. Art. 2.5 of the *TBT Agreement*: '…Whenever a technical regulation is pre-

international standards are deemed to be necessary to protect human, animal and plant life and health and are presumed to be consistent with the relevant provisions of the *SPS Agreement* and the provisions of *GATT 1994*.[93] Under the *TBT Agreement*, a similar presumption of consistency is contained in Article 2.5. In addition to the condition that a national measure be in conformity with international standards, the presumption of consistency, under the *TBT Agreement*, also requires that the national measure aims to achieve one of the enumerated legitimate objectives under Article 2.2 of the *TBT Agreement*.

The presumption of consistency as laid down in Article 3.2 of the *SPS Agreement* and Article 2.5 of the *TBT Agreement* provide WTO Members with an important advantage in cases in which they conform their measures to international standards. This presumption allows WTO Members that have put their measures into conformity with international standards to be exempted from justifying the necessity of their measures. Under Article 3.2 of the *SPS Agreement*, the presumption of consistency results in a consistency of the relevant national measure with the relevant provisions of the *SPS Agreement* and of GATT 1994. In contrast, under Article 2.5 of the *TBT Agreement*, the presumption of consistency only results in a presumption that the national measure is not an unnecessary obstacle to international trade and it is not made clear whether this entails conformity with certain provisions. It is not clear whether, in practice, the difference in formulation constitutes an important difference with regard to the consequences of the presumption of consistency under both agreements. The presumption that the national measure does not constitute an unnecessary obstacle to international trade is a presumption of the national measure's consistency with an essential obligation laid down in the *TBT Agreement*.

It has to be noted that the presumption of consistency under both agreements is rebuttable. The *TBT Agreement* is clear on this, as it uses the terms '*rebuttably presumed*'.[94] Despite the fact that the *SPS Agreement* lacks the explicit addition of the term 'rebuttably', according to the Appellate Body in *EC-Hormones*, the presumption of consistency under the *SPS Agreement* is also to be considered as being rebuttable.[95] Does the fact that the presumption is rebuttable result in an idle ad-

---

pared, adopted or applied for one of the legitimate objectives explicitly mentioned in paragraph 2, and is in accordance with relevant international standards, it shall be rebuttably presumed not to create an unnecessary obstacle to international trade.' As results from the comparison of the texts in both the *SPS Agreement* and the *TBT Agreement*, the legitimate objective of a technical regulation in the context of the *TBT Agreement* if conform with an international standard is not automatically recognised. However, in practice it is very likely that if a technical regulation aims to achieve the same objective as the international standard with which it is conform even though this objective is not explicitly enumerated in Art. 2.2 of the *TBT Agreement*, will still be considered as legitimate for the simple reason that WTO Members have in their capacity as Codex Members agreed upon the legitimacy of this objection. A rejection of the legitimacy would largely undermine the work undertaken by the Codex Alimentarius Commission.

[93] Art. 3.2 of the *SPS Agreement*.
[94] Art. 2.5 of the *TBT Agreement*.
[95] See Report of the Appellate Body, *EC-Measures Concerning Meat and Meat Products (Hormones)*, WT/DS26/AB/R, WT/DS48/AB/R (1998), para. 170: 'Such a measure enjoys the benefit of a

vantage for WTO Members that have brought their measures into conformity with international standards?[96] This depends on how difficult it is to rebut the presumption of consistency. To date, neither a panel, nor the Appellate Body has been confronted with this issue. However, several elements indicate that the balance is in favour of the WTO Member that complies with the international standard. For instance, the presumption of consistency is, somehow, reflected in the notification requirements. WTO Members that have adopted national requirements that are substantially the same as the international standard, guideline or recommendation are exempted from having to notify these requirements.[97] Furthermore, rebutting the presumption is part of the burden of proof as attributed to the complaining party. As this party claims that the presumption of consistency would not be valid in the particular circumstances, it has to prove it. Whether this burden of proof is a heavy burden will also depend on the status of whether the international standard is perceived as a necessary barrier to trade (to be discussed in Section 3.2 of this chapter).

3. THE STATUS OF CODEX MEASURES UNDER THE *SPS AGREEMENT* AND THE *TBT AGREEMENT*

The reference to the work of 'international standard-setting bodies' and the recognition of the Codex Alimentarius Commission and its work as an instrument of harmonisation in the context of both agreements raises the question of what the resulting function and status of Codex measures actually are? Most scholars agree that the 'harmonisation provisions' in the *SPS Agreement* and the *TBT Agreement* result in an increased status for Codex measures. Several scholars have referred to international standards as being as benchmarks[98] or 'safe harbour' standards.[99] Despite the general agreement amongst scholars that the references in the *SPS Agree-*

---

presumption (albeit a rebuttable one) that it is consistent with the relevant provisions of the SPS Agreement and of the GATT 1994.' The basis of the Appellate Body's conclusion (and even contained in brackets) is not clear. For a discussion on the presumption of consistency under the *SPS Agreement* see also, J.J. Barcelo, 'Product Standards to Protect the Local Environment - the GATT and the Uruguay Round Sanitary and Phytosanitary Agreement', 27 *Cornell International Law Journal* (1994), pp. 755-776.

[96] J. Scott, *supra* n. 71, pp. 325-326.

[97] Annex 5(a) of the *SPS Agreement*: 'Whenever an international standard, guideline or recommendation does not exist or the content of a proposed sanitary or phytosanitary regulation is not substantially the same as the content of an international standard, guideline or recommendation ... Members shall ... notify other Members.' Art. 2.9 of the *TBT Agreement*: 'Whenever a relevant international standard does not exist or the technical content of a proposed technical regulation is not in accordance with the technical content of relevant international standards...Members shall ... notify other Members.'

[98] D. Prévost and P. Van den Bossche, *supra* n. 16, p. 269. See also, S. Shubber, *The International Code of Marketing of Breast-Milk Substitutes. An International Measure to Protect and Promote Breast-feeding* (The Hague, Kluwer Law International 1998), p. 191.

[99] D. Roberts, *supra* n. 70, p. 379.

ment and the *TBT Agreement* entail an increased status for international standards, there is no agreement as to what this status precisely consist of.

This section examines the relevant TBT and SPS provisions and their interpretation and application by panels and the Appellate Body in order to define the increased status of Codex measures under both agreements with more precision.

This analysis is based upon the discussions and conclusions of the preceding section. It is conducted in different steps. It starts with a discussion on the reference to 'international standard-setting bodies' in both agreements, and the recognition of the Codex Alimentarius Commission as such a body. This discussion is followed by an analysis of the status of Codex measures as necessary measures as a result of the presumption of consistency as contained in Article 3.2 of the *SPS Agreement* and Article 2.5 of the *TBT Agreement*. The next sub-section considers the binding status of Codex measures as a consequence of the obligation to use international standards as a basis for national regulations. As the obligation to 'use international standards as a basis' is limited by the 'right to deviate' from these standards, as discussed in Section 2.2.1.2, the last two sub-sections concentrate on the status of Codex measures as 'appropriate' and 'effective' measures to achieve the objectives pursued by WTO Members. It examines the *TBT Agreement* and the *SPS Agreement* separately. The reason for doing this lies in the different position that the right to deviate and its criteria hold under these two agreements. As discussed in Section 2.2.1.2, Article 2.4 of the *TBT Agreement* contains an independent provision. Inconsistency with this provision entails inconsistency with the *TBT Agreement*. Conclusions on the status of Codex measures as 'appropriate' and 'effective' can thus be based upon an analysis of the Panel's and the Appellate Body's interpretation and application of the second sentence of Article 2.4. However, the structure under the *SPS Agreement* is different. The panel in *EC-Hormones* concluded that the criteria under Article 3.3 of the *SPS Agreement* required conformity with Article 5 of the selfsame agreement. Consequently, the last sub-section examines the function of international standards within the context of Article 5 of the *SPS Agreement*.

3.1 The Codex Alimentarius Commission as a recognised standard-setting body under the *SPS Agreement* and the *TBT Agreement*

The *SPS Agreement* explicitly recognises the Codex Alimentarius Commission as one of the three international standardising bodies upon which it relies for the provision of international standards for purposes of harmonisation.[100] In its Annex A,

---

[100] The preamble and Annex A.3(a), (b) and (c) of the *SPS Agreement*. The other two international standardising bodies to which the *SPS Agreement* refers are the International Office of Epizootics and the international and regional organisations operating within the framework of the International Plant Protection Convention. See L. Rosman, 'Public participation in international pesticide regulation: when the Codex Commission decides, who will listen?', 12 *Virginia Environmental Law Journal* (1993), p. 340, he mentioned one way in which the *SPS Agreement* advances harmonisation: 'It specifically names those international bodies whose standards will be considered the international standard.' He based his writing on the 'Decision by Contracting Parties on the Application of Sanitary and Phytosanitary Measures' a draft agreement, as the *SPS Agreement* had not yet been adopted).

it refers to the work of the Codex Alimentarius Commission, a reference which is directly linked to one of the three objectives recognised by the agreement, by which WTO Members have the right to adopt the necessary measures, namely, food safety.[101] It enumerates several specific issues, such as food additives, veterinary drug and pesticide residues, contaminants, methods of analysis and sampling,[102] and codes and guidelines of hygienic practice.[103] The relevance of Codex work under the *SPS Agreement* is thus restricted to these documents which are adopted for this objective. This means that the other objective of the Codex Alimentarius (the protection of fair practices of trade in food products) is not included. However, in the area of food safety standards, the Codex Alimentarius Commission has acquired a dominant position *vis-à-vis* other international standardising bodies working in this field. This is clearly laid down in Annex A, provision 3(d) that states:

'd) for matters *not covered* by the above organizations, appropriate standards, guidelines and recommendations promulgated by other relevant international organizations open for membership to all Members, as identified by the Committee.'[104]

It is, however, not clear whether the term 'not covered' refers to the mandate of the Codex Alimentarius Commission (in other words, it also covers any future measures to be adopted under this mandate) or to the actual adopted standards, guidelines or recommendations.

In contrast to the *SPS Agreement*, the *TBT Agreement* is less clear, as it does not refer to the Codex Alimentarius Commission as such. This may be due to the fact that the list of legitimate objectives, as enumerated under Article 2.2, is (contrary to the *SPS Agreement*) non-exhaustive. Be this as it may, in *EC-Sardines*, the recognition of the Codex Alimentarius Commission as an international standardising body in the context of the *TBT Agreement* was, without doubt, and was not disputed by either party to the conflict.[105] The only criterion that the *TBT Agreement* imposes

---

[101] Provision 3 under (a) of Annex A of the *SPS Agreement*. For the other two objectives, animal health and zoonoses and plant health, the *SPS Agreement* recognises the work of the International Office of Epizootics, respectively the Secretariat of the International Plant Protection Convention.

[102] Presumably these methods of analysis and sampling that are related to food safety (thus for instance excluding methods of analysis and sampling relevant to the conformity assessment procedures of trade descriptions for foods).

[103] Annex A(3) of the *SPS Agreement*.

[104] [emphasis added] It is thereby a response to concerns as expressed by some Codex Members. See, for instance, the Report of the 9th Session of the Codex Committee on General Principles, Paris, 24-28 April 1989, ALINORM 89/33, paras. 18-30. Para. 21: 'Clarification was also sought as to GATT procedures in the event of a dispute in which Codex standards and those of other international organizations had been referred to, but which differed.'

[105] Panel report in *European Communities – Trade Description of Sardines*, WT/DS231/R, 29 May 2002, paras. 7.61, 7.62 and 7.66. The recognition of the Codex Alimentarius Commission as international standardising body in both agreements (either implicit or explicit) originates from the negotiations during the Tokyo Round that preceded the adoption of the TBT Code (the former *TBT Agreement*). Its recognition was related to the decision of the negotiating parties to include agricultural products within the scope of the Code (Report of the 11th Session of the Joint FAO/WHO Codex Alimentarius

on international standard-setting bodies is that its membership is open to at least all WTO Members.[106] As held by the Panel in *EC-Sardines*, the Codex Alimentarius Commission fulfils this criterion.[107] However, as an explicit reference is lacking under the *TBT Agreement*, the recognition of Codex standards is not fixed, but has to be determined on a case-to-case basis. In general, this is unlikely to pose any problems, given the fact that the Codex Alimentarius Commission works in areas closely related to the prevention of deceptive practices, an objective that is explicitly recognised as legitimate for the adoption of technical regulations by WTO Members. Also with regard to the provisions which aim to protect human health or safety, which is a legitimate objective enumerated in Article 2.2 of the *TBT Agreement*, the Codex Alimentarius Commission functions as an international standardising body. Although the sanitary provisions contained in Codex standards fall within the scope of the *SPS Agreement*, given its prevailing scope over the scope of the *TBT Agreement*, it has to be noted that the objective of the 'protection of human health or safety' covers more than just sanitary measures as defined by the *SPS Agreement*.[108] For example, the provisions in the Codex standards on nutrition may be relevant for the purposes of the *TBT Agreement*.

Both the *SPS Agreement* and the *TBT Agreement* mention three types of documents that are recognised for the purpose of harmonisation: 1) standards, 2) recommendations, and 3) guidelines.[109] The *SPS Agreement* does not differentiate between the documents and refers to them in the same context. Article 3.1 states that Members 'shall base' their measures on *international standards, guidelines or recommendations*. The *TBT Agreement*, on the other hand, does not refer to all three types of measures in the same provision. The terminology of 'international standards' is used with regard to the provisions on technical regulations (Article 2.4), whereas 'guides' and 'recommendations', as developed by international standardising bodies, apply in the context of the requirements relating to conformity assessment pro-

---

Commission, Rome, 29 March-9 April 1976, paras. 10-13). The text of the TBT Code reveals this recognition. Art. 13.3 of the TBT Code states that 'it is understood that unnecessary duplication should be avoided between the work under this Agreement and that of governments in other technical bodies, e.g., the *Joint FAO/WHO Codex Alimentarius Commission*...' [emphasis added]. During the Uruguay Round, the regulation on non-tariff measures, which was contained in the TBT Code was separated over the *SPS Agreement* and the *TBT Agreement*. The recognition of the Codex Alimentarius Commission as international standard-setting body is found back in the context of both agreements.

[106] Annex 1.4 of the *TBT Agreement*.

[107] Panel report in *European Communities – Trade Description of Sardines*, WT/DS231/R, 29 May 2002, para. 7.66.

[108] Sanitary measures related to food products are defined in Annex A (b) as: 'to protect human or animal life or health within the territory of the member country from risks arising from additives, contaminants, toxins or disease-causing organisms in food, beverages or foodstuffs.' Sometimes the demarcation between the *SPS Agreement* and the *TBT Agreement* is very small. For instance, measures that regulate the use of an additive for the protection of human health would fall under *SPS Agreement*, whereas a similar regulation to ensure the compositional value of a product would be defined as a TBT measure. See also, D. Roberts, *supra* n. 70, p. 383.

[109] See, for a comparison of the reference to these different measures with the Codex acceptance procedure, Section 4.1 of this chapter.

cedures (Article 5.4).¹¹⁰ Thus, the *TBT Agreement* seems to distinguish 'internationals standards' on the one hand, from 'guides' and 'recommendations' on the other.

3.2 The status of Codex standards as necessary measures to protect legitimate objectives

As discussed in Section 2.2.2 of this chapter, Article 3.2 of the *SPS Agreement* and Article 2.5 of the *TBT Agreement* hold that technical regulations that are in conformity with international standards are deemed to be necessary and are not considered to be more trade-restrictive than required.¹¹¹ Consequently, the presumption of consistency grants the international standards an increased importance, including Codex measures.¹¹² It has been held that the fact that the presumption of consistency is rebuttable means that any challenge to a national measure that is in conformity with an international standard calls into question the validity of the underlying international standard.¹¹³ This would mean that the actual validity of Codex standards as necessary measures under the *SPS Agreement* and the *TBT Agreement* is at stake. However, the topic of assessment of the rebuttal is first and foremost the national measure that is challenged and not the international standard as such.¹¹⁴ This means that the validity of the Codex standard is only at stake to the extent that it may not correspond to the least-trade restrictive measure possible, given the conditions and circumstances of the WTO Member that imposes the national measure. Even in this case, it is doubtful as to whether a future panel will go as far as to question the validity of Codex standards. The reports in *EC-Hormones* and *EC-Sardines* illustrate the hesitance of WTO panels and the Appellate Body to call the validity of both the procedures preceding the adoption of the Codex standards as well as their substance into question.¹¹⁵ This corresponds to the issue emphasised by several countries during the Uruguay Round of negotiations that it was not for the dispute settlement process to question the validity of an international standard.¹¹⁶ Consequently, a rebuttal of the presumption will not be an easy

---

¹¹⁰ Art. 2.4 of the *TBT Agreement*: 'Where technical regulations are required and relevant *international standards* exist or their completion is imminent....'[emphasis added] In contrast, Art. 5.4 of the *TBT Agreement* for instance states that: 'In cases where a positive assurance is required that products conform with technical regulations or standards, and relevant *guides or recommendations issued by international standardizing bodies* exist or their completion is imminent ...'. [emphasis added]

¹¹¹ See also, J.J. Barcelo, *supra* n. 95, p. 767.

¹¹² See also, D. Prévost and P. Van den Bossche, *supra* n. 16, p. 276.

¹¹³ J. Scott, *supra* n. 71, p. 326.

¹¹⁴ Scott compares the situation with the EC situation and the softer approach of the standard of review as regards Community measures *vis-à-vis* member states measures and holds that the Appellate Body may apply a 'softer' approach when examining measures conforming international standards, J. Scott, *supra* n. 71, p. 326.

¹¹⁵ See, for more details on this hesitance in Sections 3.4 and 3.5 (on the substance) and Section 5.1.1 (on the preceding adoption procedure) of this chapter.

¹¹⁶ D. Abdel Motaal, 'The 'Multilateral Scientific Consensus' and the World Trade Organization', 38 *Journal of World Trade* (2004), p. 863.

task. The strong role accorded to Codex standards through the presumption of consistency means that they will function as benchmarks against which national measures are examined.[117] As held by Abdel Motaal, 'they set the dividing line between what the WTO is automatically willing to consider as acceptable scientific justification and what requires further discussion and evidence'.[118]

One may question whether the presumption of consistency contains another aspect, a more implicit one: namely, whether the objective of the international standard is to be perceived as legitimate. This will clearly be the case in situations in which the objective of the international standard equals the objectives enumerated in the *SPS Agreement* and the *TBT Agreement*. However, in situations where it concerns an objective that is not mentioned in either agreement, one may wonder as to whether the acceptance of the objective in the context of international standardising bodies entails legitimacy within the context of the *SPS Agreement* and the *TBT Agreement*. This may be relevant in cases where the legitimacy of the objective is still the subject of controversy.[119] For instance, it is far from certain as to how many countries will accept animal welfare as a legitimate objective for technical regulations that obstruct the trade of products. Likewise, several other types of objectives, of which the area under discussion is located in another member country (such as human rights and child labour conditions), are controversial. An automatic presumption has to be rejected on the basis of Article 2.4 of the *TBT Agreement*, which requires that the measure aims to achieve one of the objectives explicitly enumerated in Article 2.2. It is, however, not unthinkable that international standardising bodies have a role to play in the acceptance of whether or not an objective is legitimate.

3.3 The status of international standards as binding norms

Does the obligation for WTO Members to base their measures on international standards (including Codex standards) entail a binding force of these standards? The *SPS Agreement* and the *TBT Agreement* remain silent on this issue. Nor is it clear from the negotiating history of the agreements whether this was the intention of the Members. Most scholars reject a binding force of Codex standards under the *SPS Agreement* and the *TBT Agreement*.[120] Clearly, the standards have not acquired

---

[117] As regards the *SPS Agreement*, D. Prévost and P. Van den Bossche, *supra* n. 16, p. 276.

[118] D. Abdel Motaal, *supra* n. 116, p. 858. See, for a similar conclusion, L. Rosman, *supra* n. 100, p. 336.

[119] In *EC-Sardines*, the legitimacy of the objectives consumer protection, market transparency and fair competition was not put into question by Peru. See Report of the WTO Panel, *European Communities – Trade Description of Sardines*, WT/DS231/R, 29 May 2002, para. 7.122.

[120] J. Scott, *supra* n. 71, p. 330. G. Bossis, *La sécurité sanitaire des aliments en droit international et communautaire. Rapports croisés et perspectives d'harmonisation* (Bruxelles, Bruylant 2005), p. 130. R. Quick and A. Bluthner, 'Has the Appellate Body erred? An appraisal and criticism of the ruling in the WTO Hormones Case', 2 *Journal of International Economic Law* (1999), p. 613. J. Pauwelyn, 'The role of public international law in the WTO. How far can we go?', 95 *American Journal of International Law* (2001), p. 555.

a binding status on their own right. Given the fact that the agreements remain silent on the status of Codex standards, it would be premature to conclude that the standards have acquired a *de jure* binding force. However, a closer study of the interpretation and, in particular, the application of the relevant provisions both by panels and the Appellate Body illustrate that a *de facto* binding force, at least with regard to some elements of Codex measures, is not to be rejected too rapidly.

3.3.1 Explicit rejection of binding status of Codex standards by the Appellate Body

In Section 2.2.1.1 of this chapter, it was seen that the Appellate Body in *EC-Hormones* reversed the finding of the Panel that an obligation to base measures on international standards is not the equivalent of an obligation to bring measures into conformity with these standards. In doing so, it also rejected the binding force of Codex standards under the *SPS Agreement*. It concluded that:

> 'to read Article 3.1 as requiring Members to harmonize their SPS measures *by conforming those measures with international standards*, guidelines or recommendations *in the here and now*, is, in effect, to vest such international standards, guidelines and recommendations (which are by the terms of the Codex recommendatory in form and nature) with *obligatory* force and effect. The Panel's interpretation of Article 3.1 would, in other words, transform those standards, guidelines and recommendations into binding *norms*. But as already noted, the *SPS Agreement* itself sets out no indication of any intent on the part of the Members to do so. We cannot lightly assume that sovereign states intended to impose upon themselves the more onerous, rather than less burdensome, obligation by mandating *conformity* or *compliance with* such standards, guidelines and recommendations.'[121]

It is clear from the Appellate Body's reaction that, in its opinion, the *SPS Agreement* has not changed the voluntary character of the Codex standards, and that a binding status is to be rejected on the basis that no mention of such an intention was made in the *SPS Agreement*. The Appellate Body clearly hesitated to assume binding force without the expressed consent of the WTO Members themselves. Consequently, in this case of doubt, the Appellate Body applied the principle of *in dubio mitius*, in other words, it adopted an interpretation that imposed a lesser burden on national sovereignty.[122] However, the argumentation of the Appellate Body is not persuasive.

It justified the rejection of binding force by stating that:

---

[121] Report of the Appellate Body, *EC-Measures Concerning Meat and Meat Products (Hormones)*, WT/DS26/AB/R, WT/DS48/AB/R (1998), para. 165.

[122] Th. Cottier, 'Risk management experience in WTO dispute settlement', in D. Robertson and A. Kellow (eds.), *Globalization and the Environment. Risk Assessment and the WTO* (Cheltenham, UK, Edward Elgar 1999), p. 45.

'... harmonization of SPS measures of members on the basis of international standards is projected in the Agreement, as a goal, yet to be realized in the future.'[123]

However, the fact that harmonisation is a goal to be established in the future does not necessarily reduce the binding status of the existing international standards. The realisation of harmonisation depends on several elements of which the accorded status under the *SPS Agreement* is just one. For example, the realisation of harmonisation equally depends on the progress of the development and adoption of international standards covering all relevant areas that takes place within the international standardising bodies themselves.

Furthermore, another justification for the rejection of binding force is the Appellate Body's rather hasty deduction that 'conformity with' entails binding force, whereas 'based on' does not. However, the difference between an obligation of 'conformity with' and 'based on' is not simply to be reduced to the fact that the former entails binding force while the latter does not. Norms that need to be used as a basis can still have binding force, although not necessarily with regard to all their elements. For instance, directives under EC law are binding, but only with regard to the result to be achieved. They do not require complete conformity. The difference between the two terminologies says something about the method of harmonisation chosen, for instance, full harmonisation or even uniformity (conformity with) or partial harmonisation (use as a basis). To consider this logic while analysing the findings of the Appellate Body on the differences between 'conformity with' and 'based on', the following can be concluded. The Appellate Body held that a requirement to bring national measures into conformity with international standards would entail binding status of the international standards. As mentioned in Section 2.2.1.1, it found that the obligation to confirm measures with international standards means a complete embodiment of the standard. It also held that the requirement to base a measure on an international standard means that a national measure must incorporate at least some elements of the international standard. Consequently, it would not be illogical that these elements of the international standard, which, according to the Appellate Body, are to be incorporated as a consequence of the requirement to 'use as a basis' under Article 3.1 of the *SPS Agreement* and Article 2.4 of the *TBT Agreement*, have obtained a binding status. For this reason, it is very unfortunate that the Appellate Body omitted, in both *EC-Hormones* and *EC-Sardines*, to state precisely *to what extent and specifically which elements* of the international standards or its relevant parts are to be used. However, the application of Article 3.1 of the *SPS Agreement* and Article 2.4 of the *TBT Agreement* can provide clarity on these matters.

---

[123] Report of the Appellate Body, *EC-Measures Concerning Meat and Meat Products (Hormones)*, WT/DS26/AB/R, WT/DS48/AB/R (1998), para. 165.

3.3.2 *The application of the terms 'used as a basis' by panels*

As explained in Chapter II, most Codex standards are very detailed in content and do not often leave a large margin of discretion for Members. The standards applied in both *EC-Hormones* and *EC-Sardines* are clear examples of this. In *EC-Hormones*, the standards applied concerned maximum residue levels (MRLs) expressed in the form of single numerical limits.[124] In *EC-Sardines*, the provision of the standard constituted a detailed labelling requirement with regard to the labelling of the sardines and sardine-type products contained in Codex Stan 98.[125]

The Panel in *EC-Hormones* concluded that:

'... Article 3.3 stipulates the conditions to be met for a Member to enact or maintain a certain sanitary measure which are not based on international standards. It applies more specifically to measures "which result in a *higher level* of sanitary..protection than would be achieved by measures based on the relevant international standards" or measures "which result in a *level* of sanitary.. protection *different* from that would be achieved by measures based on international standards". One of the determining factors in deciding whether a measure is *based on* an international standard is, therefore, the level of protection that measure achieves.'[126]

In other words, the Panel in *EC-Hormones* defined the *level of protection* as an important element. In the point of view of the Panel, a national (or regional) measure would need to reflect the same level of protection as aimed for by the international standard in order to ensure that the measure is 'based on' the international standard. Although reversed by the Appellate Body, the Panel's selection of the level of protection as an element to be applied by WTO Members is an interesting conclusion. This would have allowed for some flexibility as the national MRL would not necessarily have to be the same numerical limit as the Codex MRL, provided that it achieved the same level of protection. Nevertheless, the Panel's application is rather strict. It compared the allowed maximum levels of residues (the one applied by the EC and the one adopted by the Codex Alimentarius Commission) and concluded that as they were not the same, the EC measure was not based on the Codex MRL.[127] As mentioned above, the Appellate Body rejected the Panel's inter-

---

[124] Report of the WTO Panel, *EC-Measures Concerning Meat and Meat Products (Hormones)*, Complaint by the United States, WT/DS26/R/USA (1997), para. 8.64 Report of the WTO Panel, *EC-Measures Concerning Meat and Meat Products (Hormones)*, Complaint by Canada, WT/DS48/R/CAN (1997), para. 9.66.

[125] Report of the WTO Panel, *European Communities – Trade Description of Sardines*, WT/DS231/R, 29 May 2002, para. 2.9.

[126] Report of the WTO Panel, *EC-Measures Concerning Meat and Meat Products (Hormones)*, Complaint by the United States, WT/DS26/R/USA (1997), para. 8.72 Report of the WTO Panel, *EC-Measures Concerning Meat and Meat Products (Hormones)*, Complaint by Canada, WT/DS48/R/CAN (1997), para. 9.75. However, the Appellate Body rejects this finding as an error in law, Report of the Appellate Body, *EC-Measures Concerning Meat and Meat Products (Hormones)*, WT/DS26/AB/R, WT/DS48/AB/R (1998), para. 168.

[127] For instance, as regards Codex standards on zeranol and trenbolone, the Codex MRL allowed a level of 2 and 10 µg/kg, whereas the EC regulations imposed a 'no residue' level. Report of the WTO

pretation of the terms 'based on' and its conclusion that the national measures had to reflect the same level of protection.[128] Unfortunately, the Appellate Body omitted to examine whether the MRL applied by the EC was based on the Codex MRL and left this question for 'another day and another case'.[129] When Codex standards are expressed in numerical limits as in the case of MRLs, it is, however, hard to establish a case in which, according to the logic of the Appellate Body, a measure would be based on a MRL, but, at the same time, would not be in conformity with it.[130] This is particularly so, given the fact that the Appellate Body rejected the level of protection as an element of linkage under the obligation to base measures on international standards.

In *EC-Sardines*, the conclusions of the Panel and the Appellate Body provided more guidance. Their application of the first part of Article 2.4 of the *TBT Agreement* illustrates that the difference between an obligation to 'base on' and an obligation to 'conform with' may actually be explained by the restricted purpose of harmonisation under the *TBT Agreement* (trade liberalisation) *vis-à-vis* the objective of the Codex Alimentarius (consumer protection *and* trade liberalisation). To illustrate this, the relevant international standard, Codex Stan 94, contains a provision that states that other sardine-types than *Sardina pilchardus*, 'shall be named "X Sardines" of a country, a geographic area, the species, or the common name of the species in accordance with the law and custom of the country in which the product is sold'.[131] Both the Panel and the Appellate Body concluded that, as the European Regulation does not allow other sardine-types than *Sardina pilchardus* to be called 'sardines', it was not based on the Codex standard.[132] An examination of whether the European Regulation does actually conform with the Codex standards (or, in other words, whether the European Regulation consists of an full incorpora-

---

Panel, *EC-Measures Concerning Meat and Meat Products (Hormones)*, Complaint by the United States, WT/DS26/R/USA (1997), para. 8.75 Report of the WTO Panel, *EC-Measures Concerning Meat and Meat Products (Hormones)*, Complaint by Canada, WT/DS48/R/CAN (1997), para. 9.78.

[128] Report of the Appellate Body, *EC-Measures Concerning Meat and Meat Products (Hormones)*, WT/DS26/AB/R, WT/DS48/AB/R (1998), paras. 166-168.

[129] Ibid., at para. 168.

[130] On the conclusion of the Appellate Body and its consequences for MRLs or ADIs see D.R. Hurst, 'Hormones: European Communities - Measures Affecting Meat and Meat Products', 9 *European Journal of International Law* (1998). The reason why Codex Alimentarius Commission itself did not allow acceptance with deviations of MRLs is that deviations would render the MRL as instrument for harmonisation useless. See, on this issue, Chapter II, Section 4.3.

[131] Provision 6.1.1(ii) of Codex Stan 94-1981 Rev. 1-1995.

[132] Report of the Appellate Body, *European Communities – Trade Description of Sardines*, WT/DS231/AB/R, 26 September 2002, para. 257: 'The effect of Article 2 of the EC Regulation is to prohibit preserved fish products prepared from the 20 species of fish other than Sardina pilchardus to which Codex Stan 94 refers-including Sardinops sagax-from being identified and marketed under the appellation "sardines", even with one of the four qualifiers set out in the standard. Codex Stan 94, by contrast, permits the use of the term "sardines" with any one of the four qualifiers for the identification and marketing of preserved fish products prepared from 20 species of fish other than Sardina pilchardus. Thus, the EC Regulation and Codex Stan 94 are manifestly contradictory. To us, the existence of this contradiction confirms that Codex Stan 94 was not used "as a basis for" the EC Regulation.'

tion of the standard) would, however, not only concentrate on the fact that the European Regulation has to allow other sardine-types to be named 'sardines'. It would also examine whether the European Regulation ensures that the naming is applied in combination with one of the additional indications; the country, the geographic area, etc. This latter element is the element in the Codex standard that actually aims to protect consumers against misleading information on the nature or origin of the sardine-types in question.

The application of both the Panel and the Appellate Body of the first part of Article 2.4 of the *TBT Agreement* is strict. When it comes to the element of the Codex provision that promotes trade liberalisation by regulating that the name of other sardine-types may include the term 'sardine', neither the Panel nor the Appellate Body leave any option for deviation. Consequently, the application of both the Panel and the Appellate Body in *EC-Sardines* of Article 2.4 of the *TBT Agreement* indicate a *de facto* binding force of the Codex standard in question with regard to the effect resulting from the incorporated provisions which aim to facilitate the international trade in food products.

3.4 The status of international standards as appropriate and effective standards under the *TBT Agreement*

As mentioned in Section 2.2.1.2 above, with the reversal of the burden of proof under Article 2.4 of the *TBT Agreement*, it is no longer the Member that deviates from the international standard that has to establish a case that the international standard is ineffective or inappropriate. Instead, the complainant has to establish a *prima facie* case. The reversal of the burden of proof brings about a reversal of the terms of the second part of Article 2.4 of the *TBT Agreement*.[133] Article 2.4 of the *TBT Agreement* is expressed in negative terms and uses the term, 'when such international standards or relevant parts would be an *ineffective or inappropriate means for the fulfilment of the legitimate objectives pursued*'. The Panel in *EC-Sardines* had defined the terms '*ineffective*' as 'not having the function to accomplish the legitimate objectives', and '*inappropriate*' as 'not being suitable for their fulfilment'. In other words, whereas '*effectiveness*' refers to the *results* of Codex Stan 94, '*appropriateness*' relates to the *nature* of the standard.[134] The reversal of the term meant that Peru had to establish a *prima facie* case that the international standards were *effective* and *appropriate*. But to what extent does the reversal of the burden of proof affect the status of international standards as appropriate and effective? Does it mean that international standards are not assumed to be appropriate and effective? In Section 2.2.1.2 above, it was concluded that the evidentiary weight to

---

[133] A.E. Appleton and V. Heiskanen, *supra* n. 46, p. 187.

[134] Panel report in *European Communities – Trade Description of Sardines*, WT/DS231/R, 29 May 2002, para. 7.116 and Appellate Body report EC-Sardines, para. 267. The *TBT Agreement* adds in its Art. 12.4 that developing countries are not to be expected to base their technical regulations, standards or conformity assessment procedures on international standards, guides or recommendations if these are not appropriate for their development, financial and trade needs.

establish a *prima facie* case by the complainant was very light. The evidentiary weight for the complainant and the status of the international standards are closely related. The only elements examined by the Appellate Body to establish the status of Codex Stan 94 as effective and appropriate were the similarity of the objective of Codex Stan 94 in comparison with the objective of the EC Regulation and the presence of a provision that requires distinct denomination for the two types of products. When it comes to demonstrating (by the defendant) that the international standards are ineffective or inappropriate, the threshold is, however, much higher.[135] According to the Appellate Body, it had to be demonstrated that consumers in *most* EC member states should *always* have associated the common name 'sardines' exclusively with *Sardina pilchardus*, in order to establish that Codex Stan 94 is, indeed, ineffective and inappropriate.[136]

This means that the reversal of the burden of proof and the reversal of the terms of the second part of Article 2.4 hardly – if at all – affect the status of international standards. The Appellate Body's hesitance to question the effectiveness and appropriateness of Codex Stan 94 is understandable given the prominent role attributed to international standards in the *TBT Agreement*. Furthermore, international standards have been subject to negotiations amongst the relevant stakeholders, including the Member countries of the standard-setting body in question. The recognition of this element is reflected in the Panel's decision. It states that:

> '*Even if we were to assume* that the consumers in the European Communities associate the term "sardines" exclusively with *Sardina pilchardus*, the concern expressed by the European Communities, in our view, was taken into account when Codex Stan 94 was adopted…'[137]

3.5 The status of international standards under Article 5 of the *SPS Agreement*

It is held that the Appellate Body's interpretation of Article 3.3 of the *SPS Agreement*, as discussed in Section 2.2.1.2 above, reduces the 'bite of international standards in the WTO'.[138] Although the Appellate Body's finding means that the provisions on harmonisation and the use of international standards cannot be examined independently from the 'regular provisions' of the *SPS Agreement*, it does not necessarily result in a reduction of the function of international standards. This will depend on the role that international standards play in the other provisions of the *SPS Agreement*, in particular, Article 5 of the *SPS Agreement*.

---

[135] See on the consequences of the reversal of the burden of proof for national sovereignty with similar conclusions, H. Horn and J.H.H. Weiler, *supra* n. 48, pp. 273-274.

[136] Report of the Appellate Body, *European Communities – Trade Description of Sardines*, WT/DS231/AB/R, 26 September 2002, para. 290, referring to para. 7.138 of the Panel report. H. Horn and J.H.H. Weiler, *supra* n. 48, pp. 273-274.

[137] Report of the WTO Panel, *European Communities – Trade Description of Sardines*, WT/DS231/R, 29 May 2002, para. 7.133. [emphasis added]

[138] J. Scott, *supra* n. 71, p. 327.

Article 5 of the *SPS Agreement* regulates the obligation to justify sanitary or phytosanitary measures on the basis of scientific assessments. The decisions of the panels and the Appellate Body reveal that even in the context of Article 5 of the *SPS Agreement*, international standards are also used as an instrument in the examination of the conformity of national measures with these provisions. Although Codex standards were only subject to examination in the *EC-Hormones* conflict, other panel and Appellate Body reports, in particular, in *Australia-Salmon*, also touch upon the role of international standards under Article 5 of the *SPS Agreement*. For this reason, they are also discussed in this section.

Within the context of three elements of Article 5, international standards are used as a reference point:

- Article 5.1 explicitly refers to 'risk assessment techniques developed by the relevant international organisations';
- In case international standard-setting bodies have conducted risk assessments in the sense of Article 5.1 of the *SPS Agreement* themselves; and
- Under Article 5.6 of the *SPS Agreement* that obliges Members to ensure that measures are not more trade-restrictive than is necessary in order to achieve the appropriate level of protection.

3.5.1 Article 5.1: Risk assessment techniques developed by the relevant international organisations

A first element that ensures the inclusion of the work undertaken by the relevant international standard-setting bodies is the explicit reference to risk assessment techniques as developed by them. According to Article 5.1, Members have to take risk assessment techniques into account, as developed by the relevant international organisations, when ensuring that their measures are based on risk assessments. According to the Panel in *Japan-Apples*, this does not entail that risk assessments need to be based on, or be in conformity with, these internationally determined risk assessment techniques.[139] It suggests that these techniques should be considered as relevant. However, the Panel also found the international risk assessment techniques to be 'a very useful guidance as to whether the risk assessment at issue constitutes a proper risk assessment within the meaning of Article 5.1'.[140]

In the past, the application of Article 5.1 of the *SPS Agreement* by different panels illustrated the necessity of this guidance, as provided by the international risk assessment techniques, in that they contribute to the acceptability of scientific reports as risk assessments. In this context, they are used in addition to the definition of risk assessment included in Annex A paragraph 4, and the factors which

---

[139] Report of the WTO Panel, *Japan-Measures Affecting the Importation of Apples*, WT/DS245/R (2003), para. 8.241.

[140] Ibid., para. 8.241.

have to be taken into account being enumerated in Article 5.2.[141] The presence of international risk assessment techniques is welcomed, as neither the definition of risk assessment, given its rather general character, nor the factors enumerated in Article 5.2 have proven sufficient to determine adequately whether the mentioned reports were risk assessments *as appropriate to the circumstances* for the purposes of Article 5.1. To illustrate this, in *EC-Hormones*, the Panel had difficulties in concluding whether the relevant scientific reports used by the European Communities actually constituted risk assessments in the sense of Article 5.1.[142] It searched for the specific requirements of what in fact constitutes a risk assessment in accordance with Article 5.[143] As no formal risk assessment technique had been adopted by the Codex Alimentarius Commission,[144] the Panel lacked an instrument to assist in filling the gap. Consequently, it concluded by *assuming* that the reports were risk assessments in conformity with Article 5.1.[145] The opinions of other panels in SPS disputes (*Australia-Salmon* and *Japan-Apples*) illustrate, however, the importance attached to the relevant risk assessment techniques developed by the relevant international organisations.[146] It has to be recalled that the relevant risk assessment techniques in these cases are not developed by the Codex Alimentarius Commission, but by the OIE (International Office of Epizootics) and the IPPC (the framework of the International Plant Protection Convention), instead. However, given the similar status accorded to these standard-setting bodies under the *SPS Agreement*, the interpretations of these panels are useful, with regard to the Codex Alimentarius Commission, in assessing the potential role that the Codex risk assessment techniques maintain under Article 5.1. In *Australia-Salmon*, the relevant risk assessment technique concerned the OIE guidelines on risk assessment.[147] The Panel used the OIE guidelines as guidance in its interpretation of the meaning of a risk assessment

---

[141] Annex A.4 of the *SPS Agreement*: 'Risk assessment – The evaluation of the likelihood of entry, establishment or spread of a pest or disease within the territory of an importing Member according to the sanitary or phytosanitary measures which might be applied, and of the associated potential biological and economic consequences; or the evaluation of the potential for adverse effects on human or animal health arising from the presence of additives, contaminants, toxins or disease-causing organisms in food, beverages or feedstuffs.' Art. 5.2 of the *SPS Agreement*: 'In the assessment of risks, Members shall take into account available scientific evidence; relevant processes and production methods; relevant inspection, sampling and testing methods; prevalence of specific diseases or pests; existence of pest- or disease-free areas; relevant ecological and environmental conditions; and quarantine or other treatment.'

[142] Report of the WTO Panel, *EC-Measures Concerning Meat and Meat Products (Hormones)*, Complaint by the United States, WT/DS26/R/USA (1997), para. 8.111 Report of the WTO Panel, *EC-Measures Concerning Meat and Meat Products (Hormones)*, Complaint by Canada, WT/DS48/R/CAN (1997), para. 9.114.

[143] Ibid., paras. 9.106-9.110.

[144] Ibid., para. 9.113.

[145] Ibid., para. 9.114.

[146] Report of the WTO Panel, *Australia-Measures Affecting the Importation of Salmon*, WT/DS18/R (1998). Report of the WTO Panel, *Japan-Measures Affecting the Importation of Apples*, WT/DS245/R (2003).

[147] Report of the WTO Panel, *Australia-Measures Affecting the Importation of Salmon*, WT/DS18/R (1998), para. 8.49.

appropriate to the circumstances.[148] These guidelines served as a tool to assess whether the scientific reports are, in fact, risk assessments. On the basis of the use of the terms '*probability*' and '*likelihood*' in the *SPS Agreement*, as well as the OIE definition of risk and risk assessment in the OIE Guidelines on Risk Assessment, the Panel concluded that a risk assessment appropriate to the circumstances needs 'to provide some evaluation or estimation of the likelihood or probability' of the relevant diseases and the related consequences.[149] Furthermore, when determining the factors to be taken into account in a proper risk assessment, the Panel also used the OIE guidelines to interpret Article 5.2 of the *SPS Agreement* better. It referred to the enumeration of risk reduction factors in the guidelines[150] and determined that the relevant risk assessments need to take into account the options that are considered to reduce the risk which might be applicable.[151] The relevant risk assessment techniques in *Japan-Apples* were the IPPC Guidelines for Pest Risk Analysis of 1995 and the IPPC Pest Risk Assessment standard of 2001. When evaluating whether the risk assessment used by Japan as justification for its measures constitutes a risk assessment in the sense of Article 5.1 and Annex A, paragraph 4, the Panel held that Japan's risk assessment should have evaluated the '*likelihood*' or the '*probability*' of the entry, or the spread of fire-blight through the importation of apple fruit. The scientific experts consulted by the Panel played an important role in the Panel's conclusion regarding whether Japan's risk assessment fulfilled this criterion. The relevant question that the experts were asked revealed the importance of the international risk assessment techniques employed by the Panel:

---

[148] Ibid., para. 8.71. The parties to the dispute recognised the guidelines for Risk assessment as developed by the OIE as containing risk assessment techniques, which need to be taken into account by Australia under Art. 5.1 of the *SPS Agreement*. It needs to be taken into account that the risk assessment procedure and criteria in this case are relevant to the spread of pest or disease-causing organisms and thus contain different risk assessment procedures and methodologies as established in the risk assessment with regard to food safety. The analysis in this paragraph therefore focuses on the use of international guidelines as such by the Panel rather than on the resulting consequences.

[149] Report of the WTO Panel, *Australia-Measures Affecting the Importation of Salmon*, WT/DS18/R (1998), paras. 8.78 and 8.80. In *EC-Hormones*, the Appellate Body had corrected the interpretation of the Panel that the terminology 'potential' would imply the evaluation of probabilities and stated that the threshold was lower and that the term 'potential' had to be read as possibilities. Report of the Appellate Body, *EC-Measures Concerning Meat and Meat Products (Hormones)*, WT/DS26/AB/R, WT/DS48/AB/R (1998), para. 184. The Panel in *Australia-Salmon* thus needed to carefully argue why it would require the evaluation of probability in this case. By arguing that the other definition of risk assessment in the context of entry, establishment, or spread of pest or disease focused on the likelihood and therefore required a determination of probabilities and thus differed in this respect from the risk assessment procedures in the context of food safety, the Panel relied heavily on the use of the OIE guidelines on risk assessment of the terms 'probabilities' (in particular OIE Code, Section 1.1 and OIE Code Art. 1.4.2.1).

[150] Report of the WTO Panel, *Australia-Measures Affecting the Importation of Salmon*, WT/DS18/R (1998), para. 8.86.

[151] Ibid., para. 8.88. It needs to be noted that Member countries are not required to use the particular risk reduction factors as enumerated in the OIE guidelines. They can determine their own relevant risk reduction factors.

'How does Japan's 1999 PRA compare with the IPPC Guidelines for Pest Risk Analysis of 1995 and the IPPC Pest Risk Assessment standard of 2001?'[152]

The answers given by two experts contributed to the Panel's conclusion that the risk assessment of Japan had some inadequacies and consequently did not meet the requirements of a risk assessment in the sense of Article 5.1 of the *SPS Agreement*.[153]

From the above, it can be concluded that the risk assessment techniques developed by international organisations are used as an interpretation tool that provides content to the more general requirements of the *SPS Agreement* on risk assessment. This means that, depending on the content of such techniques or guidelines, Codex risk assessment guidelines are liable to restrict the discretion of WTO Members to determine which scientific studies they can use as a scientific justification under Article 5 of the *SPS Agreement*.

3.5.2 *The role of the scientific basis of the international standards to interpret the scientific justification for a higher national level of protection*

As mentioned in Section 3.2, the presumption of consistency implies that it is international standards which set the dividing line as to what the WTO accepts as being scientifically-justified and what needs further discussion.[154] The presumption of consistency implies that measures that are in conformity with international standards are automatically based upon a risk assessment and in compliance with Articles 5.1-5.3.[155] However, what does this mean for the status of international standards and their scientific basis in the context of Article 5.1?

In *EC-Hormones*, the relevant international standards (Codex MRLs) were based upon the risk assessments undertaken by the Joint Expert Committee on Food Additives (JECFA). As described in Section 3.5 of Chapter 1, the JECFA's mandate is not limited to setting up methodologies and guidelines for risk assessments, it also carries out risk assessments itself. These scientific conclusions play an important role in the determination of whether there are scientific conclusions that support the measures adopted by the EC.

When examining the scientific studies as advanced by the EC as regards its specific concerns related to the use of growth hormones, the Panel evaluated whether these studies – in one way or another – invalidated the JECFA risk assessments. To be more precise, the Panel examined whether there were any concerns raised in scientific assessments, articles or opinions – as invoked by the EC – which had not been taken into account in the risk assessments of the JECFA.

---

[152] Report of the WTO Panel, *Japan-Measures Affecting the Importation of Apples*, WT/DS245/R (2003), question 33, p. 123.
[153] Ibid., para. 8.279.
[154] See also, D. Abdel Motaal, *supra* n. 116, p. 858. L. Rosman, *supra* n. 100, p. 336.
[155] D. Prévost and P. Van den Bossche, *supra* n. 16, p. 274.

For instance, the EC sought scientific justification for its measures on the use of growth hormones from the 1987 IARC Monographs relating to the risks of carcinogenicity of hormones.[156] These justifications were rejected by the Panel.[157] Besides the fact that these assessments did not specifically evaluate the potential for adverse effects arising from the presence of hormones in food, the Panel concluded that the IARC monographs had been taken into account in the JECFA reports and that they were not contradictory.[158] It stated that:

> 'We further note that, according to the scientific experts advising the Panel, the data and studies contained in these Monographs with respect to the carcinogenic potential of the hormones in dispute have been fully taken into account in the 1988 and 1989 JECFA Reports which, at several occasions, explicitly refer to these Monographs. Nowhere do the 1988 and 1989 JECFA Reports reject the conclusions reached in the 1987 IARC Monographs. On the contrary, the Monographs constitute part of the evidence on which the JECFA Reports are based. JECFA recognized that all five hormones at issue have a carcinogenic potential but concluded that this potential was linked to the hormonal effect of these hormones. Since JECFA considered that the additional residues of the three natural hormones present in treated meat are not capable of exerting any toxic effect, it decided that it was unnecessary to set ADIs and MRLs for these hormones. With respect to zeranol and trenbolone, JECFA identified a no-hormonal-effect level and adopted, on that basis, ADIs and MRLs which, if respected, would ensure the safe use of these hormones. *The IARC Monographs and JECFA Reports did not, therefore, reach contradictory but rather complementary scientific conclusions.*'[159]

Also with regard to other scientific articles and opinions invoked by the EC, which criticised the scientific methodology and conclusions of the JECFA reports and other studies which indicated that the use of hormones is not unsafe, the Panel concluded that these articles and opinions do not invalidate the conclusions of their other studies:

> 'We further note that, according to the Codex expert advising the Panel, most of the evidence contained in these articles and opinions and the potential risks addressed therein were already evaluated and taken into account in the 1988 and 1989 JECFA Reports. Indeed, in the event these articles and opinions should be considered as evidence which was, as the European Communities itself argued at the end of the Panel

---

[156] Report of the WTO Panel, *EC-Measures Concerning Meat and Meat Products (Hormones)*, Complaint by the United States, WT/DS26/R/USA (1997), paras. 8.125-8.126. Report of the WTO Panel, *EC-Measures Concerning Meat and Meat Products (Hormones)*, Complaint by Canada, WT/DS48/R/CAN (1997), paras. 9.128-9.129.

[157] Ibid., para. 9.132.

[158] See also, D. Roberts, *supra* n. 70, p. 393.

[159] Report of the WTO Panel, *EC-Measures Concerning Meat and Meat Products (Hormones)*, Complaint by the United States, WT/DS26/R/USA (1997), para. 8.128. Report of the WTO Panel, *EC-Measures Concerning Meat and Meat Products (Hormones)*, Complaint by Canada, WT/DS48/R/CAN (1997), para. 9.131. [emphasis added]

proceedings, not "new" but was *already taken into account* in the 1988 or 1989 JECFA Reports, the Lamming Report or the 1995 EC Scientific Conference, these articles and opinions would then *not invalidate or contradict* the scientific conclusions reached in these other studies, which specifically address the use of the hormones in dispute for growth promotion purposes, but rather constitute part of the evidence on which these studies are based.'[160]

The EC raised other concerns that implicitly called into question the validity and relevance of the risk assessment methodology and the risk assessment policy which formed the basis of the JECFA assessments and other well-established scientific reports that concludes that the use of hormones is safe. These concerns are derived from the appropriate level of protection as determined by the EC and the specific circumstances that would justify its measures. Examples of concerns invoked by the EC are, for example, the nature and mode of action of the hormones in dispute, the action of metabolites or combinations of, and multiple exposure to, these hormones, as well as risks arising from problems related to the detection and control of hormones.[161] The Panel examined whether the risk assessments had taken these concerns into account, in particular, with regard to the nature and mode of action of the hormones in dispute, the action of metabolites or combinations of and multiple exposure to these hormones.[162] The Panel paid special attention to the JECFA *as scientific basis of the Codex standards*. For instance, Questions 5 and 19 to which the Panel frequently referred in its reasoning, clearly illustrated this:

'The EC has identified certain potential health hazards of concern to it (*i.e.*, carcinogenicity, synergistic effects, genotoxic effects, long-term use and exposure to combination of the hormones in question). To what extent were these hazards taken into consideration in the context of the 1988 JECFA report *or in the establishment of the Codex standards for the hormones in issue*? To what extent do the risk assessments referred to by the EC assess these suggested hazards? Had the six hormones in dispute been used as animal growth-promotors over a sufficient number of years for an assessment of the long-term effects of such hormones on human and animal health to be made?' Question 19: 'Further to question 5, to what extent *do the ADIs and MRLs established by Codex for any food additive, pesticide, etc., take into account* the effects on human health of exposures to mixtures of veterinary drugs, or exposure to the hormones in question originated by other sources? To what extent *do the ADIs and MRLs established by Codex take into account* the effects on human health of exposure to mixtures of veterinary drugs, or exposure to the hormones in question originated by other sources?'[163]

Or question 6:

---

[160] Ibid., para. 9.135. [emphasis added]
[161] Ibid., para. 9.142.
[162] Ibid., para. 9.145.
[163] Ibid. [emphasis added].

'Is there any more recent scientific evidence available with respect to the effects on human or animal health of the use of any of the six hormones in dispute especially when used for growth promotion purposes, other than the evidence *already taken into account by Codex?*...Are you aware of new scientific evidence showing that the carcinogenicity, synergistic effects, genotoxic effects, long-term use and/or exposure to combinations of the hormones in question and their metabolites, if any do not only depend on the hormonal activity of the dose administered? *Would such evidence invalidate the ADIs and MRLs established by the JECFA/Codex?*'[164]

The above illustrates that the Panel sought to examine whether there was scientific evidence that invalidated the Codex MRLs or its scientific basis in the context of Article 5.1.

Consequently, it can be concluded that the risk assessments executed by the expert bodies of the Codex Alimentarius Commission are considered to be benchmarks. From the Panel report in *EC-Hormones*, it clearly appears that these risk assessments are, indeed, accepted without further discussion, and that other elements and concerns have to be justified. This means that the existence of risk assessments that are carried out in the context of the Joint FAO/WHO Food Standards Programme restrict the sovereignty of WTO Members. This restriction of sovereignty extends to both the risk assessment policy and the risk assessment methodology. Thus, international risk assessments undertaken by the JECFA are based upon certain policy decisions, the so-called risk assessment policy, which is distinct from, but closely linked to, the determination of appropriate level of protection. For instance, the risk assessment related to the Codex MRLs with regard to the use of growth hormones as veterinary drugs is based upon the concept of Good Agricultural Practices (GAP), which assumes that certain practices are respected by farmers, and is based upon a certain violation rate. Members that have adopted measures that are based upon distinct risk assessment policy decisions, such as the EC regulations with regard to the use of growth hormones (for example, the decision concerning the violation rate is distinct (see above)), will have to justify these distinct risk assessment policy decisions with scientific evidence.[165] This is not an easy task, as risk assessment conducted at an international level reflects the established scientific principles and opinions at that given moment, and few scientists, including the scientists advising the Panel in its assessment, are inclined to rebut them. As these international scientific assessments have a high standing among the members of the scientific community, its relevance is not easily brought into question.[166]

---

[164] Ibid. [emphasis added].

[165] This scientific justification is not related to the burden of proof issue. It is inherent to the main obligations resulting from the *SPS Agreement*. The burden of proof concerns only who has to make a *prima facie* case with respect to the violation of these obligations. Therefore the fact that the Appellate Body in *EC-Hormones* reversed the Panel's finding on the rule-exception relationship and puts the burden of proof under Art. 5 with the complainant instead as with the defendant (as done by the Panel) does not invalidate the examination or render the subject of the Panel's assessment irrelevant.

[166] This may be particularly difficult for cases where risk assessment policy decisions of Members trigger other risk assessment methodologies and where scientific uncertainty and controversy still dominate the international scientific debate.

3.5.3 *The role of international standards and the choice of a least trade restrictive measure achieving this level*

Article 5.6 prohibits WTO Members from adopting more trade-restrictive measures than are necessary to achieve the appropriate level of protection. It states that:

'Without prejudice to paragraph 2 of Article 3, when establishing or maintaining sanitary or phytosanitary measures to achieve the appropriate level of sanitary or phytosanitary protection, Members shall ensure that such measures are not more trade-restrictive than required to achieve their appropriate level of sanitary or phytosanitary protection, taking into account technical and economic feasibility.'

According to the Panel in *Australia-Salmon*, a measure is more trade restrictive if there is another measure that:

- is reasonably available taking into account technical and economic feasibility;
- achieves the Member's appropriate level of sanitary or phytosanitary protection; *and*
- is significantly less trade restrictive than the contested measure.[167]

If an international standard is adopted, the relevance of the international standard as the least-trade restrictive measure is triggered by the recognition that this standard is necessary for sanitary protection: a recognition that flows from the presumption of consistency under Article 3.2. This is clearly reflected in the terms 'without prejudice to paragraph 2 of Article 3' at the beginning of Article 5.6. This means that contested measures will be compared with international standards in order to define whether the measure is in conformity with Article 5.6. In *Australia-Salmon*, the Panel stated that:

'Given the repeated reference made in the SPS Agreement to the relevant international organizations, in this dispute the OIE, and the recommendations they produce (e.g. Articles 3.1 and 5.1), as well as to the more general objective of harmonization (e.g. Article 3.4 and the sixth preamble), we consider that *appropriate weight should be given to this opinion on option 5* [the option as proposed by the OIE Code]...'[168]

Although the relevant international standard in this dispute is not a Codex standard, but an international standard adopted by the OIE (see also, above, in Section 3.5.1),

---

[167] Report of the WTO Panel, *Australia-Measures Affecting the Importation of Salmon*, WT/DS18/R (1998), para. 8.167. As confirmed by the Appellate Body in Australia-Salmon, these elements have to be met cumulatively in order to conclude non-conformity with Art. 5.6. (Report of the Appellate Body, *Australia-Measures Affecting the Importation of Salmon*, WT/DS18/AB/R, para 194). See also, D. Prévost and P. Van den Bossche, *supra* n. 16, p. 320.

[168] Report of the WTO Panel, *Australia-Measures Affecting the Importation of Salmon*, WT/DS18/R (1998), para. 8.180.

this opinion illustrates that the Panel explicitly recognised the role of international standards in the context of Article 5.6.[169]

In *EC-Hormones*, the only SPS dispute in which Codex standards played a role, the Panel did not examine Article 5.6 for reasons of judicial economy.[170] However, this did not prevent the Panel from comparing the effectiveness of the EC measures with the regime as incorporated in the Codex MRLs. The Panel undertook an examination of the EC regulations in the context of Articles 5.1 and 5.2 of the *SPS Agreement*, which concentrated on the effectiveness of the EC measures as a response to the problems in inspecting natural hormones and in ensuring an appropriate level of protection. In this context, a comparison with the Codex standards was undertaken, and the discussion went as follows: the EC cited its complications in inspecting natural hormones as a risk factor and as a justification for adopting a ban on the use of added natural hormones for growth purposes. The Panel held that:

> 'We recall that, in this dispute, the factors which can be taken into account in a risk assessment under Articles 5.1 and 5.2 are limited to "available scientific evidence" and "relevant inspection, sampling and testing methods". To the extent that the problems in inspecting (sampling and testing) natural hormones would actually pose a risk and could thus, arguably, be taken into account as a risk arising from "relevant inspection, sampling and testing methods" in the sense of Article 5.2, we consider that the European Communities encounters *the same problems in inspecting for natural hormones under its current regime....*'[171]

In other words, the Panel held that the risks resulting from inspection methods are actually not effectively reduced by the measures adopted by the European Community. It went on to compare the regime adopted by the European Community with the regime of a MRL or a tolerance level, namely, a regime used by the Codex Alimentarius Commission. It stated that:

> '... The EC ban of natural hormones used for growth promotion purposes, combined with its tolerance for these hormones when used for therapeutic or zootechnical pur-

---

[169] The fact that the international standard in question played a rather limited role is due to the agreement of both the complainant and defendant and the consulted experts that the option of evisceration proposed by the OIE Code would not be effective to achieve the appropriate level of protection as sought for by Australia (see footnote 465 to paragraph 8.180 of the Panel Report). In other words, the effectiveness of the OIE standard was contested by both parties and the consulted experts. Notwithstanding, the OIE standard was used to choose another option as reference point in order to measure the effectiveness of Australian's sanitary measure.

[170] Report of the WTO Panel, *EC-Measures Concerning Meat and Meat Products (Hormones)*, Complaint by the United States, WT/DS26/R/USA (1997), para. 8.247. Report of the WTO Panel, *EC-Measures Concerning Meat and Meat Products (Hormones)*, Complaint by Canada, WT/DS48/R/CAN (1997), para. 9.250.

[171] Report of the WTO Panel, *EC-Measures Concerning Meat and Meat Products (Hormones)*, Complaint by the United States, WT/DS26/R/USA (1997), para. 8.145. Report of the WTO Panel, *EC-Measures Concerning Meat and Meat Products (Hormones)*, Complaint by Canada, WT/DS48/R/CAN (1997), para. 9.148. [emphasis added]

poses or when present endogenously in meat and other foods, would seem to cause more problems in inspecting for banned natural hormones than a regime where the use of all natural hormones would be allowed in combination with, for example, a maximum residue or tolerance level for all natural hormones in any meat regardless of the origin and use of these hormones. Indeed, only under the EC current regime does the problem of how to distinguish between endogenous and added natural hormones arise; *under a regime with an MRL or tolerance level for all natural hormones* there would be no need to distinguish endogenous from added natural hormones.'[172]

The discussion above illustrates that future panels will most probably not hesitate to compare the contested measures with relevant Codex measures. This means that, in the context of Article 5.6, Codex measures also have a role as benchmarks, as they are seen as 'necessary' to achieve sanitary protection.

4. THE CHANGED STATUS OF CODEX MEASURES AND THE ABOLITION OF THE CODEX ACCEPTANCE PROCEDURE

From the discussion in Section 3 above, it can be concluded that, to a certain extent, Codex standards have a *de facto* binding force through the *SPS Agreement* and the *TBT Agreement*. This binding force is, however, limited to some elements of the standards, and does not concern the standards as a whole. Furthermore, the binding force of the relevant elements is indirect, as it only works through the binding provisions of the WTO agreements. As mentioned in Sections 4.3 and 6 of Chapter II, the Codex acceptance procedure also aims to obtain a binding force of a large part of Codex provisions. During the 28th Session of the Codex Alimentarius Commission in 2005, this acceptance procedure was abolished for the reason that it was no longer relevant in the context of the WTO agreements.

What were the differences between the two systems? What has the impact of the entry into force of the WTO agreements on the Codex acceptance procedure been? Has the entry into force of the WTO agreements really rendered the Codex acceptance procedure obsolete? This section aims to respond to these questions with the view to considering the consequences of the entry into force of the WTO agreements and of the abolition of the Codex acceptance procedure in terms of the character of Codex measures as harmonisation instrument.

---

[172] Ibid. [emphasis added]. In para. 8.146 (9.149) the Panel continues to conclude that '... the experts advising the Panel made clear that the potential for abuse under both regimes would be comparable, some noting that abuse would probably occur more frequently under a regime where the hormones are banned compared to one allowing the controlled use of prescribed products in predetermined dosages with well-defined educational programmes, good communication between the different actors involved and appropriate penalties for misuse. In this context, we note, therefore, that banning the use of a substance *does not necessarily offer better protection of human health than other means of regulating its use.*' [emphasis added]

4.1 The obligations under the *SPS Agreement* and the *TBT Agreement* cover all documents adopted by the Codex Alimentarius Commission

As explained in Section 4.3 of Chapter II, the Codex acceptance procedure only applied to standards and MRLs. Other documents, such as recommended codes of practice and guidelines were not subject to acceptance and did not result in binding provisions.[173] This is not the case in the context of the *SPS Agreement* and the *TBT Agreement*. The obligations resulting from these agreements also extend to the guidelines and the recommended codes of practice. Although Section 3 of this chapter has mainly focused on the status of Codex standards and MRLs (as only these types of measures were relevant to the *EC-Hormones* and *EC-Sardines* disputes), the terms of both of these agreements imply that the conclusion of Section 3, namely, a *de facto* binding force of some elements of the standards applies in a similar way to the other Codex measures.

Article 3 and Annex A of the *SPS Agreement* use the terminology of 'standards, guidelines and recommendations'. Consequently, the obligation and the presumption of consistency under Article 3 clearly include other Codex measures. The SPS Committee emphasised, in its response to a question posed by the Codex Alimentarius Commission, that the *SPS Agreement* does not make a distinction between standards, guidelines and recommendations. However, it added that the actual application by the Members depends on the substantive content, rather than on the category of the adopted text.[174] From this response, it can thus be concluded that the *SPS Agreement* focuses on the *content* of the decision, rather than on the intended *status*.

The approach of the *TBT Agreement* is slightly different. It refers to the three types of documents in different sections. The term of 'standards' is used in Article 2 of the Section 'Technical regulations and standards' of the *TBT Agreement*, which regulates the harmonisation of technical regulations. 'Guides and recommendations' are referred to in Article 5 of the section entitled 'Conformity with technical regulations and standards' of the *TBT Agreement*, which contains similar provisions with regard to conformity assessment procedures. The terminology used in the *TBT Agreement* is confusing, as it does not correspond to the types of Codex measures or their nature. As indicated in Chapter II, the Codex does not exclusively use guidelines and recommendations for aspects related to conformity assessment procedures. An interpretation given by the Secretariat to the TBT Committee and the Codex Secretariat solves this complication. It refers to the response of the SPS Committee mentioned above (the substance of a document is more important than

---

[173] See, for more detail, Sections 4.3 and 6 of Chapter II.

[174] Committee on Sanitary and Phytosanitary Measures, *draft response to the Codex Alimentarius Commission,* G/SPS/W/86/Rev.1, 13 March 1998, para. 5. See also, D. Prévost and P. Van den Bossche, *supra* n. 16, p. 270. M.D. Matthee, 'L'identification et l'étiquetage des OGM: la démocratie existe-elle sur le marché international des aliments génétiquement modifiés?', in J. Bourrinet and S. Maljean-Dubois, *Le commerce international des organismes génétiquement modifiés* (CERIC, Université d'Aix-Marseille, La documentation Française 2002), p. 119.

the actual title of the document) and concludes that this answer also applies to the *TBT Agreement*.[175] Consequently, if this line of interpretation is followed, all relevant documents, which, in substance, relate to conformity assessment procedures, (often, but not always, in the form of guidelines or recommended codes of practice) will be applicable under Article 5 of the *TBT Agreement*, whereas all other relevant Codex documents with regard to the technical regulations are applicable under Article 2 of the *TBT Agreement*.

4.2 The different objective and content of the obligations resulting from Codex measures in the context of the WTO agreements

The obligations under both the *TBT Agreement* and the *SPS Agreement* and the Codex acceptance procedure are result-oriented, meaning that the implementation of the Codex measures in question needs to achieve a certain result without imposing the ways that implementation has to proceed upon the Members.

However, as the purpose of the harmonisation process of, on the one hand, the Codex acceptance procedure, and, on the other, the WTO agreements differs, the result to be achieved is also different. The obligations resulting from full acceptance in the context of the Codex acceptance procedure were directed towards two objectives:

– to ensure that a conforming product will be distributed freely and not hindered by administrative or legal provisions related to the food issues covered by the standard; and
– to ensure that products which are not in conformity with the standard in question (which do not ensure a certain level of protection) are not distributed.[176]

The objective of the obligations is thus directed to facilitate trade in food products *as well as* the protection of human health or fair practices in food trade. In contrast, as explained in Section 2.1 of this chapter, the obligations under the *SPS Agreement* and the *TBT Agreement* are directed to prevent and eliminate obstructions to international trade which arise from the use of technical standards, and sanitary and phytosanitary measures. Neither agreements contain any obligation to use Codex measures in order to ensure a minimum level of protection.

This means that the scope of Codex provisions used in the harmonisation process of the WTO agreements is limited. Whereas the obligations resulting from full acceptance under the Codex acceptance procedure covered all elements of Codex standards, the obligations under the *SPS Agreement* and the *TBT Agreement* do not necessarily cover all the elements of the measures.[177] As reasoned in Section 3.3.2,

---

[175] Status of Codex texts in the framework of the *TBT Agreement*, CX/GP 99/7, Codex Committee on General Principles, 14th Session, Paris, France 19-23 April 1999, p. 4.

[176] See Section 4.3 of Chapter II for a more detailed discussion.

[177] Report of the Appellate Body, *EC-Measures Concerning Meat and Meat Products (Hormones)*, WT/DS26/AB/R, WT/DS48/AB/R (1998), para. 163.

the obligations are limited to the elements contained in Codex measures that aim for the free circulation of food products. In the same way, it can be remarked that the obligations resulting from full acceptance relate to both imported and domestic products alike. In contrast, the obligations under the *SPS Agreement* and the *TBT Agreement* only concentrate on *imported* food products with the intention of preventing trade barriers. This can be confirmed by reading Article 12.4 of the *SPS Agreement*, which regulates the monitoring process on international harmonisation by the SPS Committee.[178] In the context of this procedure, WTO Members are requested to indicate 'those international standards, guidelines or recommendations which they apply *as conditions for import* or on the basis of which *imported products* conforming to these standards can enjoy access to their markets'.[179]

This means that, in comparison with the intentions of the Codex Alimentarius Commission as reflected in its acceptance procedure, the use of Codex measures as a harmonisation instrument in the context of the WTO agreements is clearly restricted.

4.3 Consequences related to the different membership of both institutions

Although most of WTO Members are also Members of the Codex Alimentarius Commission and *vice versa*, the issue of a different membership cannot be ignored. The membership of the WTO amounts to 148 countries, whereas the Codex Alimentarius Commission has 173 member countries. Only 6 Members of the WTO are not Members of the Codex Alimentarius Commission. In contrast, 31 Codex Members are not adhered as Member to the WTO, including member countries such as the Russian Federation. The consequence is clear. The 31 Codex Members that are not WTO Members are not bound by the obligations of the *SPS Agreement* and the *TBT Agreement*. For these 31 Codex Members, Codex measures only have a voluntary status. The situation in which the 6 WTO Members that are not Codex Members are placed is more complicated. They are bound to use Codex measures, even though, at the same time, they possess no decision-making power in the context of the Codex Alimentarius Commission.

4.4 The role of the 'explicit consent' under the acceptance procedure and under the *SPS Agreement* and the *TBT Agreement*

As explained in Sections 4.3 and 6 of Chapter II, the binding force of Codex standards under the acceptance procedure was based upon the expressed consent of countries for each standard individually. Obligations to apply Codex standards and MRLs only arose as a result of the expressed consent of Members through acceptance. Furthermore, as explained in Section 6 of Chapter II, the acceptance of

---

[178] See, for a more detailed discussion on the monitoring process on harmonisation, Section 5.2 of this chapter.

[179] [emphasis added].

Codex standards under the acceptance procedure was a unilateral action, and the resulting obligations lacked a reciprocal character. In addition, the Codex acceptance procedure impeded the uniform application of Codex standards, as it provided the Codex Members with different options in order to accept Codex standards and MRLs. They could decide upon which provisions to accept, and, consequently, which provisions were to be implemented.

It becomes clear from the decisions of the panels in *EC-Sardines* and *EC-Hormones* that this way of expressed consent of countries is no longer relevant to activate obligations resulting from Codex standards. In *EC-Sardines*, the Panel explicitly stated that the validity of Codex standards under the *TBT Agreement* does not depend on whether the standards have been accepted by its Members. In the interim report, in response to the EC argument that neither EC member states nor Peru (the complainant in this dispute) had accepted the relevant Codex standard in the context of the Codex acceptance procedure, the Panel stated:

> 'We did consider this argument but were not persuaded that this argument was relevant in determining whether Codex Stan 94 is an international standard...We recall that Annex 1.2 of the Agreement on Technical Barriers to Trade (the "TBT Agreement") defines a standard as a "document approved by a recognized body" and *does not require that the standard be accepted by countries* as part of their domestic law. Codex Stan 94 was adopted by the Codex Alimentarius Commission and we consider that this is the relevant factor for purposes of determining the relevance of an international standard within the meaning of the TBT Agreement.'[180]

In *EC-Hormones*, the Panel decided that the *SPS Agreement* imposed no conditions on the relevance of international standards for the purposes of Article 3. The relevance for Article 3 resulted from the fact that they existed.[181] Consequently, under both WTO agreements, the adoption of a standard or MRL by the Codex Alimentarius Commission is sufficient to entail obligations.

This means that it is no longer for Members to decide and choose the relevant parts of the international standards, guidelines or recommendations to be used as a basis for its national requirements. As held by the Appellate Body in *EC-Sardines*:

> 'In addition, the examination must be broad enough to address all of those relevant parts; *the regulating Member is not permitted to select only some of the "relevant parts" of an international standard*. If a "part" is "relevant", then it must be one of the elements which is "a basis for" the technical regulation.'[182]

---

[180] Report of the WTO Panel, *EC-Trade Descriptions of Sardines*, WT/DS231/R (2002), para. 6.5. [emphasis added]. Reading Art. 2.4 it can be concluded that the Panel can even go further, as even the adoption of the international standard does not seem a requirement. It reads 'or their completion is imminent' and refers to a situation before the adoption has actually taken place.

[181] Report of the WTO Panel, *EC-Measures Concerning Meat and Meat Products (Hormones)*, Complaint by the United States, WT/DS26/R/USA (1997), para. 8.69. Report of the WTO Panel, *EC-Measures Concerning Meat and Meat Products (Hormones)*, Complaint by Canada, WT/DS48/R/CAN (1997), para. 9.72. On this conclusion of the Panel, see also, Section 5.1.1 of this chapter.

[182] Report of the Appellate Body, *European Communities – Trade Description of Sardines*, WT/DS231/AB/R, 26 September 2002, para. 250. [emphasis added]

In order to be defined as a relevant international standard or a relevant part of it, the Panel in *EC-Sardines* determined that the Codex standard (or Recommended International Code of Practice (RCP) or guideline) would have 'to bear upon, relate to, or be pertinent to' a particular national or regional measure.[183] The Appellate Body adds that:

> 'In our view, the phase "*relevant parts of them*" defines the appropriate focus of an analysis to determine whether a relevant international standard has been used "as a basis for" a technical regulation. In other words, the examination must be limited to those parts of the relevant international standards that relate to the subject-matter of the challenged prescriptions or requirements.'[184]

Consequently, the international standards or provisions are recognised as relevant to the case in hand if they relate to the subject-matter of the national measure.[185]

In both the *EC-Sardines* and *EC-Hormones* decisions, the Panels selected the relevant international standard. In *EC-Sardines*, the Codex standard had a wider scope than the EC regulation in dispute, as the former covered various sardine-type products, whereas the EC regulation only explicitly regulated *sardina pilchardus*. However, this restricted scope of the EC regulation was not an obstacle for the Panel, which concluded that both the regulation and the international standard dealt with preserved sardines and contained corresponding provisions.[186]

In *EC-Hormones*, the EC argued that the Codex Maximum Residue Levels (MRLs for veterinary drugs) were not relevant as international standards, as the MRLs merely reflected a level of protection and did not really consist of a measure.[187] In other words, the MRLs do not *regulate the use* of hormone growth promoters as the EC regulation does. As held by the EC, the only Codex standard relevant to the EC regulation is the Codex Code of Practice for Control of the Use of Veterinary Drugs.[188] However, the Panel held that the MRLs were the relevant international standards.[189] It based this finding upon the mere fact that the standards actually

---

[183] Panel report in *European Communities – Trade Description of Sardines*, WT/DS231/R, 29 May 2002, para. 7.68, quoting Webster's New World Dictionary (William Collins & World Publishing Co., Inc. 1976), p. 1199.

[184] Report of the Appellate Body, *European Communities – Trade Description of Sardines*, WT/DS231/AB/R, 26 September 2002, para. 250. [emphasis added]

[185] In its analysis, the Appellate Body rejects any relevance that the Codex acceptance procedure or an incorrect application of the Codex standard-setting procedure may have on the determination whether an international standard is a relevant standard. See for a more detailed discussion on this issue, sections 4.4 and 5.1.1 of this chapter.

[186] Panel report in *European Communities – Trade Description of Sardines*, WT/DS231/R, 29 May 2002, para. 7.69.

[187] Report of the WTO Panel, *EC-Measures Concerning Meat and Meat Products (Hormones)*, Complaint by the United States, WT/DS26/R/USA (1997), para. 4.79 Report of the WTO Panel, *EC-Measures Concerning Meat and Meat Products (Hormones)*, Complaint by Canada, WT/DS48/R/CAN (1997).

[188] Ibid.

[189] Ibid., para. 9.73.

existed. According to the Panel, this is the only condition imposed by Article 3.1 ('where they [international standards, recommendations and guidelines] exist') in conjunction with paragraph 3 of Annex A of the *SPS Agreement* ('international standards ... for food safety, the standards.. established by the Codex Alimentarius Commission relating to...veterinary drugs...residues...').[190] Unfortunately, due to this rather narrow interpretation of the SPS provisions, the Panel did not provide any further motivation for its rejection of the Codex Code of Practice for Control of the Use of Veterinary Drugs as a relevant international recommendation.

As already briefly indicated in Section 2.2.1 of this chapter, the terms of Article 2.4 of the *TBT Agreement* bring along another restriction as regards the consent of Codex Members. Article 2.4 equally refers to relevant draft international standards, which completion is imminent. This also means that draft international standards can become relevant within the context of the obligation of Article 2.4, even prior to their adoption by the Codex Alimentarius Commission. So far, no definition or explanation on the meaning of the term 'imminent' has been given. It is clear that a wide interpretation and application of this term would be problematical. The reasons are as follows: the final consent of Codex Members during the standard-setting procedure is crucial, in particular now that they have no longer the possibility of expressing their consent during the acceptance procedure. Furthermore, for reasons of legal certainty, WTO Members will, at least, have to rely on the fact that the content of the draft international standards will not change (besides the possibility of some editorial modifications) between the time that the completion of the draft standards is 'imminent' and the time that the draft standards are adopted as international standards.

The new status of Codex standards is a considerable change when compared with the status previously accorded under the former acceptance procedure of the Codex Alimentarius Commission. It already results from the final adoption or even when completion is imminent in the case of Article 2.4 of the *TBT Agreement*. As the binding force is indirect and works through the WTO agreements, the resulting obligations are similar to the resulting obligations of WTO provisions, which are bilateral in nature.[191] The fact that the Codex Members have lost the possibility of explicitly consenting to the standard after adoption has taken place also means that the different harmonisation levels, as reflected in the flexible acceptance procedure, disappear. As discussed in Section 5 of Chapter II, the possibilities of deviating from Codex standards were mainly incorporated in the differentiated ways of acceptance. Consequently, WTO Members are obliged to apply the relevant parts of Codex measures, which are, especially, in the case of Codex standards, expressed in a mandatory way which does not leave much room for flexible application at domestic level. This means that the possibilities of deviating from Codex measure

---

[190] Ibid., paras. 9.72-9.73.
[191] See, on the bilateral nature of WTO obligations, J. Pauwelyn, 'A typology of multilateral treaty obligations: are WTO obligations bilateral or collective in nature?', 14 *European Journal of International Law* (2003), p. 907-951.

are mainly examined through the interpretation of the relevant WTO provisions that allow for deviations. Thus, the incorporation of different harmonisation levels – as envisaged by the Codex Alimentarius Commission – to a large extent disappears.

4.5 Consequences of the abolition of the Codex acceptance procedure

Despite the increased status of Codex measures, it can be concluded from the discussion in Section 4.2 and Section 4.3 above that the entry into force did not make the Codex acceptance procedure completely obsolete.[192] The WTO agreements are only concerned with Codex measures as an instrument of harmonisation to reduce trade barriers, but do not, however, cover their harmonisation function to ensure a minimum level of consumer protection. This means that, with the abolition of the acceptance procedure, the Codex Alimentarius Commission has reduced the second function of their measures as instruments of harmonisation, the application of Codex standards as minimum platform standards. Codex provisions which aim to ensure a minimum level of protection remain merely recommendatory in nature. Although it is true that the acceptance procedure had never really functioned well, is this sufficient reason to remove the only 'compliance' mechanism which the Codex Alimentarius Commission had to promote the application of Codex provisions in order to ensure a certain level of consumer protection in its member countries?

5. THE INTER-INSTITUTIONAL RELATIONSHIP OF THE WTO – CODEX ALIMENTARIUS COMMISSION

The authority to elaborate international standards has not been attributed to the WTO. In line with the earlier approach taken under the TBT Code (the former *TBT Agreement*),[193] the WTO relies on other existing international standardising bodies

---

[192] See also, concerns expressed by Suppan: S. Suppan, *Governance in the Codex Alimentarius Commission,* (Consumers International 2005), pp. 22-24.

[193] Report of the 10th Session of the Joint FAO/WHO Codex Alimentarius Commission, Rome, 1-11 July, 1974, Appendix III (Statement by GATT representative to the 10th Session of the Joint FAO/WHO Codex Alimentarius Commission), para. 10: 'The GATT has taken the view that it should not itself get into the standards writing process, recognizing that other organizations have the technical competence and experience in this field....The GATT Code would contribute to the harmonization of standards by laying an obligation on signatories to play an full part in the work of appropriate international standards-writing bodies and to adopt international standards as a basis for their own mandatory standards except in cases where these are inappropriate for them. The intention is therefore to make a contribution towards the strengthening of existing standards-writing bodies, such as the Codex Alimentarius Commission.' The fact that the GATT was not a *de jure* international organisation at the time the TBT Code was negotiated may have contributed to the decision that the GATT itself would not be turned into an international standard-setting body for the purposes of establishing common measures.

and recognises the expertise and competence that these bodies have.[194] In doing so, it was chosen to strengthen these international standardising bodies, instead of attributing new competencies to the WTO. As discussed in Section 3.1 of this chapter, the Codex Alimentarius Commission is included in these international standardising bodies. This means that the inter-institutional relationship between the WTO and the Codex Alimentarius Commission is characterised by a separation of powers with regard to the harmonisation process.[195] The Codex Alimentarius Commission is responsible for the 'legislative acts', while the WTO is responsible for ensuring the application of the standards.[196]

The WTO does not leave this separation of powers untouched. For the purpose of harmonising measures, both the *SPS Agreement* and the *TBT Agreement* require WTO Members to participate fully, within the limits of their resources, in the decision-making procedure of the international standardising bodies.[197] Furthermore, the *SPS Agreement* stresses that its Members must '*promote* ... the development and the periodic review of international standards, guidelines or recommendations'.[198] Likewise, under both agreements, the participation of developing countries in the international standard-setting bodies is particularly encouraged.[199] In another effort to promote participation in international standard-setting bodies, the TBT Committee has adopted the Principles for the Development of International Standards, Guides and Recommendations.[200] This decision of the TBT Committee

---

[194] L. Eicher, 'Technical Regulations and standards', in J.M. Finger and A. Olechowski, *The Uruguay Round. A Handbook on the Multilateral Trade Negotiations* (Washington D.C., The World Bank 1987), p. 140. G. Marceau and J. Trachtman, *supra* n. 74, p. 840.

[195] See also M.A. Livermore, 'Authority and Legitimacy in Global Governance: Deliberation, Institutional Differentiation, and the Codex Alimentarius', 81 *New York University Law Review* (2006), p. 790.

[196] The *SPS Agreement* defines harmonisation as 'the *establishment*, *recognition* and *application* of common [sanitary and phytosanitary] measures by different Members'. (Annex A.2 of the *SPS Agreement* [emphasis added]). From this definition the separation of powers between the two institutions is clear. Although mentioning the need for horizontal co-ordination with other intergovernmental organisations (Sutherland Report, p. 35), the Sutherland Report does not address this important type of inter-institutional relationship that the WTO has with the Codex Alimentarius Commission and with other standard-setting bodies.

[197] Art. 2.6 of the *TBT Agreement* and Art. 3.4 of the *SPS Agreement*.

[198] Under the *SPS Agreement*, Members are thus not solely required to fully participate, they are indeed obliged to put effort in the actual development of common standards. The *TBT Agreement* restricts the requirement to a full participation in the standard-setting procedure related to these standards that concern products for which the individual Members have either adopted, or expect to adopt technical regulations. This in contrast with the obligation under the *SPS Agreement* that covers *all* aspects of sanitary or phytosanitary measures.

[199] Art. 10.4 of the *SPS Agreement* indicates that active participation of developing countries should be encouraged and facilitated by the other Members. Art. 12.5 and Art. 12.6 of the *TBT Agreement* are more specific and require that Members shall ensure that international standard-setting bodies operate in such a manner, which facilitates active and representative participation of all Members, in particular developing countries. Furthermore, they are to examine the possibility of preparing international standards related to products of a particular interest to developing countries, if the developing countries so require.

[200] Decision of the Committee on Principles for the Development of international Standards, Guides and Recommendations with relation to Arts. 2, 5 and Annex 3 of the Agreement, in Decisions and

holds that the following principles and procedures should be observed: transparency, openness, impartiality and consensus, and the development dimension.[201] Despite its appearance, the Principles do not aim to dictate how to develop international standards to international standard-setting bodies, but to encourage the participation of WTO Members in these standard-setting bodies, instead.[202]

However, there are also aspects which are related to the institutional structure of the WTO that may obstruct the functioning of the Codex Alimentarius Commission in its capacity as standard-setting body. In this context, both the attributed mandate and the operation of several WTO bodies require particular attention: the panels and the Appellate Body that function as adjudicative bodies responsible for the interpretation of Codex measures, and the SPS Committee with its function to monitor the harmonisation process.

5.1 The dispute settlement mechanism

As already mentioned above in Section 2.1, the *SPS Agreement* and the *TBT Agreement* were adopted as an integral part of the WTO Agreement. This means that all WTO Members can bring disputes relating to rights and obligations under the *SPS Agreement* and the *TBT Agreement* under the procedure of the dispute settlement mechanism. Since the entry into force of the WTO Agreement, panels which function as adjudicative bodies have obtained an increased status.[203] Whereas, under the GATT system, the decisions from panels were adopted by consensus, the *Dispute Settlement Understanding* (*DSU*) annexed to the WTO Agreement indicates that Panel decisions (as reversed, modified or upheld by the Appellate Body) can only be *rejected* by consensus.[204] Furthermore, with the establishment of the WTO, the dispute settlement mechanism has been strengthened with an Appellate Body, which has the task of hearing appeals from panel cases.[205] Both bodies can interpret and apply the rights and obligations of the *SPS Agreement* and *TBT Agreement*. Given the references to Codex measures contained in these agreements, panels and the Appellate Body also act as adjudicating bodies for disputes which arise over the interpretation and application of these measures. And it is here that the complica-

---

Recommendations adopted by the Committee since 1 January 1995, Note by the Secretariat, G/TBT/1/Rev. 8, 23 May 2002. This decision of the TBT Committee does not have binding force.

[201] Decision of the Committee on Principles for the Development of international Standards, Guides and Recommendations with relation to Arts. 2 and 5 and Annex 3 of the Agreement, in Decisions and Recommendations adopted by the Committee since 1 January 1995, Note by the Secretariat, G/TBT/1/Rev. 8, 23 May 2002.

[202] G. Marceau and J. Trachtman, *supra* n. 74, p. 840.

[203] See, on the WTO dispute settlement mechanism, M. Matsushita, et al., *The World Trade Organization* (Oxford, Oxford University Press 2003), pp. 17-51. P. Van den Bossche, *supra* n. 13, pp. 172-306. R. Hudec, 'The New WTO Dispute Settlement Procedure', 8 *Minnesota Journal of Global Trade* (1999), pp. 1-53. E.U. Petersmann, *The GATT/WTO Dispute Settlement System: International Law, International Organisations and Dispute Settlement* (Dordrecht, Kluwer Law International 1998).

[204] Art. 16.4 of the *DSU*.

[205] Art. 17.1 of the *DSU*.

tions arise. As mentioned in the Sutherland Report, given the fact that it is 'self-contained' in its jurisdictional responsibilities, the dispute settlement system offers little legal space for co-operation with other international organisations.[206] The *DSU* merely concentrates on the interpretation and application of WTO provisions.

This section examines whether the current rules of the *DSU* enable panels and the Appellate Body to deal adequately with Codex matters within the context of the WTO Agreements. It focuses, on the one hand, on the judicial review of the Codex standard-setting procedure, and, on the other, on the interpretation and application of Codex measures in the context of the WTO agreements.

5.1.1 Lack of judicial review to ensure legitimacy of the Codex procedures

Panels established under the WTO dispute settlement mechanism do not have the explicit authority to review the legitimacy of Codex procedures and the resulting measures.[207] The *DSU* does not foresee this. Furthermore, the panels and the Appellate Body in *EC-Hormones* and *EC-Sardines* declined to assume such a responsibility.

In *EC-Hormones*, the EC argued that the Codex standards in question had been adopted by 33 votes in favour and 29 against, while the number of 7 abstentions indicated that this was actually a minority of the Members present.[208] The EC argument concentrated on the unusual fashion in which the MRLs relating to hormones had been adopted.[209] Despite the EC arguments about the elaboration procedure preceding the standard at issue, both the Panel and the Appellate Body in *EC-Hor-*

---

[206] Report by the Consultative Board to the WTO Director-General Supachai Panitchpakdi on 'The Future of the WTO: Addressing Institutional Challenges in the New Millennium' (the Sutherland Report), 2005, p. 39.

[207] See, on the competences of the International Court of Justice as regards judicial review, K.H. Kaikobad, *The International Court of Justice and Judicial Review. A Study on the Court's Powers with Respect to Judgments of the ILO and UN Administrative Tribunals* (The Hague, Kluwer Law International 2000).

[208] The adoption of these standards have had a long and problematic past. In 1991, at a first attempt to adopt the standards, the adoption was rejected by a vote of 12 in favour of adoption, 27 against and 9 abstentions. See Report of the 19th Session of the Joint FAO/WHO Codex Alimentarius Commission, Rome, 1-10 July 1991, ALINORM 91/40, para. 161. In 1995, thus on a second occasion, the standards were adopted. See Report of the 21st Session of the Joint FAO/WHO Codex Alimentarius Commission, Rome, 3-8 July 1995, ALINORM 95/37, paras. 44-45: 'After a lengthy debate on the issue in relation to whether to base a decision on currently available scientific evidence or to take into account factors other than health concerns, the Delegation of Spain, on behalf of the member countries of EU, proposed a roll-call vote on the adjournment of debate on the adoption of certain growth-promoting hormones at Step 8. The motion for adjournment failed.' A majority of Member countries voted to proceed by the use of a secret ballot, as requested by the Delegation of the United States. As a result of the secret ballot, the Commission adoption the MRLs for growth-promoting hormones (33 votes in favour of adoption, 29 votes against adoption, and 7 abstentions).' See also, Section 3.1.1 of Chapter V.

[209] A voting procedure has only rarely been used by the Codex Alimentarius Commission to adopt its standards. Normally, the Codex Alimentarius Commission adopts its standards by consensus. (see Section 3.1.1 of Chapter V).

mones did not verify whether the procedure had been conducted in accordance with the internal rules of the Codex Alimentarius Commission. The Panel declined all responsibility to examine the legitimacy of the standard-setting procedure:

> 'No other conditions are imposed in the SPS Agreement on the relevance of international standards for the purposes of Article 3. Therefore, as a panel making a finding on whether or not a Member has an obligation to base its sanitary measure on international standards in accordance with Article 3.1, we only need to examine whether such international standard exist. For these purposes, we need not consider….(ii) whether these standards have been adopted by consensus of by a wide or narrow majority…'[210]

The procedural rules of the Codex Alimentarius Commission were also invoked in *EC-Sardines*. Here, the EC relied on the Codex internal rule Consideration 5 of the 'Guide to the Consideration of Standards at Step 8' and argued that:

> 'With regard to the elaboration procedure of Codex Stan 94, ... an editorial change, and not a substantial change, was made at step 8 of the procedure. If a substantive amendment had been made at this stage, it would have been necessary to refer the text back to the relevant committee for comments before its adoption. However, if a substantive change had nevertheless been made at step 8 of the Codex elaboration procedure, the European Communities claims that Codex Stan 94 would, in this case, be *rendered invalid* and could not, therefore, be considered a relevant international standards within the meaning of Article 2.4 of the TBT Agreement.'[211]

However, the Panel did not address this issue. Instead, it relied on the definition of 'standard' as defined in Annex 1.2 of the *TBT Agreement*. As the definition of 'standard' neither includes this procedural requirement related to the adoption of international standards, nor contains any reference to the Procedural Manual of the Codex Alimentarius Commission, judicial review of the procedural correctness of the adoption of Codex Stan 94 was not conducted. For another argument, the EC relied on the principle of consensus contained in the 'Principles for the Development of International Standards, Guides and Recommendations' as adopted by the TBT Committee, and argued that the Codex measure had not been adopted by consensus.[212]

---

[210] Report of the WTO Panel, *EC-Measures Concerning Meat and Meat Products (Hormones)*, Complaint by the United States, WT/DS26/R/USA (1997), para. 8.69. Report of the WTO Panel, *EC-Measures Concerning Meat and Meat Products (Hormones)*, Complaint by Canada, WT/DS48/R/CAN (1997), para. 9.72.

[211] Report of the WTO Panel, *European Communities – Trade Description of Sardines*, WT/DS231/R, 29 May 2002, para. 4.34. [emphasis added]

[212] Report of the WTO Panel, *European Communities – Trade Description of Sardines*, WT/DS231/R, 29 May 2002, para. 4.33: 'According to the European Communities, another reason for not considering Codex Stan 94 as a relevant international standard is that it was not adopted in accordance with the principle of consensus set out by the TBT Committee in the Decision of the Committee on Principles for the Development of the International Standards, Guides and Recommendations with Relation to Articles 2, 5 and Annex 3 of the Agreement (the "Decision"). In support of its claim, the European

The Panel rejected this argument as well, and declined to use the criteria of consensus as laid down in the TBT Committee's decision as an interpretation instrument to review the relevance of the Codex standard in the context of the *TBT Agreement*:

> 'The Decision to which the European Communities refers is a policy statement of preference and not the controlling provision in interpreting the expression "relevant international standard" as set out in Article 2.4 of the TBT Agreement.'[213]

Upon the appeal relating to the Panel's conclusions in *EC-Sardines*, the Appellate Body also denied that it was responsible for reviewing the correctness of the Codex elaboration procedure of Codex stan 94:

> 'This conclusion [the definition of 'standard' in Annex 1.2 to the TBT Agreement does not require that standards are adopted by consensus] is relevant only for the purposes of the TBT Agreement. It is not intended to affect, in any way, the internal requirements that international standard-setting bodies may establish for themselves for the adoption of standards within their respective operations. In other words, the fact that we find that the TBT Agreement does not require approval by consensus for standards adopted by the international standardization community should not be interpreted to mean that we believe an international standardization body should not require consensus for the adoption of its standards. *That is not for us to decide.*'[214]

The explicit rejection of the responsibility for reviewing Codex measures judicially has been the subject of criticism.[215] The main criticism is the lack of legal security and the fact that illegally-adopted measures can become relevant in the context of the WTO agreements. One example is the opinion of Chile, as expressed during the dispute settlement meeting which adopted the Panel report in *EC-Sardines*:

> '... the finding reached by the Panel and the Appellate Body to the effect that Annex 1.2 to the TBT Agreement allowed standards that were not based on consensus. In this respect, Chile agreed with the EC that only standards adopted by consensus by an international organization could be "relevant" for the purposes of Article 2.4. This was confirmed by the Decision of the Committee on Principles for the Development of In-

---

Communities submits the following: (a) According to Rule VI:2 of the Rules of Procedure of the Codex Alimentarius Commission, decisions can be taken by a majority of the votes cast; even if it is not recorded whether Codex Stan 94 was elaborated and adopted by means of a formal vote, it is clear that it was adopted in circumstances in which dissenting members could have been outvoted and, therefore, may have decided not to express their disagreement, i.e., by not insisting on a vote. This is especially so, since the General Principles of the Codex Alimentarius make clear that Codex standards are recommendations that need to be accepted by governments and that their acceptance can be unconditional, conditional or with deviations.'

[213] Report of the WTO Panel, *European Communities – Trade Description of Sardines*, WT/DS231/R, 29 May 2002, para. 7.91.

[214] Report of the Appellate Body, *European Communities – Trade Description of Sardines*, WT/DS231/AB/R, 26 September 2002, para. 227.

[215] H. Horn and J.H.H. Weiler, *supra* n. 48, p. 255. R. Muñoz, *supra* n. 8, p. 483.

ternational Standards, Guides and Recommendations in relation to Articles 2, 5 and Annex 3 of the Agreement; a decision that the Appellate Body had not even mentioned in its Report. *It appeared that the analysis by the Appellate Body made it possible to consider as relevant standards which had not been approved in accordance with the decision-making process followed in international standardization organizations.* Chile hoped that, in future, this point would be clarified by the Appellate Body.'[216]

The lack of any judicial review of the elaboration procedure places more responsibility on the Codex Alimentarius Commission to ensure that standards are properly elaborated. However, the concern expressed by Chile remains, as standards which have not been adopted in conformity with the rules of the Procedural Manual of the Codex Alimentarius Commission can still be considered as international standards in the context of the *TBT Agreement* and the *SPS Agreement*. Both Article 3.1 of the *SPS Agreement* and Article 2.4 of the *TBT Agreement* imply that only 'relevant' international standards are taken into account in the context of these agreements. In line with the restricted competence of WTO panels to interpret only the provisions of the WTO agreements, these articles could have formed a basis upon which to consider whether international standards that have not been elaborated in accordance with the procedural rules are actually 'relevant' international standards.[217] Unfortunately, the decision of the Appellate Body in *EC-Sardines* does not provide sufficient room for future panels to examine the validity of Codex measures under these articles.[218]

5.1.2 WTO panels and the interpretation of Codex measures

Due to the reference to Codex measures in the *SPS Agreement* and the *TBT Agreement*, the meaning of the relevant Codex measure has become a WTO issue. When, during the disputes, one of the parties invokes the relevant articles of the *SPS Agreement* or the *TBT Agreement*, the panels will have to interpret the relevant Codex measures in question. However, the question remains as to whether their mandate actually allows the panels to ensure an interpretation and application of Codex measures which is coherent with their original meaning, purpose and function within the overall structure of the Codex Alimentarius.

---

[216] Dispute Settlement Body, 23 October 2002, WT/DSB/M/134, p. 12, para. 41. [emphasis added]

[217] The EC's reliance in *EC-Sardines* on the TBT Committee's decision is an interesting one. Would the Panel have approved of the relevance of these principles to the interpretation of the definition of international standards, it would have recognised a competence to judge the legitimacy of a standard-setting procedure of another international body. It still would not have had the competence to nullify the action taken by the Codex Alimentarius Commission. However, it could have used these criteria to reject the relevance of the standard in question as reference point under the *TBT Agreement*. By means of the TBT decision, it could have responded to several concerns related to the increased status of the Codex standards and the importance of controlling the legitimacy of the procedure.

[218] In contrast to what has been held here, some authors believe that the Appellate Body decision in *EC-Sardines* has created a possibility to question the validity of Codex measures in future cases, see M.A. Livermore, *supra* n. 195, p. 793.

Several provisions of the *DSU* promote an objective execution by the panels of their function, as laid down in Article 11 of the *DSU*. Article 3.2 of the *DSU* requires panels to interpret the provisions of the WTO agreements in accordance with customary rules of interpretation. As held by the Appellate Body in *US-Gasoline* and *Japan-Alcoholic Beverages II*, Articles 31 and 32 of the Vienna Convention on the Law of Treaties constitute a codification of some customary rules of interpretation,[219] and, so far, the 'customary rules' that have been applied by panels have been limited to these articles.[220] The Vienna Rules of Interpretation aim to ensure that the interpretation of treaties is in line with their object, purpose and context.[221] For this reason, the Vienna Rules of Interpretation can be an important instrument to ensure a coherent interpretation of Codex measures.

Another instrument which allows panels to seek information and consult experts is contained in Article 13 of the *DSU*. This article provides for the possibility that

---

[219] Report of the Appellate Body, *United States-Standards for Reformulated and Conventional Gasoline*, WT/DS2/AB/R, 29 April 1996, p.16-17: 'The "general rule of interpretation" set out above has been relied upon by all of the participants and third participants, although not always in relation to the same issue. That general rule of interpretation has attained the status of a rule of customary or general international law. As such, it forms part of the "customary rules of interpretation of public international law" which the Appellate Body has been directed, by Article 3(2) of the DSU, to apply in seeking to clarify the provisions of the General Agreement and the other 'covered agreements' of the Marrakesh Agreement Establishing the Word Trade Organization.' Report of the Appellate Body, *Japan-Taxes on Alcoholic Beverages*, WT/DS8/AB/R, WT/DS10/AB/R, WT/DS11/AB/R, 4 October 1996, pp. 9-11.

[220] It is questionable whether other provisions of the Vienna Convention are likely to be taken into consideration by Panels as customary rules of interpretation. This rather limited scope of articles relevant to the interpretation of treaties is due to the fact that the substantive jurisdiction of WTO panels is limited to invocation of violation with provisions of WTO agreements. For this reason, several codified rules of interpretation are not relevant to WTO Panels, such as Art. 30. Palmeter and Mavroidis emphasise the fact that Panels did not consider Art. 30 of the Vienna Convention, D. Palmeter and P.C. Mavroidis, 'The WTO Legal System: Sources of Law', 92 *American Journal of International Law* (1998), p. 412.

[221] Art. 31 of the Vienna Convention reads:
'Article 31: General rule of interpretation
1. A treaty shall be interpreted in good faith in accordance with the ordinary meaning to be given to the terms of the treaty in their context and in the light of its object and purpose.
2. The context for the purpose of the interpretation of a treaty shall comprise, in addition to the text, including the preamble and annexes:
(a) any agreement relating to the treaty which was made between all the parties in connexion with the conclusion of the treaty;
(b) any instrument which was made by one or more parties in connexion with the conclusion of the treaty and accepted by the other parties as an instrument related to the treaty.
3. There shall be taken into account, together with the context:
(a) any subsequent agreement between the parties regarding the interpretation of the treaty or the application of its provisions;
(b) any subsequent practice in the application of the treaty which establishes the agreement of the parties regarding its interpretation;
(c) any relevant rules of international law applicable in the relations between the parties.
4. a special meaning shall be given to a term if it is established that the parties so intended.'

the assistance of the Codex Alimentarius Commission or one of its subsidiary bodies be sought for the interpretation of Codex measures.

However, it appears from the interpretation and application of these provisions (the use of the customary rules of interpretation, and the possibility of seeking information and consulting experts) that they both contain some shortcomings that may seriously undermine the interpretation of Codex measures.

5.1.2.1 The scope of the obligation to use customary rules of interpretation

The obstacle to the application of the customary rules of interpretation to the interpretation of Codex measures is the scope of the obligation to use the customary rules under Article 3.2 of the *DSU*. When Article 3.2 of the *DSU* mentions the 'customary international rules of treaty interpretation', it does so in the context of clarifying the provisions of the WTO agreements, which implies that the rules of interpretation solely concern the interpretation of these provisions. Formally speaking, Codex standards do not form an integral part of the WTO agreements, and are, therefore, not directly enforceable by WTO panels. According to Peru, in an argument raised before the Appellate Body in *EC-Sardines*, it is for this reason that the interpretation of Codex standards is to be regarded as an assessment of facts, instead of being regarded as an interpretation of law.[222] Consequently, the obligation to use the customary rules of interpretation would not apply to Codex measures.

WTO case-law illustrates that the scope of the obligation to use the rules of interpretation is not as restricted as may appear at first sight. The panels' application of the rules of interpretation extends beyond the WTO provisions, as panels have also applied the rules of interpretation to 'outside' provisions. For instance, in *EC-Bananas III*, the applicability of the Lomé Convention to the dispute was questioned, as it was not included in the terms of reference of the established panel and was not part of the covered agreements. The Panel held that:

> 'We note that since the GATT CONTRACTING PARTIES incorporated a *reference* to the Lomé Convention into the Lomé waiver, the *meaning* of the Lomé Convention became a GATT/WTO issue, *at least to that extent*. Thus, we have no alternative but to examine the provisions of the Lomé Convention ourselves in so far as it is *necessary to interpret* the Lomé waiver.' [223]

---

[222] Report of the Appellate Body, *European Communities – Trade Description of Sardines*, WT/DS231/AB/R, 26 September 2002, para. 75: 'In respect of the European Communities' argument that the Panel incorrectly interpreted Codex Stan 94, Peru asserts that the European Communities mistakenly treats this alleged error as an error in interpretation, rather than a failure to conduct an objective assessment of fact. According to Peru, the Codex standard is not a covered agreement within the meaning of Article 1.1 of the DSU, nor is it a treaty or another source of international law. Peru thus contends that, like municipal law, the Codex standard must be treated by an international tribunal as a fact to be examined, not as law to be interpreted.'

[223] Reports of the WTO Panel, *European Communities - Regime for the Importation, Sale and Distribution of Bananas*, adopted on 25 September 1997, WT/DS27/R, para. 7.98. [emphasis added]

In *US-Section 211*, too, the Panel recognised, in the context of the *TRIPs Agreement*, the applicability of the rules of interpretation to the provisions of the Paris Convention.[224]

Be that as it may, neither the panels nor the Appellate Body have explicitly recognised Codex measures as applicable law under the WTO agreements and it has to be admitted that the current formulation of Article 3.2 *DSU* is not an incentive to do so. However, it must also be stated that the Appellate Body does not reject Codex measures as applicable law, either. In *EC-Sardines*, the Appellate Body did not address Peru's argument that Codex measures were not to be considered as law to be interpreted. It remains silent on this point. Nevertheless, the Appellate Body did conduct an assessment of the meaning of the Codex standard. In fact, as the competences of the Appellate Body are restricted to addressing questions of law, this action points to a *de facto* recognition of the Codex standard as *applicable law*.[225] These situations in which panels and the Appellate Body interpret Codex measures, while not being bound by the rules of interpretation, as was clearly the case in *EC-Sardines*, need to be avoided. The absence of applicable rules of interpretation creates the risk that interpretations of Codex measures may run counter to the meaning, purpose and function of the measures within the system of the Codex Alimentarius as intended by Codex Members. Furthermore, the absence of such applicable rules makes it difficult to control the legitimacy of the panels' and the Appellate Body's conduct.

---

[224] Report of the WTO Panel, *United States – Section 211 Omnibus Appropriations Act of 1998*, WT/DS176/R, 6 August 2001, para. 8.16: '… We will apply the principles enunciated by the Appellate Body in the United States-Gasoline [Article 31 and 32 of the Vienna Convention on the Law of Treaties] to interpret the relevant provisions of the TRIPs Agreement throughout the report, *including the provisions of the Paris Convention (1967)* incorporated into the Agreement.'

[225] Art. 17.13 of the *DSU* states that 'the Appellate Body may uphold, modify or reverse the *legal* findings and conclusions of the panel.'[emphasis added]. Furthermore, Art. 17.6 of the *DSU* provides that 'an appeal shall be limited to issues of law covered in the panel report and legal interpretations developed by the panel.' Art. 17.12 of the *DSU*: 'The Appellate Body shall address each of the issues raised in accordance with paragraph 6 during the appellate proceedings.' This does not mean that the competence of the Appellate Body is restricted to the review of the interpretation and the application of law and that the panels' discretion to assess facts is unrestricted (see P. Van den Bossche, 'Appellate Review in WTO Dispute Settlement', in F. Weiss (ed.), *Improving WTO Dispute Settlement Procedures. Issues and Lessons from the Practice of Other International Courts and Tribunals* (London, Cameron May 2000), p. 308). In *EC-Hormones*, the Appellate Body confirmed that the question of whether or not the Panel had made an objective assessment of the facts is a legal question and so is the examination on 'the consistency or inconsistency of a given fact or set of facts with the requirements of a given treaty provision is … a legal characterization issue. It is a legal question.' (Report of the Appellate Body, *EC-Measures Concerning Meat and Meat Products (Hormones)*, WT/DS26/AB/R, WT/DS48/AB/R (1998), para. 132). However, it considered that the finding of facts itself it not subject to review by the Appellate Body, as it does not constitute a question of law (Report of the Appellate Body, *EC-Measures Concerning Meat and Meat Products (Hormones)*, WT/DS26/AB/R, WT/DS48/AB/R (1998), para. 132). It gives an example of a question of fact which is the determination of whether or not a certain event did occur in time and space. It continues by stating that whether or not Codex has adopted an international standard, guideline or recommendation on MGA is a factual question: as it is an event that occurred in time and space.

5.1.2.2 The competence to seek information from 'outside' sources

Article 13.1 of the *DSU* states that a panel has 'the right to seek information and technical advice from any individual or body which it deems appropriate'. Article 13.2 of the *DSU* provides that panels may 'seek information from any relevant source and may consult experts to obtain their opinion on certain aspects of the matter'. Consequently, as mentioned above, Article 13 of the *DSU* provides panels with the possibility of consulting the Codex Alimentarius Commission or its subsidiary bodies with regard to the meaning and validity of Codex measures. In this sense, this article is one of the rare examples in the *DSU* that allows panels to 'co-operate' with other international organisations.[226] In fact, in previous decisions, panels have used this article as a basis to consult international organisations. However, from the previous applications of Article 13 of the *DSU* by panels, it becomes apparent that the consultation of international organisations remains restricted to the request for factual information and does not involve the legal interpretation of the relevant provisions. In *India-Quantitative Restrictions*, the Panel concluded that Article 13.1 of the *DSU* provided a basis for the Panel to consult with the International Monetary Fund (IMF) in order to obtain any relevant information relating to India's monetary reserves and its balance-of-payments situation.[227] Likewise, in *US-Section 211*, the Panel consulted the WIPO to requests factual information on the provisions of the Paris Convention on the basis of Articles 13.1 and 13.2 of the *DSU*.[228]

In *EC-Hormones*, the Panel decided, on the basis of both Article 13 of the *DSU* and Article 11.2 of the *SPS Agreement*, to consult six experts.[229] Although the majority of the experts were scientific experts, one of the experts was a representative of the Codex Secretariat, who informed the Panel on the history of the international standards and the functioning of the procedures.[230] He also gave some explanations

---

[226] It has to be noted that for instance the International Court of Justice has the capacity and even the obligation to ask for legal interpretation if it is to decide upon constitutional documents of international organisations. In these cases it is the political organs that decide on the appropriate interpretation and not the administrative organs. Schermers and Blokker advance several reasons why policy-making organs may be more suitable than courts. See H.G. Schermers and N.M. Blokker, *International Institutional Law*, 4th edn. (The Hague, Martinus Nijhoff Publishers 2003), pp. 852-853.

[227] Report of the WTO Panel, *India-Quantitative Restrictions on Imports of Agricultural, Textile and Industrial Products*, WT/DS90/R, 6 April 1999, para. 5.12.

[228] Report of the WTO Panel, *United States – Section 211 Omnibus Appropriations Act of 1998*, WT/DS176/R, 6 August 2001, paras. 8.12 and 8.13.

[229] Report of the WTO Panel, *EC-Measures Concerning Meat and Meat Products (Hormones)*, Complaint by the United States, WT/DS26/R/USA (1997), para. 8.8. Report of the WTO Panel, *EC-Measures Concerning Meat and Meat Products (Hormones)*, Complaint by Canada, WT/DS48/R/CAN (1997), paras. 9.8-9.9.

[230] Transcript of the Joint Meeting with Experts, held on 17 and 18 February 1997, Annex to the Report of the WTO Panel, *EC-Measures Concerning Meat and Meat Products (Hormones)*, Complaint by the United States, WT/DS26/R/USA (1997) and the Report of the WTO Panel, *EC-Measures Concerning Meat and Meat Products (Hormones)*, Complaint by Canada, WT/DS48/R/CAN (1997), paras. 26 and 27.

on the concept of MRLs and ADIs, and whether or not these could be considered as measures. The example of consulting the representative of the Codex Secretariat illustrates that Article 13 of the *DSU* can be very useful in assuring that the interpretation of panels are in line with the original meaning and purpose of Codex measures. However, Article 13 is not without flaws. First, the competence to consult international organisations is not mandatory and is left to the discretion of panels. This led, for instance, to the unfortunate situation in which the Panel in *EC-Sardines* refused to consult the Codex Alimentarius Commission on the meaning of the relevant Codex standard as it did not consider it necessary.[231] However, the Appellate Body concluded that the Panel had not exceeded its discretion by refusing:

> 'We also reject the EC claim regarding the fourth instance of supposed impropriety, which related to the decision of the Panel not to seek information from the Codex Commission. Article 13.2 of the DSU provides that "panels may seek information from any relevant source and may consult experts to obtain their opinion on certain aspects of the matter". This provision is clearly phrased in a manner that attributes discretion to panels, and we have interpreted it in this vein. Our statements in *EC-Hormones*, *Argentina-Textiles and Apparel*, and *US-Shrimp*, all support the conclusion that, under Article 13.2 of the DSU, panels enjoy discretion as to whether or not to seek information from external sources. *In this case, the Panel evidently concluded that it did not need to request information from the Codex Commission, and conducted itself accordingly. We believe that, in doing so, the Panel acted within the limits of Article 13.2 of the DSU...*'[232]

Second, the competence to seek information or consult experts covers both the consultation of the representatives of the relevant institutions *and* the consultation of scientific experts. The following discussion illustrates that Article 13 of the *DSU* and the other relevant articles have shortcomings that lead to an important concern.[233] These articles do not distinguish between the two types of consultations, even though the functions of the representatives from international organisations on the one hand, and those of the scientific experts on the other, are very distinct. The function of the former group aims to provide the background information that assists panels to interpret the case in question better. The scientific experts are consulted for their *scientific* expertise and to assist panels on the difficult scientific questions that arise in the context of the dispute. Due to this lack of a clear distinction between the two groups, a blend of responsibilities may result in a situation in which the very information or advice which the individuals give falls outside the

---

[231] Report of the WTO Panel, *European Communities – Trade Description of Sardines*, WT/DS231/R, 29 May 2002, para. 6.8.

[232] Report of the Appellate Body, *European Communities – Trade Description of Sardines*, WT/DS231/AB/R, 26 September 2002, para. 302. [emphasis added]

[233] Art. 13 of the *DSU*, Art. 11.2 of the *SPS Agreement* and the 14.2 of the *TBT Agreement*. Art. 14.2 of the *TBT Agreement* reads: 'At the request of a party to a dispute, or at its own initiative, a panel may establish a technical expert group to assist in questions of a technical nature, requiring consideration by experts.'

scope of the capacity for which they have been selected. The application of Article 13 of the *DSU* by panels, in particularly in the context of the *SPS Agreement*, illustrates the importance attached to the consultation of scientific experts. The decisions of panels with regard to SPS disputes reflect the enormous influence that the opinions of the scientific experts have on the conclusions of the panels. It has even been held that the consultation sought from scientific experts actually gives the impression that panels partially delegate their adjudicatory responsibility to the experts.[234] The lack of a clearly described mandate in these cases has, in fact, led to the scientific experts giving advice on the interpretation of international standards, instead of this being done by the representatives of the international standard-setting bodies.

That this concern is not without foundation is clearly demonstrated by the following examples. In *EC-Hormones*, the scientific experts gave information on the aspects that had been taken into account in the report of the Joint FAO/WHO Expert Committee on Food Additives (JECFA) and in the establishment of Codex standards.[235] However, this task should be reserved for an appointed representative of the Codex Alimentarius Commission. Likewise, in *Australia-Salmon*, two scientific experts gave their advice on the interpretation of the OIE guidelines.[236] In *Japan-Apples*, the scientific experts were asked by the Panel: 'How does Japan's 1999 PRA compare with the IPPC Guidelines for Pest Risk Analysis for 1995 (Exhibit JPN-30) and the IPPC Pest Risk Assessment standard of 2001 (Exhibit USA-15)?'[237] Consequently, the experts were asked for their opinion not only on the meaning of the IPPC Guidelines, but also on the compatibility of Japan's assessment with [whether Japan's assessment was compatible with the guidelines] the guidelines.

---

[234] C. Button, *supra* n. 19, p. 52: 'One concern relates to the risk that, because of their lack of expertise, panels will ultimately delegate responsibility for assessing issues like whether there is a scientific basis for a health measure to experts ... In asking its experts this question, the panel risked giving the impression that it was delegating its adjudicatory responsibility to its experts.'

[235] For instance, Report of the WTO Panel, *EC-Measures Concerning Meat and Meat Products (Hormones)*, Complaint by the United States, WT/DS26/R/USA (1997), para. 7.58 and the Report of the WTO Panel, *EC-Measures Concerning Meat and Meat Products (Hormones)*, Complaint by Canada, WT/DS48/R/CAN (1997) para. 8.57. It has to be remarked that the information was provided on the basis of a Panel's question: 'to what extent were these hazards [the potential hazards identified by the EC] taken into consideration in the context of the 1988 JECFA report or in the establishment of the Codex standards for the hormones at issue?'

[236] Report of the WTO Panel, *Australia-Measures Affecting the Importation of Salmon*, WT/DS18/R (1998), paras. 6.151-6.155.

[237] See also, discussion in Section 3.5.1 of this chapter. Report of the WTO Panel, *Japan-Measures Affecting the Importation of Apples*, WT/DS245/R (2003), question 33, p. 123. The IPPC Guidelines for Pest Risk Analysis are recognised international scientific assessment methods as developed by the IPPC, which are relevant under Art. 5.1 of the *SPS Agreement* and will have to be taken into account by WTO Members.

5.2 The SPS Committee: Monitoring the process of international harmonisation

The *SPS Agreement* assigns the SPS Committee, as a 'regular forum for consultation',[238] with the task of developing a procedure to monitor the process of international harmonisation and the use of international standards.[239] This procedure is to work as a supplement to the obligations that already exist for Members in the context of the notification procedure,[240] and gives Members the right to request reasons for sanitary measures which have trade distorting effects and are not based on the existent international standard.[241]

The monitoring function of the Committee contains elements which overlap with the role of the Codex Alimentarius Commission or its Secretariat. Given the different objectives of the WTO, in comparison to the Codex Alimentarius Commission, a clear separation of responsibilities needs to be promoted. The monitoring procedure, as described in Article 12.4 and as developed by the SPS Committee, concentrates on inviting WTO Members to inform the Committee on their use and non-use of international standards, recommendations and guidelines that have major trade effects.[242] This element of the monitoring procedure resembles the 'notification dimension' of the former acceptance procedure of the Codex Alimentarius Commission. With the abolition of the Codex acceptance procedure, this means that the

---

[238] Art. 12.1 of the *SPS Agreement*.

[239] Art. 12.4 of the *SPS Agreement*. See, on the mandate of the SPS Committee, D. Prévost and P. Van den Bossche, *supra* n. 16, p. 343, G. Marceau and J. Trachtman, *supra* n. 74, p. 840, G. Bossis, *supra* n. 120, pp. 122-126. G. Stanton, 'A review of the operation of the Agreement on Sanitary and Phytosanitary Measures', in M. Ingco and A. Winters (eds.), *Agriculture and the New Trade Agenda* (Cambridge, Cambridge University Press 2004), pp. 101-110.

[240] Annex B of the *SPS Agreement*. See Section 2.2.1.2 of this chapter for further discussion on the notification procedure and the right of WTO Members to request justification, as laid down in Art. 5.8 *SPS Agreement*.

[241] Art. 5.8 of the *SPS Agreement*.

[242] Committee on Sanitary and Phytosanitary Measures, Procedure to monitor the process of international harmonisation. Decision of the Committee, G/SPS/11, 22 October 1997. In 1997, this procedure has been adopted on a provisional basis for 18 months and has been prolonged three times and has been revised in 2004. At its meeting in June 2006, the SPS Committee decided to extend the provisional procedure indefinitely and review it every four years (Committee on Sanitary and Phytosanitary Measures, Decision to modify and extend the provisional procedure to monitor the process of international harmonisation, G/SPS/40, 5 July 2006). The procedure, as adopted by the SPS Committee is as follows. At least 10 days prior to each regular meeting of the SPS Committee, Members submit concrete examples of standards, guidelines and recommendations that cause problems with a significant trade impact, which is either related to the use or non-use of these international standards, guidelines or recommendations. These complaints are submitted to the Secretariat, which compiles the comments and send them to all Members as much in advance of the regular Committee meeting as possible to give other Members the opportunity to prepare comments on the use or non-use of the international standards at issue. This information is used by the Committee to establish a list of international standards, guidelines and recommendations, which is to be reviewed at each meeting of the Committee. Furthermore, each year, the Committee prepares an annual report on the list of identified international standards, guidelines and recommendations and the major trade impacts they cause due to the use, non-use or non-existence. This report is sent to the relevant international standard-setting bodies.

monitoring of the notifications related to Codex measures has become the prime responsibility of the SPS Committee.

Furthermore, a large part of the procedure is dedicated to the monitoring of the establishment procedure of international standards, guidelines and recommendations, as one of the aims of the procedure developed by the SPS Committee is the identification, 'for the benefit of the relevant international organisations, where standards, guidelines or recommendations were needed or were not appropriate for their purpose'.[243] The procedure is linked with the Committee's competence to ask the Codex Alimentarius Commission to examine a particular aspect of a standard, recommendation or guideline, and has been used in the light of a given Member's reasons for the non-use of a particular standard.[244] For instance, during the SPS Committee meeting of 15-16 October 1997, it was noted that:

> 'In its submission, the Philippines noted that benzoic acid was an anti-microbial preservative used in sauces which required a shelf-life of 6 to 12 months, and therefore used in products marketed for export. The lack of an international standard allowed countries to discriminate in the use of the additive. An Acceptable Daily Intake (ADI) of 5 mg/kg body weight had been determined for benzoic acid. This had been used as a reason for restricting its use, as high levels of consumption of foods containing the additive could create a health risk. However, as there were no internationally agreed methodologies for assessing risk due to dietary exposure to food additives, the potential for discrimination in usage existed. The use of benzoic acid in sauces was included in the draft Codex General Standard for Food Additives (GSFA). *It was important that the GSFA be finalized soon* because the lack of an international standard not only harmed international trade in sauces, but had the potential to affect trade in other processed food products where preservatives were used. Other Members expressed their views in the subsequent Committee discussions, and in their responses to G/SPS/W/100. The United States, Norway, Thailand, Cyprus, the European Communities, Hungary and the Czech Republic *agreed that there was a need for an international standard*. ..... Norway observed that *the standard should take into consideration* the frequent use of benzoic acid in foods together with the knowledge of low ADI and potential risks of allergic reactions.'[245]

For this moment, risks of negative consequences resulting from this overlap are negligible. Article 12 of the *SPS Agreement* contains several provisions that aim to avoid duplication with the work undertaken by the international standard-setting bodies. For instance, Article 12.3 states that 'the Committee shall maintain close

---

[243] Decision of the SPS Committee, Procedure to monitor the process of international harmonization, G/SPS/11, 22 October 1997, para. 1.

[244] This competence is laid down in Art. 12.6 of the *SPS Agreement*.

[245] Committee on Sanitary and Phytosanitary Measures, Procedure to monitor the process of international harmonisation, First Annual Report, adopted by the Committee on 8 July 1999, G/SPS/13, 12 July 1999, paras. 23-24. Another more recent example is the issue raised as regards the lack of a Codex standard maximum residue level of surphur dioxide in cinnamon. The lack of such a standards has caused trade problems for Sri Lanka. Committee on Sanitary and Phytosanitary Measures, Summary of the meeting of 29-30 March 2006, G/SPS/R/40, 26 May 2006, para. 106.

contact with the relevant international organisations.....in order to ensure that unnecessary duplication of effort is avoided'.[246] This is executed, e.g., by the attendance of the Codex Secretariat at SPS Committee meetings and the presence of a representative of the WTO at the Codex Alimentarius Commission meetings.[247] Furthermore, Article 12.5 provides the Committee with the possibility to 'use the information generated by the procedures, particularly for notification, which are in operation in the relevant international organizations'.

In addition, the monitoring procedure, in conjunction with the competence to request international organisations to examine a particular aspect of a standard, recommendation or guideline, has no formal implications for these organisations. Although there is a possibility for the relevant WTO Committees to submit recommendations to the Codex procedure, this competence lacks any formal hierarchical structure, as it is not binding on the Codex Commission.[248] Informally, though, the SPS Committee expects its Members, through their participation in international standardisation bodies, to take this information into account when establishing the work priorities.[249]

6. Conclusions

One of the instruments used in the *SPS Agreement* and the *TBT Agreement* to further trade liberalisation is the instrument of harmonisation. To achieve the harmonisation of national measures, both agreements use international standards that have been developed in 'outside' standard-setting bodies as a harmonisation tool. The instrument of harmonisation is expressed in an encouragement to conform national measures to these international standards and in an obligation to use them as a basis for national measures. Through the reference to international standards in these provisions and through the explicit recognition of Codex standards as international standards, Codex measures have acquired an increased status which consists of the following.

First, the encouragement to conform national measures to international standards, also referred to as the presumption of consistency, accords an important advantage for those WTO Members that conform their measures to international standards. These Members are exempted from having to justify the necessity of their measures. The presumption of consistency means that international standards are automatically considered to be necessary trade barriers that are justified under the *TBT Agreement* and the *SPS Agreement*. This strong role accorded to Codex standards through the presumption of consistency means that they function as benchmarks against which national measures are examined.

---

[246] See also, M. Echols, *supra* n. 8, p. 139.
[247] Ibid., p. 139.
[248] A contrary opinion has been expressed by G. Bossis, *supra* n. 120, p. 124.
[249] Decision of the SPS Committee, Procedure to monitor the process of international harmonization, G/SPS/11, 22 October 1997, para. 10.

Second, the requirement to use international standards as a basis for national measures is clearly obligatory in nature. This leads to the question as to whether Codex standards have acquired a binding force under the *SPS Agreement* and the *TBT Agreement*. Although the Appellate Body has rejected a binding force of Codex standards, its argumentation is not persuasive. Furthermore, the rather strict application of the obligation to use international standards as a basis by both panels and the Appellate Body to the cases in hand indicates a *de facto* binding force of the standards, at least to some extent. In this context, it has to be noted that the *de facto* binding force of the Codex standards is not direct as it only applies through the binding provisions of the WTO agreements. Furthermore, the binding force does not concern Codex standards as a whole, but is restricted to some elements or provisions of the standards. Although the Appellate Body does not specify upon which elements of the standards national measures have to be based, the *EC-Sardines* decisions provide some indication that the restricted nature of the scope of the binding elements may actually be linked with the restricted nature of the objective of the *SPS Agreement* and the *TBT Agreement* itself (*vis-à-vis* the objectives of the Codex Alimentarius Commission). In other words, the scope of the provisions of Codex standards that have acquired binding force may be restricted to the elements of the standards that aim to further trade liberalisation, and thus the scope does not cover the elements that envisage the protection of human health or fair practices in the food trade.

The explicit and autonomous right to deviate under Article 3.1 in conjunction with Article 3.3 of the *SPS Agreement*, and the second sentence of Article 2.4 of the *TBT Agreement* grant WTO Members the permanent possibility of deviating from the international standards. However, neither under the *SPS Agreement*, nor under the *TBT Agreement* is this right to deviate absolute or unconditional. It is restricted by the ability of the international standards in question to achieve a level of sanitary or phytosanitary protection or other legitimate objectives pursued at a national or regional level. Furthermore, the notification procedure under both agreements requires WTO Members to notify deviations from international standards and to justify these deviations upon request. Whether the autonomous right to deviate from international standards actually reduces the status of Codex standards as containing binding elements, under the agreements depends upon how easily panels and the Appellate Body accept the justification for these deviations.

Under the *TBT Agreement*, deviations from international standards are accepted if the international standards are an inappropriate or ineffective means of fulfilling their legitimate objectives. In spite of the reversal of the burden of proof, which now rests upon the complainant that challenges the deviation from the international standard, the threshold of evidentiary weight to justify the deviations still remains high, as the decisions in *EC-Sardines* illustrate.

Under the *SPS Agreement*, measures can only be consistent with Article 3.3 of the Agreement if they are in conformity with Article 5 of the Agreement. In spite of this 'subordinate' position of Article 3.3 *vis-à-vis* Article 5, even in the context of the latter article, international standards are taken as reference point to examine conformity of the measures. The reports in *EC-Hormones* indicate that the scientific reports, the risk methodology and the risk assessment policy of the JECFA (which forms the basis of the relevant Codex standards) have played a significant role as a reference point in the examination under Article 5 of the *SPS Agreement*. Also in the context of Article 5.6, which prohibits WTO Members from adopting more trade-restrictive measures than necessary to achieve the level of protection, Codex standards have a role as reference point. This position results from the recognition of Codex standards as necessary trade barriers, which flows from the presumption of consistency under Article 3.2 of the *SPS Agreement*.

To conclude, under both agreements, it is not easy to justify deviations from Codex standards, which only confirms the *de facto* binding force.

The status of Codex standards under the *SPS Agreement* and the *TBT Agreement* is an increased status if one compares it with the initial status accorded under the Codex acceptance procedure, under which, Codex standards could only obtain binding force upon explicit acceptance by countries. In contrast, under the *SPS Agreement* and the *TBT Agreement*, the status of Codex standards no longer depends on the consent of countries, but results from the mere fact that Codex standards have been adopted as a result of the Codex standard-setting procedure. This also means that the different harmonisation levels, as reflected in the flexible Codex acceptance procedure, disappear. Furthermore, it can be noted that, as both the *SPS Agreement* and the *TBT Agreement* refer to international standards, guides *and* recommendations, all these three types of Codex measures are covered by the obligatory harmonisation provisions. This is in contrast with the Codex acceptance procedure, which only applied to standards and MRLs. In the context of the WTO agreements, the acceptance procedure was no longer relevant. However, this did not mean that it had become obsolete. Although both the obligations under the WTO agreements and the obligations which resulted from the Codex acceptance procedure are result-oriented, the objectives pursued are partly different. Within the context of the *SPS Agreement* and the *TBT Agreement*, the obligations primarily serve the objective of trade liberalisation, but do not cover the objectives of the protection of human health and fair practices in the food trade as envisaged by the Codex Alimentarius. This means that, in the context of the *SPS Agreement* and the *TBT Agreement*, Codex measures function as a ceiling and not as a floor. Consequently, with the abolition of the acceptance procedure, the Codex Alimentarius Commission has done away with the mechanism for promoting the use of Codex measures as minimum-platform standards to ensure a minimum level of consumer protection.

The WTO is not a standardising body and relies on 'outside' international standard-setting bodies, such as the Codex Alimentarius Commission, for the adoption of harmonisation measures. This means that the relationship between the WTO and the Codex Alimentarius Commission can be characterised by a separation of powers: the Codex Alimentarius Commission being responsible for the 'legislative' acts, and the WTO being responsible for the enforcement of the application of harmonisation measures. Two aspects relating to the institutional rules of the WTO may potentially obstruct the functioning of the Codex Alimentarius Commission. First, the rules regulating the WTO dispute settlement mechanism as laid down in the *DSU* do not ensure that panels and the Appellate Body function adequately as adjudicating bodies when it comes to disputes on the proper interpretation and application of Codex measures. The *DSU* does not foresee in the possibility of judicial review of Codex measures that have been adopted in conflict with the internal rules of the Codex Alimentarius Commission. This creates legal uncertainty in situations where Codex measures have been unlawfully adopted, but must be used as a basis for national measures in the context of the WTO agreements. Furthermore, through the reference to international standards in the *SPS Agreement* and the *TBT Agreement*, the interpretation of Codex measures has become a WTO issue. The fact that panels have not used the Vienna rules of interpretation, and the fact that the competence to consult 'outside sources' is not mandatory and not clearly de-limited under the *DSU*, means that the interpretation of Codex measures may run counter to their meaning, purpose and function within the *Codex Alimentarius*.

Second, the function of the SPS Committee, as 'regular forum for consultation' also overlaps with the mandate of the Codex Alimentarius Commission. However, at this moment, the overlap does not lead to negative consequences, as the *SPS Agreement* contains provisions that provide for inter-institutional co-operation, and because there is no hierarchal structure between the SPS Committee and the Codex Alimentarius Commission.

Chapter V
THE LEGITIMACY OF THE CODEX ALIMENTARIUS, THE STANDARD-SETTING PROCEDURE AND THE INSTITUTIONAL FRAMEWORK

1. INTRODUCTION

As held in Chapter I, despite its subsidiary position *vis-à-vis* the FAO and the WHO, the Codex Alimentarius Commission has been attributed with important powers: independent normative powers and the task of acting as the promoter of the co-ordination of international food standard activities. It was explained that the 'host country construction' has allowed the Codex Alimentarius Commission to establish a strong institutional structure of subsidiary committees.[1] Furthermore, co-operation initiatives with other international standard-setting bodies have been established, by submitting their standards for final adoption under the Codex elaboration procedure. Both this strong institutional framework and the assistance of the expert bodies as risk assessors have enabled the Codex Alimentarius Commission to give content, *ratione materiae*, to its mandate. It was also explained that the Codex Alimentarius Commission has developed a complex system of food requirements that reflect a high level of harmonisation. Agreement by Codex Members over the often very detailed provisions was made possible through the establishment of the acceptance procedure. Under the former acceptance procedure, Codex standards did not have binding force unless Codex Members accepted them as binding.

As demonstrated in Chapters III and IV, this last aspect has changed considerably. Through the entry into force of the WTO agreements, in particular, the *SPS Agreement* and the *TBT Agreement*, the legal status of Codex measures has increased. Even within the EC legal order, Codex measures have also gained in importance due to the reference contained in Regulation 178/2002. Consequently, the Codex Alimentarius Commission has acquired increased authority as a standard-setting body.[2]

The increased status has raised questions regarding the legitimacy of the Codex Alimentarius. For instance, the reaction of the European Parliament on the WTO Panel decision in *EC-Hormones* indicated concern about the undemocratic and obscure procedural rules.[3] Likewise, on several occasions, the EC Commission has

---

[1] The financial dependence of the Codex Alimentarius Commission on its parent organisations has led to the use of the 'host country' construction to establish subsidiary bodies. This means that lack of financial resources is solved by the financial and administrative commitment of Codex Members as regards the organisation of the Codex Committees. See Section 3.4.4 of Chapter I.

[2] J. Braithwaite and P. Drahos, *Global Business Regulation* (Cambridge, Cambridge University Press 2000), pp. 405-406.

[3] EP resolution, OJ 1997, C222/53.

been called upon to initiate a discussion on the legitimacy of the Codex procedure, given the modified status of Codex measures. A clear example of this can be found in the (Council) Resolution on the Commission Green Paper which:

'... calls on the Commission also to make every effort to amend the Codex Alimentarius decision-making process so as to bring it more into line with the real impact of Codex standards since the WTO was established.' [4]

Likewise, questions relating to the legitimacy of the Codex Alimentarius Commission have also been raised at an academic level.[5]

Several evaluations have been made to examine the functioning of the bodies involved in the Codex standard-setting procedure. For instance, an evaluation of the independence of the Joint Meeting on Pesticides Residues (JMPR) has been conducted. Likewise, in 2002, a Joint FAO/WHO Evaluation of the Codex Alimentarius, commissioned by the Directors-General of the FAO and the WHO, was also conducted. This evaluation examined the work of both the Codex Alimentarius Commission and its subsidiary bodies.[6]

Taking the recently conducted evaluations and the resultant amendments to the internal rules into account, this chapter examines the legitimacy of the Codex Alimentarius Commission. It starts by analysing the institutional legitimacy of the Codex Alimentarius Commission (Section 2). In other words, whether the execution of powers by the Codex Alimentarius Commission corresponds with its attributed powers and whether the delegation of its powers to other bodies raises concerns, and how the Codex Alimentarius Commission respond to these concerns. In a next

---

[4] See Resolution on the Commission Green Paper on the general principles of food law in the European Union, A4-0009/98, para. 74. See also, question by Marianne Thyssen, E-2929/96, 8 November 1996: 'Since the start of the World Trade Organization (WTO) the degree of enforceability of certain standards in the framework of the Codex Alimentarius has increased. As the Commission is aware, in the Codex Alimentarius decisions are taken by a simple majority of votes cast. As these standards have an indirect influence on the content of European legislation, it would seem appropriate for the Commission to recommend the Member States to press in the Codex Alimentarius for a change in the decision-making procedure so that it would relate to the actual impact which the Codex standards will have in future. Does the Commission intend to take steps in this direction?'

[5] T. Makatsch, *Gesundheitsschutz im Recht der Welthandelsorganisation (WTO). Die WTO und das SPS-Übereinkommen im lichte von Wissenschaftlichkeit, Verrechtlichung und Harmonisierung* (Berlin, Duncker & Humblot 2004). R. Romi, 'Codex Alimentarius: de l'ambivalence à l'ambiguité', *Revue Juridique de l'Environnement* (2001), pp. 201-213. R. Muñoz, 'La Communauté entre les mains des normes internationales: les conséquences de la décision Sardines au sein de l'OMC', 4 *Revue du Droit de l'Union Européenne* (2003), pp. 457-484.

[6] The term of reference of the evaluation team consisted of examining and provide recommendations on: the evolving context and challenges surrounding international food standards and their relevance in ensuring food safety, consumer protection, trade and economic development; expectations of governments as to the validity, acceptability and institutional mechanisms for food standard setting within Codex; the particular interests and expectations of developing countries and of producers, industry and civil society; the effectiveness of existing institutional arrangements, management, methods of work and resources for international food standard setting. The evaluation is based on consultations with Codex Members, INGOs and other international standard-setting bodies.

section (Section 3), the legitimacy of the standard-setting procedure, in particular, with regard to consensus, impartial participation and transparency, is analysed. In the last section (Section 4), the substantive legitimacy of the Codex is considered, or, in other words, the legitimacy of the content of the adopted Codex measures, in particular, the level of harmonisation that they reflect.

2. QUESTIONS OF LEGITIMACY RELATED TO THE INSTITUTIONAL STRUCTURE

When it comes to international organisations, the scope of their attributed powers is a fundamental concern with regard to their legitimacy. The scope of the competence of international organisations is limited, and depends upon the transfer of powers by governments. As explained in the introduction to this research, the question of institutional legitimacy examines the powers executed by the international organisation and compares them with the scope of the attributed powers. If an international organisation exceeds its attributed powers, this can lead to a legitimacy-crisis of the institution itself. Furthermore, institutional legitimacy also depends on whether the division of powers amongst the organs of the institution and the other external organs involved is perceived as being right by the participating member countries.

The question of institutional legitimacy in the context of this section relates to the scope of the powers delegated to the Codex Alimentarius Commission, its subsidiary bodies and the other relevant bodies that are involved in the standard-setting procedure. It concentrates on the mandate of the Codex Alimentarius Commission with regard to the scope of its normative competence as laid down in Article 1 of the Statutes. In addition, it analyses the position of Codex committees and task forces, expert bodies and the other international organisations within the institutional structure responsible for the standard-setting procedure.

Three types of bodies are involved in the standard-setting procedure, each of which has a different type of function and each of which faces different restrictions with regard to their mandate:

- Codex committees and task forces;
- Expert bodies; and
- International organisations.

The first type of bodies owe their competence to the direct delegation of powers from the Codex Alimentarius Commission. Their function is to prepare draft standards and other documents to be submitted for approval by the Codex Alimentarius Commission. The second type of bodies (expert bodies) work directly under the auspices of the Food and Agriculture Organisation (FAO) and the World Health Organisation (WHO) and are bound by the internal procedures established by these organisations.[7] Their relationship with the Codex Alimentarius Commission and

---

[7] See Section 3.5 of Chapter I.

its subsidiary bodies is based on the separation of powers between risk assessors and risk managers.[8] Their function is advisory and they execute risk assessments by, for example, the mandate of the Codex Alimentarius Commission and its subsidiary bodies. The third type of bodies consists of international organisations that have a similar function as the Codex committees and task forces within the standard-setting procedure, in that they are involved in the preparation of draft proposed standards and other documents. However, they are not subsidiary bodies of the Codex Alimentarius Commission. Consequently, these organisations are not bound by the internal rules established by the Codex Alimentarius Commission.

2.1 The normative competence of the Codex Alimentarius Commission

In the late 1960s, Dobbert addressed the question of the legal basis of the normative competence of the Codex Alimentarius Commission.[9] At that time, he concluded that a flexible application of the competence of both parent organisations to submit recommendations to its members had led to the normative competence of the Codex Alimentarius Commission.[10] One of the reasons that Dobbert gave for the fact that Codex Members had never objected to the flexible application of this competence was that the adoption of standards, being recommendations, did not have mandatory force.[11] However, as a result of the entry into force of the WTO agreements, the adoption of international standards, guidelines and recommendations is no longer a mere act of submitting recommendations to governments. This development shows the importance of reconsidering the legal basis of the normative competence of the Codex Alimentarius Commission again.

The mandate of the Codex Alimentarius Commission, as contained in Article 1 of the Statutes of the Codex Alimentarius Commission does not mention a competence to adopt *binding* provisions. As indicated in Section 3.1.1.2 of Chapter I, Article 1 only implies that the Codex Alimentarius Commission is to finalise standards and publish them in the Codex Alimentarius.[12] Does the mandate of Article 1 contain the implied powers to adopt binding provisions? The response to this question also depends on the powers that have been attributed to its parent organisations, the FAO and the WHO, by their members. In accordance with the doctrine of attributed competences (competence d'attribution),[13] the FAO and the WHO cannot delegate more powers to the Codex Alimentarius Commission than they themselves have acquired from their members. Consequently, examining the scope of their

---

[8] See Sections 3.1.1.2 and 3.5 of Chapter I.

[9] J.P. Dobbert, 'Le Codex Alimentarius vers une nouvelle méthode de réglementation internationale', 15 *Annuaire Français de Droit International* (1969), pp. 677-717.

[10] Ibid., pp. 707-712. Art. IV.3 of the Constitution of the FAO and Art. 23 of the Constitution of the WHO.

[11] J.P. Dobbert, *supra* n. 9, p. 709.

[12] Art. 1(d) of the Statutes.

[13] See, on the doctrine of attributed competences, H.G. Schermers and N.M. Blokker, *International Institutional Law*, 4th edn. (The Hague, Martinus Nijhoff Publishers 2003), pp. 155-157.

normative competence can be of assistance for the interpretation of Article 1 of the Statutes of the Codex Alimentarius Commission.

Article XIV of the Constitution of the FAO allows the Conference of the FAO to adopt *conventions* and *agreements*.[14] A first question would be to ask whether Codex measures could be considered as being conventions or agreements in the sense of Article XIV of the FAO Constitution. However, even if the terminology of conventions and agreements may be considered as being broad enough to include standards and other measures, as adopted by the Codex Alimentarius Commission, the entry into force still depends on a provision contained in the convention or agreement in question.[15] This is in conflict with Article 3.3 of the *SPS Agreement* and Article 2.4 of the *TBT Agreement*, as these articles are applicable to all standards, guidelines and recommendations as adopted by international standard-setting bodies, and do not require the consent of the member countries of these bodies.[16]

When it comes to the WHO Constitution, Article 21 allows the Health Assembly of the WHO to adopt *regulations*.[17] According to Article 22 of the WHO Constitution these regulations:

'shall come into force for all Members after due notice has been given of their adoption by the Health Assembly except for such Members as may notify the Director-General of rejection or reservations within the period stated in the notice.'

Although the entry into force is more closely connected to the adoption of the regulations, members still have the possibility of preventing the regulations from having binding force. Like the competence of the FAO Conference, the normative competence of the WHO Health Assembly remains subordinate to a final consent of member countries.

From the above, it can be concluded that there is a lack of a legal basis in the mandate of both the Codex Alimentarius Commission and the constitution of its parent organisations to adopt measures (containing elements) that have binding force upon adoption. Despite this lack of a legal basis, there is no trace within the reports of the Codex Alimentarius Commission and its subsidiary bodies of the legitimacy of the normative powers of the Codex Alimentarius Commission being questioned by its Members.[18] The recognition of Codex measures in the *SPS Agreement* and the *TBT Agreement* seem, for the moment, to be sufficient. However, this reference is a rather weak basis upon which to guarantee long-term legitimacy, as it does not provide Codex Members with legal security with regard to the status of the decisions of the Codex Alimentarius Commission.

---

[14] Basis texts of the Food and Agriculture Organization of the United Nations, Vol. I, 2004 edn. Emphasis added.

[15] Art. XIV.4 of the FAO Constitution. Basis texts of the Food and Agriculture Organization of the United Nations, Vol. I, 2004 edn.

[16] See Section 5 of Chapter IV.

[17] Emphasis added.

[18] This in contrast with many aspects that are related to the normative powers, such as procedural aspects, e.g., decision-making, participation and transparency (see Section 3 of this chapter).

2.2 Delegation of tasks to Codex Committees: a question of decentralisation

Financial independence provides a solid basis for the operation of Codex Committees. It has ensured their continuity. It also allows them to meet regularly; in most cases, more often than the Codex Alimentarius Commission has done. Furthermore, as already mentioned in Section 3.4.3 of Chapter I, Codex Committees, in particular, General Codex Committees and Task Forces are characterised by a high level of expertise. This is ensured by Rule XI.4 of the Rules of Procedure, which requires the representatives of the members of the subsidiary bodies to be specialists in the relevant field of the body in question. In addition, Codex Committees have an important function within the standard-setting procedure. As explained in Section 4 of Chapter II, they are responsible for the preparation of the (proposed) draft standards which are submitted to the Codex Alimentarius Commission for approval. Likewise, they maintain the contacts with the expert bodies and undertake the main part of the risk management functions.[19] These aspects strengthen the position of the Codex Committees in the standard-setting procedure. The influence of the recommendations of the Codex Committees is great, and the high number of Committees' proposals that is accepted by the Commission without any further comments illustrates this.[20] This brings about a tendency towards the decentralisation of the standard-setting procedure. This decentralisation is characterised by a structure of Codex Committees, each of which has their own specific field of expertise and their own specific perspective and way of approaching issues, as is exemplified by a statement during the 14th Meeting of the CCGP:

> 'It was also pointed out that the application of these concepts [the concept of Food Safety Objectives and the concept of the determination of the Appropriate Level of Protection] at a technical level would probably differ in the different fields covered by specialist Codex Committees.'[21]

---

[19] See, for further discussion, Section 2.3 of this chapter.

[20] Although it is hard to demonstrate the exact influence of the Committees in the standard-setting procedure, some indication may be given by the adoption of standards at Step 8. See, for instance, the Report of the 28th Session of the Codex Alimentarius Commission, Rome 4-9 July 2005, ALINORM 05/28/41, paras. 45-88: the Codex Alimentarius Commission adopted all three codes of practice as proposed by the Committees (without further amendments); one out of the two guidelines proposed by Committees (the other one was adopted with a small amendment as proposed by the Codex Alimentarius Commission); one amended draft standard as proposed by the Committee on Oils and Fats; the Draft General Standard for Fruit Juices and Nectars with the amendment as proposed by the Committee on Methods of Analysis and Sampling, with just one clarification inserted in an additional footnote as proposed by the Codex Alimentarius Commission; 216 out of 217 proposed MRLs for pesticides and 70 Interim MRLs for pesticides as proposed by the Codex Committee for Pesticides Residues.

[21] See Report of the 14th Session of the Codex Committee on General Principles, ALINORM 99/33A, 1999, para. 8. See also, D.L. Leive, *International Regulatory Regimes* (Lexington, Lexington Books 1976), pp. 436-437.

Their terms of reference are defined in general terms and often provide the Codex Committees with a basis that allows them to undertake a wide range of activities related to the specific field in question.[22] These broadly-defined mandates, in conjunction with the competence(s) of Codex Committees to propose the initiation of new work, strengthen the position of the Codex Committees and may impede the coherent operation of the Codex Alimentarius Commission and its subsidiary bodies as an integral institution if this is not carefully co-ordinated. In 2005, a team of consultants examined the structure and the mandates of the Codex Committees. During the 55th Session of the Executive Committee, they expressed their concerns with regard to this rather autonomous character of the Codex Committees:

'...They (the consultants) were under the impression that Codex was composed of quasi-autonomous committees which were setting their own agenda and were receiving little oversight from the Commission ...'.[23]

The question of institutional legitimacy as regards the scope of their delegate powers consists of two principal concerns:

1. the restrictions imposed on the proposals for new work (definition of the scope of their activities *vis-à-vis* the Codex Alimentarius Commission); and
2. the definition of the inter-Committee relations.

2.2.1   *Concerns related to the position of the Codex Committees*

2.2.1.1   Codex Committees and the initiation of new work

A decision to initiate new work can be taken by a Codex Committee.[24] However, this decision needs to be submitted for approval to the Codex Alimentarius Com-

---

[22] See Terms of Reference of the Codex Committees as contained in the Procedural Manual, 16th edn., 2006, pp. 124-158. For example, the Terms of Reference of the Codex Committee on Food Hygiene state: '(a) to draft basic provisions on food hygiene applicable to all food; (b) to consider, amend if necessary and endorse provisions on hygiene prepared by Codex commodity committees and contained in Codex commodity standards, and (c) to consider, amend if necessary, and endorse provisions on hygiene prepared by Codex commodity committees and contained in Codex codes of practice unless, in specific cases, the Commission has decided otherwise, or (d) to draft provisions on hygiene applicable to specific food items or food groups, whether coming within the terms of reference of a Codex commodity committee or not; (e) to consider specific hygienic problems assigned to it by the Commission; (f) to suggest and prioritize areas where there is a need for microbiological risk assessment at the international level and to develop questions to be addressed by the risk assessors, (g) to consider microbiological risk management matters in relation to food hygiene and in relation to the risk assessment of FAO and WHO.' (Procedural Manual, 16th edn., 2006, pp. 128-129).
[23] See Report of the 55th Session of the Executive Committee, Rome, 9-11 February 2005, ALINORM 05/28/3, para. 27.
[24] Step 1 of the 'Uniform Procedure for the elaboration of codex standards and related texts', Procedures for the Elaboration of Codex standards and related texts, Procedural Manual, 16th edn., 2006, p. 25. See also, Section 4.1 of Chapter II.

mission.²⁵ The majority of the proposals to initiate the procedure arise at Committee level and are submitted for approval to the Codex Alimentarius Commission or the Executive Committee. Some Codex Committees, such as the Codex Committee on Pesticides Residues (CCPR), the Codex Committee on Residues of Veterinary Drugs in Foods (CCRVDF) and the former Codex Committee on Food Additives and Contaminants (CCFAC), now divised into the Codex Committee on Food Additives (CCFA) and the Codex Committee on Contaminants in Foods (CCCF) play a primary role in the priority-setting of the work undertaken by the expert bodies (in particular, the JECFA and the JMPR).²⁶ During the 28th Session of the Codex Alimentarius Commission, this position was strengthened, as it was endorsed that:

> 'all requests from the Codex subsidiary bodies for JECFA advice on additives and contaminants should be routed exclusively through the Additives and Contaminants Committees and requests for JECFA advice on residues of veterinary drugs through the Committee on Residues of Veterinary Drugs in Food.'²⁷

In fact, prior to 1993 (the alignment of the different elaboration procedures into one uniform procedure), the CCPR and the CCRVDF did not require the approval of the Codex Commission to initiate the elaboration procedure for the establishment of Maximum Residue Levels (MRLs) and had a direct link with the joint secretaries of the expert bodies.²⁸

The strong role of Codex Committees in the initiation of new Codex work is reinforced by the inclusion of the task of establishing their own working priorities in the general terms of reference of Codex Committees (as expressed in the Guidelines for Codex Committees).²⁹ With regard to some of the Codex Committees, this

---

²⁵ Ibid.

²⁶ See, for instance, Report of the 13rd Session of the Codex Alimentarius Commission, Rome 3-14 December 1979, ALINORM 79/38, para. 70.

²⁷ Report of the 28th Session of the Codex Alimentarius Commission, Rome 4-9 July 2005, ALINORM 05/28/41, para. 145.

²⁸ See also, L. Salter, *Mandated Science, Science and Scientists in the Making of Standards* (Dordrecht, Kluwer Academic Publishers 1988), p. 75. Moreover, these two Codex Committees are now required to submit their proposals for decision to the Codex Alimentarius Commission. Report of the 20th Session of the Codex Alimentarius Commission, ALINORM 93/40, para. 143: 'The Chairman of the Committee [on Pesticides Residues] *informed* the Commission that priority lists had been established by the 24th and 25th Sessions of the Committee...The Commission noted the new Codex Procedures adopted under Agenda Item 15 and agreed to authorize the Codex Committee on Pesticide Residues to commence work on the elaboration of MRLs for pesticides referred to in the above lists.' [emphasis added] or para. 176 and 177: 'The Commission was informed that a Priority List had been established by the Committee [on Residues of Veterinary Drugs in Foods]... In the light of the adoption by the Commission of new Codex Procedure under Agenda Item 15, the Chairman of the Committee requested the Commission's authorization to commence work on the elaboration of MRLs for veterinary drugs on the Priority List. The Commission endorsed the Priority List as established by the Committee.' See, on the history of the Codex standard-setting procedure, Section 4 of Chapter II.

²⁹ See duty (a) of the Duties and Terms of Reference included in the Guidelines of Codex Committees and *Ad Hoc* Intergovernmental Task Forces, Procedural Manual, 16th edn., 2006, p. 48.

task is even included in their specific terms of reference.[30] Consequently, several Codex Committees have developed, or are in the process of developing, their own criteria for working priorities.[31]

In the past, the great influence of the proposals of Codex Committees on the overall initiation of new work has been a topic of concern as it might jeopardise the coherence of any actions taken by Codex Committees:

'... It had appeared that, up until now, initiatives to undertake new work were largely dependent upon the decisions of individual Codex Committees as statements of overall priorities to achieve consistency with the general objectives of the Commission and its parent organisations had not been defined. The Executive Committee indicated that it was the role of the Commission to establish overall priorities in the light of the programmes set by the FAO and WHO so as to orient the work of its subsidiary bodies accordingly. The Commission would then be able to ensure that the work undertaken by individual committees was in accordance with their (medium- or long-term) objectives ...'.[32]

This concern is relevant for another matter. As mentioned in Section 5.2 of Chapter IV, the SPS Committee examines the process of standardisation taking place in international standard-setting bodies. Members of the SPS Committee can request information from the Codex Secretariat with regard to decisions taken to elaborate new standards. Monitoring the initiation of new work within the Codex Alimentarius Commission would strengthen the communication with the SPS Committee and enhance the legitimacy of its work within the context of the WTO agreements.

2.2.1.2   The definition of inter-Committee relationships

Some issues which fall under the terms of reference of one Codex Committee touch upon issues of other Codex Committees as well. For this reason, it frequently

---

[30] This task is laid down in for instance, the Codex Committee on Food Hygiene is to 'suggest and prioritise areas where there is a need for microbiological risk assessment at the international level and to develop questions to be addressed by the risk assessors.' (Terms of Reference of the Codex Committee on Food Hygiene, para. (f), Procedural Manual, 16th edn., 2006, p. 129). Likewise, the Codex Committee on Pesticide Residues is to 'prepare priority lists of pesticides for evaluation by the Joint FAO/WHO Meeting on Pesticide Residues' as is the Codex Committee on Residues of Veterinary Drugs in Foods to 'determine priorities for the consideration of residues of veterinary drugs in foods' (Terms of Reference of the Codex Committee on Pesticide Residues, para. (c), Procedural Manual, 16th edn., 2006, p. 134 and the Terms of Reference of the Codex Committee on Residues of Veterinary Drugs in Foods, para. (a), Procedural Manual, 16th edn., 2006, p. 134).

[31] The Codex Committee on Food Hygiene launched a drafting group to among other things propose a mechanism that would allow CCFH to establish its work priorities on an ongoing basis (Report of the 35th Session of Codex Committee on Food Hygiene, Orlando, Florida, United States, 27 January-1 February 2003, ALINORM 03/13A, para. 175). Likewise, the CCPR has established a working group on work priorities and is in the process of elaborating specific criteria itself applicable for the elaboration of pesticides MRLs, see Report of the 37th Session of the Codex Committee on Pesticide Residues, The Hague, the Netherlands, 18-23 April 2005, ALINORM 05/28/24, pp. 252-256.

[32] See Report of the 20th Session of the Joint FAO/WHO Codex Alimentarius Commission, Geneva, 28 June-7 July 1993 ALINORM 93/40, para. 75.

occurs that a (proposed) draft standard or other related text will have to pass two, or even more, different Codex Committees before arriving at Commission level.

The inter-relationship between the Codex Committees has different dimensions. A first type of inter-committee relationship is the relationship between General Committees and Commodity Committees, both of which are involved in the preparation of Commodity standards. In this case, a Commodity standard – as prepared by a Commodity Committee – is sent to a General Committee in order to regulate a specific provision. For instance, the draft standard for salted Atlantic herring and salted sprat was sent from the Codex Committee for Fish and Fish Products to the Codex Committee on Food Hygiene to endorse the hygiene provisions of the draft standard.[33] Another type of inter-Committee relationship is the one between two General Committees. For instance, the Single Laboratory Validated Methods of Analysis was proposed and drafted by the Codex Committee on Methods of Analysis and Sampling (CCMAS), which forwarded it at its 24th Session to the Codex Committee on Pesticides Residues (CCPR). The Criteria for selection of Single Laboratory Validated Methods of Analysis were amended by the CCPR, whose amendment was again considerably revised by the CCMAS.[34] After the consideration of the CCMAS, it will also pass the Codex Committee on General Principles before it arrives at Commission level. Thus, the question arises as to who is responsible for which part of the standard or related texts (or revision thereof) and which committees are involved in the procedure?

Questions with regard to a clear division of powers between these bodies were raised by Codex Members during the Joint FAO/WHO Evaluation of the Codex Alimentarius Commission. For instance, the evaluation team states that:

> '... many of the comments and criticisms of Codex that we heard during country visits relate to these issues (a clearer definition of Committee and Task Force Working Procedure). We believe that a clearer division of the work between horizontal and vertical committees will go some way to reducing overlaps between committees, streamlining and increasing efficiency in the use of scarce resources ...'.[35]

Likewise, eighty percent of the responses received from governments have pointed out that eliminating inconsistencies of the work of the Codex committees has to be a very high priority for the future work of the Codex Alimentarius Commission.[36]

---

[33] See Report of the 36th Session of the Codex Committee on Food Hygiene, ALINORM 04/27/13, April 2004, para. 14.

[34] Report of the 25th Session of the Codex Committee on Methods of Analysis and Sampling, ALINORM 04/27/23, paras. 14-18: '... After some discussion regarding the applicability of the criteria, the Committee concluded that the proposed criteria should be of a general nature and should be incorporated into the Procedural Manual. It therefore amended the first bullet (i) by deleting the specific reference to the CCPR Guideline on Good Laboratory Practice.'

[35] Report of the evaluation of the Codex Alimentarius and other FAO and WHO food standards work, ALINORM 03/25/3, para. 116. Addition inserted.

[36] Ibid., para. 110.

2.2.2 Instruments to supervise and co-ordinate the activities of Codex Committees

The structure of the two readings of the elaboration procedure allows the Codex Alimentarius Commission to co-ordinate and control the work undertaken by the large number of Codex Committees. At three stages in the procedure, it has the opportunity to correct and direct the work undertaken by Codex Committees: at Step 1 (the start of the elaboration procedure), at Step 5 (the moment that draft proposed Codex standards are submitted for approval), and at Step 8 (the moment that proposed Codex measures are submitted for final adoption). Prior to 2004, the Executive Committee had the competence to approve proposals of Codex Committees for new work at Step 1 and to approve draft proposed measures at Step 5 of the procedure.[37] Given the fact that the Codex Alimentarius Commission is under pressure in terms of time to deal with all its items on the agenda, in most cases, it has been the Executive Committee that has taken on this task.[38]

This section examines the instruments that have been developed by the Codex Alimentarius Commission in order to address the concerns mentioned above, and examines the application of these instruments during the relevant steps of the standard-setting procedure. The Reports of the Codex Alimentarius Commission and of the Executive Committee illustrate the practice of supervising and co-ordinating the work of Codex Committees.

2.2.2.1 The 'Criteria for Establishing Work Priorities', the 'Medium-Term Plan' and their application by the Codex Alimentarius Commission and the Executive Committee

Even by 1969, the Codex Alimentarius Commission adopted 'Criteria for establishing work priorities' (hereafter referred to as the 'Criteria'), which most of the Codex Committees had to take into account when initiating new work.[39] However, the criteria only contained the criteria applicable for the initiation of commodity standards. In 1993, the Executive Committee emphasised the need to establish overall priorities in order to orient the work of the subsidiary bodies in accordance with the medium- and long-term objectives of the Codex Alimentarius Commission. Since 1997, the 'Criteria for Establishing Work Priorities', in addition to the already adopted checklist for commodity standards, also contain a checklist of the criteria appli-

---

[37] See also, Section 3.2.1 of Chapter I.

[38] See, for instance, as regards the task of the Executive Committee to decide on the initiation of new work, the Report of the 24th Session of the Codex Alimentarius Commission, Geneva, 2-7 July 2001, ALINORM 01/41, para. 216: 'The Commission was unable to complete its review of ... proposals for the elaboration of new standards and related texts. It requested the Directors-General to convene an extraordinary session of the Executive Committee at an early date to consider these matters on its behalf, so that progress at Committee level would not be impeded.'

[39] Report of the 6th Session of the Codex Alimentarius Commission, Geneva, 4-14 March 1969, ALINORM 69/67, para. 44.

cable to general subjects.[40] Furthermore, the revised 'Criteria' state that Codex Committees should also take into account the Medium-Term Plan and other strategic projects of the Codex Alimentarius Commission.[41]

When it comes to the actual application of the 'Criteria' by the Codex Alimentarius Commission or the Executive Committee, it can be stated that experience over the past years illustrates that the Commission has had problems completing the Medium-Term Plan in time.[42] This reduces the usefulness of the criteria quite considerably.

Furthermore, prior to 2000, the consideration of the criteria of the two 'checklists' by the Codex Alimentarius Commission or the Executive Committee was not clearly present in the reports.[43] In the majority of cases, the Commission and the Executive Committee approved proposed new work without making comments. In

---

[40] 'Criteria for the establishment of work priorities', Procedural Manual, 16th edn., 2006, p. 66. The Criteria contains two checklists of aspects to be taken into account by the Codex Committees when establishing their work priorities: one that is applicable to general subject standards and one that is applicable to commodity standards. Criteria for Commodity standards:
'a. Volume of production and consumption in individual countries and volume and pattern of trade between countries;
b. Diversification of national legislations and apparent resultant or potential impediments to international trade;
c. International or regional market potential;
d. Amenability of the commodity to standardisation;
e. Coverage of the main consumer protection and trade issues by existing or proposed general standards;
f. Number of commodities which would need separate standards indicating whether raw, semi processed or processed;
g. Work already undertaken by other international organisations in this field.'
'Criteria applicable to general subjects:
a. Diversification of national legislations and apparent resultant or potential impediments to international trade;
b. Scope of work and establishment of priorities between the various sections of work;
c. Work already undertaken by other international organizations in this field.'

[41] See Report of the 22nd Session of the Joint FAO/WHO Codex Alimentarius Commission, Geneva, 23-28 June 1997, ALINORM 97/37, Appendix II.

[42] Ibid., para. 177: 'The Commission noted the progress towards achieving the current medium-term objectives (1993-1998). It endorsed the general directions indicated in the Medium-Term Plan 1998-2002 and the proposals of the Secretariat to transmit the outline of the plan to Member Governments for comment, finalization by the Executive Committee at its 45th Session, and *approval by the Commission at its 23rd Session in 1999*.' [emphasis added] or see Report of the 51st Session (Extraordinary) of the Executive Committee, Geneva, 10-11 February 2003, ALINORM 03/25/2, para. 19: 'The Executive Committee recalled that the Medium-Term Plan was currently under consideration and that it was scheduled for finalization by the 26th Session of the Commission. However, the Executive Committee noted that the recommendations of the Evaluation might significantly affect several elements of the Medium-Term Plan, and agreed that it would be premature to finalize it before the Commission had completed its review of the Evaluation. The Executive Committee therefore proposed that the Commission postpone its consideration of the Medium-Term Plan until its 27th Session.' The 27th Session of the Codex Alimentarius Commission takes place in July 2004, which is one and a half year after the expiry of the previous Medium-Term Plan.

[43] See, for a detailed analysis of the Step 1 (and equally the rest) of the step procedure of the Codex Alimentarius Commission thirty years ago, D.L. Leive, *supra* n. 21, p. 442.

several cases, however, they indicated the criteria that had been taken into account. In these cases, the comments were related to the scope or the title of the document to be developed,[44] concerns about the workload of the responsible Codex Committee,[45] or comments relating to the necessity to co-operate with other international organisations.[46] The following criteria were recorded to justify either the rejection or the approval of the proposals in question, whether the work falls under the mandate of the Codex Alimentarius Commission,[47] whether the proposal is in line with the Medium-Term Plan (although this reference is only found back in the report of the 21st Session of the Codex Alimentarius Commission in 1995),[48] and whether there is general agreement between the Members to proceed with the proposed initiation.[49] However, as from 2000, the reports of the Executive Committee illustrate the development of a new practice, which strengthens the controlling mechanism. It encouraged the practice, established by some of the Codex Committees, of developing discussion papers or position papers that examine the merits of proceeding with new work. The Executive Committee emphasised, however, that these

---

[44] The title of the proposed Code on Aflatoxins in tree nuts was modified to ensure that the scope was not limited to the reduction of, but would equally deal with the prevention of aflatoxins. See Report of the 50th Session of the Executive Committee, Rome 26-28 June 2002, ALINORM 03/3A, para. 66. Likewise, reservations were expressed with regard to the use of the term 'precautionary principle' in the proposed draft Guidelines for the Control of *Listeria monocytogenes* in Foods. See Report of the 49th Session (Extraordinary) of the Executive Committee, Geneva, 26-27 September 2001, ALINORM 03/3, para. 23.

[45] With regard to the approval of the initiation of work on proposed draft code of hygienic practice for precut fruits and vegetables, proposed draft code of hygienic practice for primary production, harvesting and packaging of fresh produce and the proposed draft Annex to the Recommended International Code of Practice, the Executive Committee commented that careful attention needed to be paid to the effect of the increasing workload of the Committee on Food Hygiene. See Report of the 45th Session of the Executive Committee, Rome, 3-5 June 1998, ALINORM 99/3, Appendix III. Also, with regard to the development of the Proposed Draft Guidelines for the Control of *Listeria monocytogenes* in Foods, concerns were expressed on the increasing workload of the CCFH. See Report of the 49th Session (Extraordinary) of the Executive Committee, Geneva, 26-27 September 2001, ALINORM 03/3, para. 23.

[46] Report of the 23rd Session of the Joint FAO/WHO Codex Alimentarius Commission, Rome, 28 June - 3 July 1999, ALINORM 99/37, para. 206. It concerned the initiation of the Codex Standards for Apples, Table Grapes and Tomatoes to be undertaken in collaboration with UNECE.

[47] The proposal of India to initiate the establishment of a data base on importing country legislation was rejected on the basis that this type of work did not fall within the mandate of the Codex Alimentarius Commission but within the mandate of its parent organisations. See Report of the 23rd Session of the Joint FAO/WHO Codex Alimentarius Commission, Rome, 28 June-3 July 1999, ALINORM 99/37, para. 201.

[48] Report of the 21st Session of the Joint FAO/WHO Codex Alimentarius Commission, Rome, 3-8 July 1995, ALINORM 95/37, para. 85: 'The Commission endorsed the proposals made by the Executive Committee on the elaboration of new standards and or related texts at step 1, which were found to meet the Medium-Term Objective of the Commission and to be within the mandate of the Commission.'

[49] The amendment of the General Standard for the Labelling of Prepackaged Foods to include provisions for the labelling of country of origin was simply not approved due to the lack of consensus between the members of the Executive Committee. See Report of the 49th Session (Extraordinary) of the Executive Committee, Geneva, 26-27 September 2001, ALINORM 03/3, para. 25.

papers do not substitute the actual decision.[50] A first example is the Executive Committee's rejection of the proposal for the initiation of a standard for unifloral honey and the completion of Part Two of the Standard for Honey, which covers industrial uses.[51] It stated that more adequate justification was required, 'including the necessity to develop a standard for industrial uses, which is not usual practice in Codex'. In addition, with regard to the proposals for new work, the Executive Committee made the general observation that 'justification for new work in terms of the Criteria for the Establishment of Work Priorities should be well documented when making such proposals'.[52]

2.2.2.2 The definition of inter-relationship between Codex Committees

The Procedural Manual only regulates the relationship between the Commodity Committees and the General Committees, laid down in the 'Relations between Commodity Committees and General Committees'. This document indicates, for instance, the obligation for the Commodity Committees to refer the draft standard to a general committee and mentions the precise stage in the process at which this should, preferably, take place. Likewise, with regard to the section on food additives in a draft commodity standard, any proposed standard is to be endorsed by the Codex Committee on Food Additives and Contaminants after the standard has been advanced to Step 5, or before it is considered by the Commodity Committee at Step 7.

The relationships between General Committees are not regulated by the Procedural Manual. However, the reports of sessions of the Committees indicate that these relationships are regulated in internal documents.[53] Several reports indicate

---

[50] See Report of the 47th Session of the Executive Committee, Geneva 28-30 June 2000, ALINORM 01/3, para. 44.

[51] See Report of the 47th Session of the Executive Committee, Geneva 28-30 June 2000, ALINORM 01/3, Annex III.8. Already during its 23rd Session in 1999, the Codex Alimentarius Commission requested the Codex Co-ordinating Committee for Asia to examine the feasibility of a Codex standard on instant noodles. However, this request was a response to the direct suggestion of Japan during the Codex Commission to elaborate a new standard. For this occasion a discussion paper was developed by Japan which studied the feasibility of elaborating a codex standard. Before it was subject of a discussion within the Codex Co-ordinating Committee for Asia, four other Asian countries had already submitted their comments. The discussion led to a wide-spread consensus on the necessity of such a standard, given that there was a sharp and continuous increase of the production and trade within Asia and other parts of the world and there were strong differences between the national regulations for instant noodles (Report of the 12th Session of the Codex Co-ordinating Committee for Asia, Chiang Mai, Thailand, 23-26 November 1999, ALINORM 01/15, para. 11). The proposal of the Codex Co-ordinating Committee for Asia was approved by the Executive Committee during its meeting in 2000 and decided that the standard was to be developed by the Codex Co-ordinating Committee for Asia itself in combination with the Codex Committee on Cereals, Pulses and Legumes.

[52] Report of the 47th Session of the Executive Committee, Geneva 28-30 June 2000, ALINORM 01/3, para. 44.

[53] Report of the 36th Session of the Codex Committee on Pesticides Residues, New Delhi, India, 19-24 April 2004, ALINORM 04/27/24.

that the chairpersons of the subsidiary bodies meet on an informal basis.[54] Although some reports demonstrate that either the Commission or the Executive Committee have used the Step Procedure to refer the standard back to a previous step in the procedure because they have not received the input of all relevant committees, this application can hardly be considered to be a systematic approach.[55] The point remains that the division of the responsibilities between the General Committees, a type of relationship that has increased in importance as a result of the horizontal approach, has not been laid down in the Procedural Manual.[56] Consequently, the definition of the relationship(s) between General Committees as directed by the informal practices or by the internal documents which specify the conduct of their meetings, or by the chairpersons of the Codex Committees, lacks overarching provisions that can be used by the Codex Alimentarius Commission as criteria to manage and co-ordinate the activities between these committees. Leaving the determination of the inter-Committees' relationships to the Committees themselves may result in a standard-setting procedure which functions incoherently.

2.2.3 Amendments of the 'Procedures for the Elaboration of Codex Standards and Related Texts' and the new function of the Executive Committee

As a result of the Joint FAO/WHO Evaluation of the Codex Alimentarius Commission, some fundamental changes have been introduced in the Procedural Manual. The Executive Committee has acquired a new function, and no longer has the competence to decide upon proposals for new work or to advance (proposed) draft standards to Step 6. Instead, it has acquired the task of conducting an on-going critical review of the proposals for new work and has to give its advisory opinion to the Commission.[57]

The new provisions of the 'Procedures for the Elaboration of Codex Standards and Related Texts' explicitly require the Executive Committee, in this capacity, to take the Criteria and the strategic plan of the Codex Commission into account when conducting the critical review of proposals for new work. Furthermore, the Executive Committee practice of requesting more information to justify a proposal for new work has also been formally codified. Provision 1 of Part 2 of the 'Procedures for the Elaboration of Codex Standards and Related Texts', states that each proposal must be accompanied by a project document. The project document must, for

---

[54] Report of the 15th Session of the Codex Committee on General Principles, Paris 10-14 April 2000, ALINORM 01/33, para. 70.

[55] For instance, at the 22nd Session of the Codex Alimentarius Commission all draft revised standards proposed by the Codex Committee on Milk Products were referred back to Step 6 awaiting the development of relevant issues by the Committee on Food Hygiene and Food Labelling. See Report of the 22nd Session of the Joint FAO/WHO Codex Alimentarius Commission, Geneva, 23-28 June 1997, ALINORM 97/37, para. 96 and further.

[56] See, for more detail on the horizontal approach of the Codex Alimentarius Commission, Section 2.1 of Chapter II.

[57] Provisions 2 and 3 of Part 2 of the Procedures for the Elaboration of Codex Standards and Related Texts, Procedural Manual, 16th edn., 2006, p. 21.

example, contain an assessment performed against the Criteria, an indication of the relevance of the proposal to the strategic objectives of the Codex, and information on the relationship of the project with other existing Codex documents.

With regard to the co-ordination of work between the Committees, the Executive Committee may advise the establishment of *ad hoc* cross-committee task forces if new work falls within the mandate of several committees.[58]

Although the new function of the Executive Committee responds to the need to supervise the position of its subsidiary bodies, several of the concerns discussed above have not been fully addressed. For instance, the requirement for Codex Committees to justify their proposal to initiate new work by submitting a project document is an important tool in the hands of the Codex Alimentarius Commission and the Executive Committee, which enables them to supervise their work. However, new work or the revision of individual MRLs for pesticides, veterinary drugs, food additives and MLs for contaminants seem to be excluded from the scope of obligation, and the procedures of the relevant Committees remain applicable. The question arises as to whether the existence of separate procedures, as developed by the relevant Committees, will impede the possibility of ensuring a unified approach in the area of standards development. Although 'Criteria for Prioritization Process of Compounds For Evaluation by JMPR' have been adopted and included in the Procedural Manual,[59] it is not clear from either the Criteria, or from the 'Procedures' how they relate to the critical review of the Executive Committee.[60]

Amendments relating to the supervision of the inter-Committee relationships seem to be limited to the possibility of the Codex Alimentarius Commission establishing *ad hoc* cross-committee task forces.[61] This competence to decide on the allocation of tasks is not combined with more explicit provisions to take corrective action during the standard-setting procedure in order to ensure the co-ordination of the work undertaken by different Codex Committees. Although the Executive Committee may propose that the work be undertaken by another committee than the one to which it was originally entrusted, the reasons for proposing such action is related to the progress of the development of the standard against the time-frame as ac-

---

[58] Provision 3 of Part II of the Procedures for the Elaboration of Codex Standards and Related Texts, Procedural Manual, 16th edn., 2006, p. 21. See also, the recommendations of the Consultants involved in the evaluation of the Codex Alimentarius Commission, Report of the evaluation of the Codex Alimentarius and other FAO and WHO food standards work, ALINORM 03/25/3, para. 109: '… with issues that involve several committees, an *ad hoc* cross-committee task force could reduce overlap and increase efficiency in work on a standard.'

[59] Procedural Manual, 16th edn., 2006, p. 67.

[60] During the 36th Session of the CCPR, attention was drawn by the Codex Secretariat to the fact that the prioritisation criteria for pesticides residues should be consistent with the 'Criteria of Establishing Work Priorities'. See Report of the 36th Session of the Codex Committee on Pesticide Residues, ALINORM 04/27/24, 2004, para. 215: 'The Codex Secretariat recalled that the criteria for prioritization of pesticides should be consistent with the Criteria for the Establishment of Work Priorities, especially if they were intended for inclusion in the Procedural Manual.'

[61] Besides, it is doubtful whether this possibility contributes to the transparency of the already complicated procedure. See, for discussion on the issue of transparency, Section 3.3 of this chapter.

corded for the development of the standard in question.[62] It does not deal with the question of correcting inter-committee relationships. During the discussions on the evaluation on the Codex Committee structure and the mandate of Codex Committees and Task Forces, several Codex Members proposed that the Codex Chairpersons' meeting be formalised.[63] This may contribute to the supervision of the inter-committee relationships if the purpose of these meetings is not limited to that of a discussion forum between committees, but becomes an instrument both to direct and to orient the work of the Codex Committees by the Codex Alimentarius Commission and the Executive Committee.

2.3 The mandate of expert bodies

As explained in Section 3.5 of Chapter I, expert bodies have an advisory role. In the context of the Codex Alimentarius, they work upon the request of the Codex Alimentarius Commission and its subsidiary bodies. The tasks of the expert bodies, with regard to the requested scientific advice, are defined on the basis of a mandate as given by the Codex Alimentarius Commission and its subsidiary bodies. Their advice aims to serve the elaboration of Codex measures. In this context, it falls under the concept of 'mandated science', as used by Liora Salter to refer to the role of science and scientists for regulatory purposes.[64] By distinguishing the different circumstances of 'mandated science' from the circumstances of 'normal science' (science conducted in the context of research), she indicates that, when conducting evaluations and providing advice for regulatory purposes, scientists are subject to public pressure.[65] This pressure is seen as a risk that puts the scientific integrity of the risk assessments conducted by the expert bodies at stake.

In order to ensure the scientific integrity of the risk assessment, the relationship between the expert bodies on the one hand, and the Codex Alimentarius Commis-

---

[62] See Provision 6 of Part II of the Procedures for the Elaboration of Codex Standards and Related Texts, Procedural Manual, 16th edn., 2006, p. 22 and Section 3.1 of this chapter on consensus-building.

[63] See Report of the 55th Session of the Executive Committee, Rome, 9-11 February, 2005, ALINORM 05/28/3, para. 26: 'Several members expressed the view that the meeting of Codex Committee chairs should be formalized in one way or another, since this meeting could provide useful guidance to the Executive Committee in its critical review function. The interaction between Codex chairs could also facilitate the coordination between Codex Committees, which were jointly involved in the development of a standard. The resource implication of formalising the meeting of Codex Chairs should however been fully analysed.'

[64] L. Salter, *supra* n. 28, p. 1.

[65] Other sociologists have examined these different situations and the consequences for the functioning of scientists for regulatory purposes. See, for instance, A. Kantrowitz, 'Controlling Technology Democratically', *American Scientist* (1975), pp. 505-509. S. Jasanoff, *The Fifth Branch* (Cambridge, Massachusetts, Harvard University Press 1990). M.A. Hermitte, 'L'Expertise scientifique à finalité réflexions sur l'organisation et la responsabilité des experts', 8 *Justices* (1997), pp. 79-103. National Research Council, *Risk Assessment in the Federal Government: Managing the Process* (Washington, National Academy Press, 1983). R. von Schomberg, *Argumentatie in de context van een wetenschappelijke controverse. Een analyse van de discussie over de introductie van genetisch gemodificeerde organismen in het milieu* (Delft, Uitgeverij Eburon 1997).

sion and its subsidiary bodies on the other hand, is characterised by a separation of powers. The expert bodies have the task of acting as the risk assessors, while the Codex Alimentarius Commission and its subsidiary bodies function as the risk managers. As stated in Consideration 9 of the Working Principles for Risk Analysis for Application in the framework of the Codex Alimentarius:

> 'There should be a functional separation of risk assessment and risk management, in order to ensure the scientific integrity of the risk assessment, to avoid confusion over the functions to be performed by risk assessors and risk managers and to reduce any conflict of interest ...'.[66]

Another aspect that reflects the aim of ensuring scientific integrity is that members of the expert bodies give their advice in their personal capacity as scientists.[67] They do not represent the institution or institutions with which they are connected.[68]

Several concerns that relate to the fear that the expert bodies or the members of the expert bodies may exceed their competences as scientific advisors have been expressed. The mandate given to expert bodies is not always clearly defined and has resulted in the blurring of scientific and political considerations.[69] Furthermore, the independence of the scientists taking place in the expert bodies has not been adequately ensured in the past.[70]

2.3.1 Relation risk assessor – risk manager

The capacity to execute a mandate strongly depends on the way the mandate has been defined.

As explained in Section 3.1 of Chapter II, the food safety provisions of Codex measures are often trans-scientific in nature. They cover issues and deal with questions that cannot be addressed by scientists alone. In order to provide recommendations that are useful for the decision-making process that precedes the adoption of Codex measures, risk assessments are based upon certain assumptions, and on certain decisions that enable scientific experts to arrive at conclusions. These assumptions and decisions are often political in nature. In contrast to a situation of 'normal science' (science conducted in the context of research) in which it is the scientists themselves who ask the questions, in a situation of scientific assessment for regulatory purposes, the questions are presented to the scientists by the decision-mak-

---

[66] Procedural Manual, 16th edn., 2006, p. 104.

[67] See FAO/WHO, preparatory meeting on the elaboration of a common framework for the functioning of joint FAO/WHO expert bodies and consultations, Rome, 3-4 September, 2001, p. 6. See also, Section 3.5.2 of Chapter I.

[68] Ibid.

[69] See Consultant's report, Review of the Working Procedures of the Joint FAO/WHO Meeting on Pesticide Residues (JMPR), 2002, p. 38.

[70] Committee of Experts on Tobacco Industry Documents, Tobacco Company Strategies to Undermine Tobacco Control Activities at the World Health Organization, July 2000, p. 1.

ers.[71] Consequently, it is the task of Codex Committees, as the risk managers and requesters of scientific advice to define the assumptions and decisions so that they can serve as a basis for the assessment undertaken by the expert bodies.

The definition of these assumptions and policy decisions that are used as a basis for the conduct of in a risk assessment is also referred to in the Procedural Manual of the Codex Alimentarius Commission as 'risk assessment policy'. According to the 'Definitions for the Purposes of the Codex Alimentarius', risk assessment policy is defined as consisting of:

'the documented guidelines for scientific judgement and policy choices to be applied at appropriate decision points during risk assessment.'[72]

The importance of a clear definition of the risk assessment policy and the interaction between the risk assessors and risk managers has been emphasised in recent years.[73] However, a consultation report on the review of the working procedure of the Joint FAO/WHO Meeting on Pesticide Residues (JMPR) has pointed out that risk assessment policy is not always determined by the relevant Codex Committee, but by the expert bodies instead:

'On some occasions, the JMPR has ventured into the area of risk assessment policy. One high profile example was in the consideration of the MRL for DDT in meat that was undertaken by 1995 JMPR. The meeting considered as wealth of monitoring data that were supplied by New Zealand. However, since these data formed a very wide distribution, the Extraneous Maximum Residue Limit (EMRL) that could have been proposed by the JMPR was crucially dependent on the "violation rate" on which the EMRL should be based. Different "violation rates" result in different EMRL values. The 1996 JMPR decided to use a violation rate of 0.2% leading to an EMRL of 5mg/kg. *The point here is not that 0.2% is too high or too low, it is that "violation rate" is not a scientific decision and should have been left to the CCPR.* To be fair to the JMPR, they were aware of this issue and an invitation to the CCPR was placed in the simultaneous report asking for "... the views of governments on the level of violation that are considered acceptable." In the absence of clear advice from the CCPR, the 2000 JMPR correctly amended the advice to the CCPR by giving a number of EMRL options depending on which violation rate is used.'[74]

---

[71] S. Jasanoff, *Risk Management and Political Culture* (United States of America, Russell Sage Foundation 1986), p. 39.

[72] Procedural Manual, 16th edn., 2006, p. 44.

[73] See Report of the evaluation of the Codex Alimentarius and other FAO and WHO food standards work, ALINORM 03/25/3, para. 116 and Report of a Joint FAO/WHO Workshop, 'Provision of Scientific Advice to Codex and Member countries', WHO Headquarters, Geneva, 27-29 January, 2004, p. 15.

[74] See Consultant's report, Review of the Working Procedures of the Joint FAO/WHO Meeting on Pesticide Residues (JMPR), 2002, p. 38 [emphasis added]. For clarification, EMRLs differ from regular MRLs as the former are taken in a case where the use of pesticides or chemicals are actually banned world-wide.

Consequently, on various occasions, the scientists have been left to make the political decisions on the selection of scientific data or on the assessment of scientific data.

At its 26th Session in 2003, the Codex Alimentarius Commission adopted the 'Working Principles for Risk Analysis for Application in the Framework of the Codex Alimentarius' (hereinafter referred to as the Working Principles).[75] The Working Principles constitute a step forward, as they provide for a definition of the responsibilities of both the risk assessors and the risk managers, which was absent before their adoption. The Working Principles clearly specify that the determination of risk assessment policy is a component of risk management.[76] Consequently, the role of Codex Committees as risk managers is not limited to a simple request followed by the integration of the risk management recommendations of the expert bodies. As risk managers, they also have the task of interacting with the expert bodies and of clearly restricting their tasks and responsibilities by defining the risk assessment policy. Consideration 14 of the Working Principles states that the determination of risk assessment policy should take place in advance of risk assessment, in consultation with risk assessors and all other interested parties.[77] Consideration 15 emphasises that the mandate given to risk assessors should be as clear as possible.[78] Furthermore, Consideration 24 limits the responsibilities of the risk assessors, as it states that 'risk assessments should be based on realistic exposure scenarios, with consideration of different situations being defined by risk assessment policy.'[79] It addresses the concern expressed in the Consultation report related to the JMPR, as mentioned above.

---

[75] These Principles are intended for internal use in the context of the Codex Alimentarius and not as guidelines for Codex member countries. The Principles lay down general provisions with regard to the process of risk analysis. Detailed provisions are to be worked out by the Codex Committees as regards their specific working area (an example of such detailed provisions are the proposed draft risk analysis principles as developed by the Codex Committee on Food Additives and Contaminants. Several other initiatives have been taken to deal with the relationship between risk assessors and risk managers. For instance, the Codex Committee on Residues of Veterinary Drugs in Foods (CCRVDF) is in the process of drafting a discussion paper on risk management methodologies, including risk assessment policies in which the interaction between risk assessors and risk managers and the role of various parties with responsibilities in risk assessment and risk management are important elements (ALINORM 03/31A, para. 89). Equally within the context of the JEMRA a consultation relating to the interaction between risk assessors and risk managers has taken place (Report of a WHO Expert Consultation in collaboration with the Institute for Hygiene and Food Safety of the Federal Dairy Research Center and the Food and Agriculture Organization, 'The Interaction between Assessors and Managers of Microbiological Hazards in Food', Kiel, 21-23 March 2000).

[76] Consideration 13, Procedural Manual, 16th edn., 2006, p. 104.

[77] Consideration 14 of the Working Principles, Procedural Manual, 16th edn., 2006, pp. 104-105. See also, for discussion, Report of the 14th Session of the Codex Committee on General Principles, ALINORM 99/33A, 1999, para. 25.

[78] Consideration 15 of the Working Principles, Procedural Manual, 16th edn., 2006, p 105. See also, for discussion, Report of the 14th Session of the Codex Committee on General Principles, ALINORM 99/33A, 1999, para. 25.

[79] Consideration 24 of the Working Principles, Procedural Manual, 16th edn., 2006, p 106.

Although the Working Principles respond, in general terms, to a need for a clearer separation of the responsibilities of the risk assessors and of the risk managers, they may not be sufficient. Several comments indicate that this definition of responsibilities is not always applied in practice and that 'there is a pressing need for mechanisms to be set up within the risk management process to ensure that risk assessment policy issues can be properly handled ...'.[80] It has been suggested that a more active and a more direct interaction between the Codex Committees and the expert bodies needs to be established.[81] For instance, this could be done by the submission of the preliminary reports of the consulted expert body to the Codex Committees in order for them to be discussed during a joint meeting of the Codex Committee and the expert body in question.

2.3.2 *Working procedures to ensure independence of the expert bodies*

As mentioned in Section 3.5 of the Chapter I, expert members and expert advisors of the Joint FAO/WHO expert bodies and consultations do not receive financial benefits for their work as expert members or temporary advisors.[82] Only their travel expenses are reimbursed. Experts are often employed or have another financial relationship with a national regulatory or academic institution. However, as stated above in the context of the Joint FAO/WHO expert bodies or consultations, they do not act as the representatives of their institutions, but as scientific experts. Conflicts of interests on the part the experts might undermine the scientific integrity of the advice given by the expert body and thus challenge its primary task. Until 1999, the scientific integrity of the expert bodies was hardly questioned, and rules to prevent

---

[80] Report of the FAO/WHO electronic forum on the provision of scientific advice to Codex and Member countries, February 2004, pp. 18-19 and p. 38.

[81] Report of a Joint FAO/WHO Workshop, 'Provision of Scientific Advice to Codex and Member countries', WHO Headquarters, Geneva, 27-29 January 2004, p. 20: '... The Group considered that there was a need for clear articulation of the questions from Codex and Member countries. In the context of Codex, the expert bodies could not rely solely on text in the official reports. Improved communication with representatives from Codex committees was needed between the annual meetings of these committees. It was important to negotiate/clarify the questions between the representatives of the expert bodies and the Codex Committees. Questions should be relevant and within the scope of the respective body. This needed the active involvement of the expert body, the requesting Codex committees and the FAO/WHO secretariat. FAO/WHO should consider means of enhancing such interactions, such as scheduling back-to-back meetings between expert body meetings and Codex committee meetings. They should also consider making members of expert bodies available at the Codex committees in expert meetings. FAO/WHO should also establish means for increased interaction among expert bodies in matters/questions that impact multiple committees (e.g., considerations involving risk-risk trade-offs).'

[82] Even if the financial resources of the FAO and the WHO were sufficient to cover payments for experts, still the provision of payment may not always be possible as some national organisations do not allow their employees to receive payment (FAO/WHO, 'Selection of experts to avail of best available expertise', Discussion Paper 2c, FAO/WHO Electronic Forum on the provision of scientific advice to Codex Alimentarius and member countries, 1 October-14 November 2003, available at: <www.fao.org/es/esn/proscad/index_en.stm>.

conflicts of interests hardly existed.[83] Only the temporary advisor (the expert who is responsible for the preparation of draft documents to be submitted to the expert members of the expert body for discussion during the meeting) had to sign a statement of ethics in which he or she promised that the information received would not be used for national purposes.[84]

In 2000, however, it became apparent that the tobacco industry had been able to infiltrate the proceedings of the scientific assessments of the JMPR EBDC pesticides which had also been used for the cultivation of tobacco plants.[85] The Director-General of the WHO, Gro Harlem Brundtland, assembled a committee of experts to evaluate tobacco company documents, which had become public through lawsuits against the tobacco industry in the United States. The reason for this action was an internal report that had provided evidence that the tobacco companies had made 'efforts to prevent the implementation of a healthy public policy and efforts to reduce the funding of the tobacco control within UN organisations'.[86] The experts' report disclosed that the tobacco industry funded the work of a temporary scientific advisor to the JMPR,[87] who was in a position to influence the conclusions of the JMPR strongly.[88] The scientist's assistance consisted of summarising international decisions on the EBDC pesticides, which was used to assist another advisor of the WHO with the drafting of a working paper, which served as the basis of the 1993 JMPR meeting.[89] During the re-evaluation of the pesticide in question at

---

[83] One of the exceptions questioning the scientific integrity of the JMPR prior to 1999 is L. Salter, *supra* n. 28, p. 78: 'According to an industry spokesman, however, the JMPR experts and its temporary advisors consult extensively with industry in preparing the draft documentation. The members of the JMPR told us that their contact with industry is limited. This is a sensitive issue, as one might imagine. The situation was described to us as follows: the implication is that reviewers who do the work for WHO lack credibility and integrity, when it is said that we can't allow you to talk to industry because they will compromise you. I would find that insulting if I were a reviewer. These guys are pros…they don't need someone else to say you be careful and don't talk to industry. And because we have the data…its not a matter of (us) being unwilling to give anyone the data.'

[84] L. Salter, *supra* n. 28, p. 78. See, on the function of the temporary advisor, Section 3.5.2 of Chapter I.

[85] Committee of Experts on Tobacco Industry Documents, Tobacco Company Strategies to Undermine Tobacco Control Activities at the World Health Organization, July 2000, p. 1.

[86] Ibid.

[87] Before his retirement in 1988, this advisor had been WHO Executive and Technical Secretary of the JMPR and the JECFA, Committee of Experts on Tobacco Industry Documents, *Tobacco Company Strategies to Undermine Tobacco Control Activities at the World Health Organization*, July 2000, p. 33.

[88] Committee of Experts on Tobacco Industry Documents, *Tobacco Company Strategies to Undermine Tobacco Control Activities at the World Health Organization*, July 2000, p. 15. According to the Expert Committee the exact impact of the advisor's opinion stays unclear due to lack of documentation of the decision-making procedure at the JMPR.

[89] The scientist used his previously conducted reports, which had been informally shared with WHO, FAO and EPA, and 'were now provided to the WHO for official use'. The scientific issues at stake concerned the genotoxic character of the Ethylenethiourea (ETU) and whether there was a threshold level under which the use of the pesticide could be considered as safe. The U.S. Environmental Protection Agency (EPA) had concluded that ETU was carcinogen and that such as threshold did not exist. The report of the scientific advisor and the conclusions of the JMPR in contrast stated that ETU was not

the 1993 JMPR meeting, he officially served as a temporary advisor, in the role of an independent scientist.

The experts' report has had an important influence on the restructuring of the use of experts within the WHO.[90] The direct result of the report on the role of the tobacco industry was the initiative taken by the FAO and the WHO to revise their procedure for the selection of experts for the JMPR and to require that all attendees sign a 'declaration of interest form' in order to identify any potential sources of bias or conflict of interest. Furthermore, the CCPR and the JMPR have undertaken studies and meetings on how to improve the organisation of the JMPR and its use of experts.[91]

An additional change in the rules of procedure of the expert bodies is related to the interaction of the temporary advisor and the sponsor (the provider of scientific data, often the producer of the pesticides, additives or veterinary drugs).[92] Copies of all exchanges of information and correspondence now have to be sent to the Secretariat.[93] The temporary advisor is explicitly discouraged from sending the draft working paper to the sponsor.[94] To avoid both influence and direct contact, the draft working paper, excluding the evaluation section undertaken by the WHO Core group member, is sent to the sponsor by the Joint Secretary of the WHO, instead, with a request to verify its accuracy.[95] The sponsor may not contact the temporary advisor directly, unless it is to inform him or her of any additionally available information.[96] Furthermore, the temporary advisor is encouraged to report any pressure from the industry to the Joint Secretary of the WHO, while any abuse of rules will be reported to the Global Crop Protection Federation (GCPF).[97]

---

a genotoxic and that a threshold level existed. These conclusions of the JMPR allows the Codex Alimentarius Commission to adopt ADI for EBDC (the substances were considered to be safe enough and a threshold ensuring safety could be determined). See Committee of Experts on Tobacco Industry Documents, *Tobacco Company Strategies to Undermine Tobacco Control Activities at the World Health Organization,* July 2000, p. 172.

[90] The report emphasizes the importance of ensuring integrity of the 'scientific phase' of the standard-setting by the Codex Alimentarius Commission the adopted standard 'has become part of international trade law.' See Committee of Experts on Tobacco Industry Documents, *Tobacco Company Strategies to Undermine Tobacco Control Activities at the World Health Organization,* July 2000, p. 156.

[91] See Consultant's Report, Review of the Working Procedures of the Joint FAO/WHO Meeting on Pesticide Residues (JMPR) and the 34th Session of the Codex Committee on Pesticide Residues, The Hague, the Netherlands, 13-18 May 2002.

[92] WHO Procedural Guidelines for the Joint FAO/WHO Meeting on Pesticides Residues, Geneva, January 2001, Section 6. See also, similar provisions in FAO procedural guidelines for the Joint FAO/WHO Expert Committee on Food Additives, Rome, September 2002, Section 6. WHO procedural guidelines for the Joint FAO/WHO Expert Committee on Food Additives, Geneva January 2001, Section 6. See, on the role of the sponsor, Section 3.5.2 of Chapter I.

[93] WHO Procedural guidelines for the Joint FAO/WHO Meeting on Pesticides Residues, Geneva, January 2001, Section 6.

[94] Ibid., at Section 5.2.

[95] Ibid.

[96] WHO Procedural guidelines for the Joint FAO/WHO Meeting on Pesticides Residues, Geneva, January 2001, Section 6.

[97] Ibid.

Similar provisions which aim to ensure the integrity of scientific expertise are included in the guidelines of the other WHO core groups and FAO Panels.

Despite these changes, some concerns about whether the application of the rules is adequately applied have been raised. For instance, although the declaration of interest is an important instrument to ensure the impartiality of the experts, the meaning of 'interest' has not been specified. Experience has demonstrated that experts limit the meaning of interest to financial interest.[98] This interpretation by the experts can lead to incomplete declarations, as other conflicts of interest, such as research conducted, may not have been included.[99] Furthermore, there is no indication of how the FAO and the WHO deal with the declared interests.[100]

2.4 The role of other international institutions in the elaboration procedure

Other international organisations also play a role in the Codex standard-setting procedure. Not only do other international organisations and non-governmental organisations have the right to participate as observers during the standard-setting procedure, they may also be delegated specific tasks in their capacities as standard-setting bodies.[101] Due to its subsidiary character, the Codex Alimentarius Commission does not formally have the competence to enter into agreements with other international organisations. Therefore, the option to co-operate in the context of the standard-setting procedure is a useful tool for establishing strong relations with other relevant organisations.[102] This section examines the basis for the delegation of tasks to other international organisations, the scope of these tasks, and the concerns expressed with regard to their role.

As explained in Section 2 of Chapter I, one of the reasons behind the establishment of the Joint FAO/WHO Food Standards Programme was to co-ordinate the international initiatives already undertaken by other international (non-)governmental

---

[98] See also, FAO/WHO, 'Selection of experts to avail of best available expertise', Discussion Paper 2c, FAO/WHO Electronic Forum on the provision of scientific advice to Codex Alimentarius and member countries, 1 October-14 November 2003, available at: <www.fao.org/es/esn/proscad/index_en.stm>. See also, Report of the FAO/WHO electronic forum on the provision of scientific advice to Codex and Member countries, February 2004, p. 31.

[99] See also, FAO/WHO, 'Selection of experts to avail of best available expertise', Discussion Paper 2c, FAO/WHO Electronic Forum on the provision of scientific advice to Codex Alimentarius and member countries, 1 October – 14 November 2003, available at: <www.fao.org/es/esn/proscad/index_en.stm>.

[100] Report of the FAO/WHO electronic forum on the provision of scientific advice to Codex and Member countries, February 2004, p. 31.

[101] The aspect of participation shall be discussed in Section 3.2 of this chapter. It should be distinguished from the discussion in this section. The role of international non-governmental organisations as participants focuses on their capacity to represent an interest group. In contrast, the call upon international governmental organisations and international non-governmental organisations in the context of this section results from their experience and expertise as standard-setting body.

[102] See, for discussion, Report of the 17th Session of the Codex Committee on General Principles, ALINORM 03/33, 2002, para. 93.

organisations. This has resulted in the task of the Codex Alimentarius Commission – as defined in Article 1(b) of its Statutes – becoming that of promoting the 'co-ordination of all food standards work undertaken by international governmental and non governmental organizations'. The Codex Alimentarius Commission has exercised this power by adopting a standard-setting procedure in which other international organisations can be, and have been, attributed several tasks. Thus, from the beginning of its existence, the Codex Alimentarius Commission has relied upon the work of other international organisations to elaborate Codex standards.

Concerns have been expressed with regard to the role of other international organisations in the Codex standard-setting procedure. For instance, during the 17th Meeting of the Codex Committee on General Principles, one delegation (supported by many other delegations) stated that:

> 'In relation to the interpretation of Article 1(b) of the Statutes of the Commission, the Delegation expressed the view that this might be construed to mean that the work undertaken by other bodies might be taken into account as references in the development of Codex standards. It should not be construed as allowing work to be undertaken by bodies other than Codex subsidiary bodies, not even at Step 2 of the Procedure. The Delegation drew attention to the conclusion of the Commission that it was the main body responsible for elaborating international food standards ...'.[103]

The delegation expressed the concern that the delegation of tasks to other international organisations might entail diminishing the supervision of the Codex Alimentarius Commission as the responsible body. Supervision by the Codex Alimentarius Commission is, indeed, complicated by the fact that, in contrast to Codex subsidiary bodies, these organisations are only bound by the internal rules contained in the Procedural Manual, to the extent that the rules relate to their observer status within the Codex Alimentarius Commission. Furthermore, other international organisations have different aims, a different type of membership, different working procedures and different priorities. The following examples of the allocation of tasks illustrate this.

The allocation of tasks to other international organisations is exercised both on an *ad hoc* basis as well as on a structural basis, and include the task of preparing a 'proposed draft standard', during Step 2, to be submitted to a Codex Committee at Step 3, and the task of providing recommendations regarding the content of the standard. An example of a structural allocation of tasks can be found in the explicit provision contained in the terms of reference of the Codex Committee on Fresh Fruits and Vegetables to co-operate with the Working party on Agricultural Quality

---

[103] Ibid., para. 94. The report states in the same section that: '.... Many other delegations supported these views and also drew attention to the importance of inclusiveness, openness and transparency in the process of elaboration of a proposed draft text and to the fact that the first drafting of a standard was of importance in terms of the source and manner of its preparation, its content and the orientation of further debates.' See also, Report of the 18th Session of the Codex Committee on General Principles, ALINORM 03/33A, 2003, para. 104.

Standards of the United Nations Economic Commission for Europe.[104] It states that the Working Party may recommend the initiation of a worldwide Codex standard for fresh fruits and vegetables, it may prepare 'proposed draft standards' and may perform specific tasks in relation to the elaboration of standards for fresh fruits and vegetables at the request of the Codex Committee on Fresh Fruits and Vegetables.[105] This Working Party is a subsidiary organ of a regional commission of the United Nations (the UN Economic Commission for Europe (UNECE)). The purpose of this organisation is different from the purpose of the Codex Alimentarius Commission, as the former concentrates on the economic progress of the European region.[106] Furthermore, it has to be noted that, in contrast to the membership of the Codex Alimentarius Commission, its membership is limited to the European Members of the United Nations, the US, Canada, Switzerland and Israel.[107] Restricted membership has been one of the main reasons why many non-member countries of the UNECE have objected to the proposal to submit the standards of the UNECE directly to the Codex Committees at Step 3.[108]

Another example can be found in the special task allocated to the International Dairy Federation (IDF). According to the 'Procedures for the Elaboration of Codex Standards and Related Texts', the recommendations of the IDF are to be distributed by the Codex Secretariat during Step 2 of the elaboration of Codex measures related to milk and milk products. Although the 'Procedures' remain silent as to what type of recommendation the IDF is to give, it has to be noted that the reference to the recommendations of the IDF is made in the same paragraph as the reference to the expert bodies consulted by the Codex Committees, such as the JECFA and the JMPR. The position of the IDF has been quite influential. For instance, the Report of the 'Evaluation of the Codex Alimentarius and other FAO and WHO Food Standards Work' indicate that, in the case of the elaboration of the maximum level for aflatoxin M1 in milk, a proposed maximum level submitted by the IDF has been used as a justification not to request the scientific advice of the JECFA.[109] The CCFAC submitted the proposed maximum level directly to the Commission for approval at Step 5. Only after a divergence of opinions had not been resolved two years later, when the standard was to be adopted at Step 8, did the CCFAC decide to submit the draft maximum level to the JECFA for examination.

---

[104] Procedural Manual, 16th edn., 2006, p. 145.

[105] Ibid., footnote 39, pp. 145-146.

[106] Art. 1 of the Terms of Reference of the Economic Commission for Europe, United Nations, E/ECE/778/Rev.: '... (a) initiate and participate in measures for facilitating concerted action for the economic reconstruction of Europe, for raising the level of European economic activity, and for maintaining and strengthening the economic relations of the European countries both among themselves and with other countries of the world ...'.

[107] Terms of Reference of the Economic Commission for Europe, United Nations, E/ECE/778/Rev. 3.

[108] Report of the 24th Session of the Codex Alimentarius Commission, Geneva, 2-7 July 2001, ALINORM 01/41, paras. 19-23.

[109] Report of the evaluation of the Codex Alimentarius and other FAO and WHO food standards work, ALINORM 03/25/3, pp. 52-53.

The IDF is a non-governmental organisation composed of representatives of the dairy industry. Its aim is to give scientific and technical advice while, at the same time, representing the dairy sector's interests in various decision-making procedures. Given its composition and its objectives, one may question the function of the IDF in the standard-setting procedure. The IDF owes this importance to the strong position that it held in the Joint FAO/WHO Committee of Government Experts on the Code of Principles concerning Milk and Milk Products, the predecessor of the Codex Committee on Milk and Milk Products. The establishment of this Joint FAO/WHO Committee was undertaken as a result of a proposal by the International Dairy Federation, concerned with the standardisation of dairy products.[110] This Committee, not being a subsidiary body of the Codex Alimentarius Commission, functioned in an autonomous way within the Codex system and had its own procedures.[111] In 1993, the IDF maintained its role when the Codex Committee on Milk and Milk Products was established to replace the Joint FAO/WHO Committee on Milk and Milk Products, and when the different standard-setting procedures were aligned into one unified procedure.[112]

At its 28th Session, in 2005, the Codex Alimentarius Commission adopted 'Guidelines on Co-operation with Other International Intergovernmental Organisations' (hereinafter the Guidelines on Co-operation).[113] However, these guidelines do not apply to international non-governmental organisations, such as the IDF. The guidelines distinguish two types of co-operation: co-operation at the initial drafting stages of a Codex standard or related text, and co-operation through the mutual exchange of information and participation in meetings.[114] These provisions also imply that the first type of co-operation includes the delegation of tasks to other intergovernmental organisations. Provision 7 of the Guidelines on Co-operation indicates that the Commission, or a subsidiary body of the Commission, may entrust the initial drafting of a proposed draft standard to an international intergovernmental organisation. It explicitly refers to the international organisations as mentioned in Annex A of the *SPS Agreement*: the International Office of Epizootics (OIE) and the international organisation operation within the framework of the International Plant Protection Convention. Furthermore, it requires the international organisations

---

[110] See D.L. Leive, *supra* n. 21, p. 377.

[111] Ibid., at p. 388.

[112] Report of the 20th Session of the Joint FAO/WHO Codex Alimentarius Commission, Geneva, 28 June-7 July 1993 ALINORM 93/40, para. 92. There were some objections to the inclusion of the reference. However, these objections have been rejected on the basis that the issue was discussed in the Codex Committee on General Principles and had been decided in favour of including the reference to the IDF given its input in this area. See, on the history of the Codex standard-setting procedure, Section 4 of Chapter II.

[113] Report of the 28th Session of the Codex Alimentarius Commission, Rome 4-9 July, 2005, ALINORM 05/28/41, paras. 43-44 and Appendix IV.

[114] Provision 4 of the Guidelines on co-operation between the Codex Alimentarius Commission and international intergovernmental organisations in the elaboration of standards and related texts, see Report of the 28th Session of the Codex Alimentarius Commission, Rome 4-9 July 2005, ALINORM 05/28/41, Appendix IV.

to adhere to the same principles of membership (meaning that membership is open to all members and associate members of the FAO and the WHO) and to have equivalent principles of standard-setting.

The adoption of the guidelines raises some questions. Firstly, how do these guidelines relate to the explicit reference to the Working Party of the UNECE and to the IDF? The first organisation is an intergovernmental organisation. However, it does not meet the criteria of having the same principles of membership. The second organisation is not an intergovernmental organisation. Thus, the reference to the Working Party of the UNECE and to the IDF seems to be in contradiction with the adopted guidelines. Given the ambiguous role of the IDF and the restricted membership of the UNECE, it is questionable whether an explicit reference is still desirable.

Secondly, the Guidelines on Co-operation do not include any criteria for the submission of 'proposed draft standards' that may assist the Codex Alimentarius Commission or its subsidiary bodies in the supervision of the execution of the delegated tasks.

3. Procedural legitimacy

Before the entry into force of the WTO agreements, the attribution of normative powers to the Codex Alimentarius Commission was restricted by the strong consensual approach reflected in the acceptance procedure.[115] The binding force of Codex standards only resulted from the explicit consent of governments.[116] As demonstrated in Section 4.4 of Chapter IV, the increased status already results from the adoption of Codex measures and no longer necessitates the explicit consent of governments. Practice has demonstrated that Codex Members are aware of this change. It has often been held that the entry into force of the WTO agreements have led to a 'politicisation' of the standards-setting procedure.[117] Agreement by consensus on

---

[115] See Section 4.3 of Chapter II.
[116] See Section 6 of Chapter II.
[117] Awareness of the Codex Members on the consequences of the new *SPS Agreement* and *TBT Agreement* was already present during the Uruguay negotiations. See Report of the 19th Session of the Joint FAO/WHO Codex Alimentarius Commission, Rome, 1-10 July 1991, ALINORM 91/40, paras. 154-161. See also, C.C. van der Tweel, *Een nieuw GATT-Verdrag en de Codex Alimentarius. Een bestuurskundig onderzoek naar internationale beleidsvorming en nationale beleidsruimte* (Leiden, Rijksuniversiteit Leiden 1993), p. 33 on the 1991 vote on growth hormones. On the politicisation of the Codex procedure, see D. Roberts, et al., 'Sanitary and phytosanitary barriers to agricultural trade: progress, prospects, and implications for developing countries', in M. Ingco and A. Winters (eds.), *Agriculture and the New Trade Agenda* (Cambridge, Cambridge University Press 2004), p. 341. S. Poli, 'The European Community and the Adoption of International Food Standards within the Codex Alimentarius Commission', 10 *European Law Journal* (2004), pp. 613-630. J. Braithwaite and P. Drahos, *supra* n. 2, p. 403. F. Veggeland and S.O. Borgen, 'Negotiating International Food Standards: The World Trade Organization's Impact on the Codex Alimentarius Commission', 18 *Governance: An International Journal of Policy, Administration, and Institutions* (2005), pp. 675-708. A. Cosbey, *A*

the adoption of Codex measures has become harder to achieve.[118] For instance, Roberts, Orden and Josling mention three events that characterise this politicisation of the standards-setting procedure:

- The controversial debate over the Codex statements of principle concerning the role of science in the Codex decision-making process and the extent to which other factors are taken into account;
- The 1995 vote on Codex MRLs for the use of hormones; and;
- The 1997 vote on the revised standard on mineral waters.

Codex Members have become keener to ensure that adopted Codex standards fully respond to their concerns, in particular, if they already have a regulatory measure in place at national or regional level.[119]

It is, in particular, the reactions of the Codex Members in cases where the final adoption of a standard has occurred by vote that reflect the legitimacy-deficit of the standard-setting procedure. For instance, after the adoption of the Revised Standard on Mineral Waters, the United States pointed out that 'it was regrettable that such a decision has been taken by voting', and made the following statement:

'With respect to the decision of the Commission, the US cannot support this Commission action because we object to several provisions of the adopted standard.'[120]

Force Evolution? The Codex Alimentarius Commission, Scientific Uncertainty and the Precautionary Principle (Canada, International Institute for Sustainable Development 2000), p. 9.

[118] F. Veggeland and S.O. Borgen, *supra* n. 117, pp. 675-708. Veggeland and Borgen describe the pre-1995 Codex as a gentlemen's club, meaning that Codex Members would refrain from obstructing the progress of standard-setting. The operation of the Codex as gentlemen's club was possible due to the voluntary nature of the results of the standard-setting procedure. However, the authors state that this has changed since the adoption of the WTO agreements, as the Codex Members are less conducted by their belief of what is appropriate and more by the consequences resulting from the link with the WTO agreements.

[119] Often the adoption of national or regional legislation has been subject to procedures balancing the concerns at stake and the outcome reflects a delicate compromise between the stakeholders. Particularly, in the European Communities EC food legislation has already been subjected to a negotiation process between the different EC member states. Where a national or regional measure has already been adopted the position of the concerned Codex Member will be concentrating on ensuring that the Codex recommendation reflects the national or regional measure as much as possible. As illustrated in Chapter III, the request of the Commission of the European Communities is a response to the increased importance of the standard-setting procedure of the Codex and in order to ensure that the concerns of the EC are fully taken into account. On the pressure of Codex Members that already have regulatory measures in place, See also, S. Henson, et al., *Review of Developing Country Needs and Involvement in International Standards-setting Bodies* (University of Reading 2001). See also, D. Victor, 'WTO Efforts to Manage Differences in National Sanitary and Phytosanitary Policies', in D. Vogel and R. Kagan, *Dynamics of Regulatory Change: How Globalization Affects National Regulatory Policies* (University of California Press 2002). D. Roberts, et al., *supra* n. 117, p. 341.

[120] Report of the 22nd Session of the Joint FAO/WHO Codex Alimentarius Commission, Geneva, 23-28 June 1997, ALINORM 97/37, para. 91.

Numerous other delegations also expressed their reservations about the way that the decision had been taken.[121] It is clear from the statement of the United States that, in this case, the Codex standard-setting procedure had not ensured that the interests defended by the United States were reflected in the outcome. Here, the United States had been unable to give its 'consent', either through remaining silent and not raising objections, or by an explicit expression regarding the content of the standard. This indicates that the possibilities of expressing consent during the standard-setting procedure have assumed greater importance.[122]

This section examines the legitimacy of the Codex standard-setting procedure. As indicated in the introduction to this research, it concentrates particularly on 'horizontal' democracy aspects of the procedure. In other words, it focuses on the equal rights and opportunities for states to ensure that their interests are taken into account. For this purpose, three procedural elements that influence the possibilities of influencing the decision-making process are discussed in this section:

1. consensus as the way of decision-making;
2. *de jure* and *de facto* possibilities of participating in the procedure; and
3. the transparency of the procedure and the decisions.

3.1 Consensus

Decision-making by consensus can be divided into two phases: 1. the process of negotiations preceding the adoption of measures, also referred to as consensus-building or 'active consensus', and 2. the actual moment of final decision-making, often in the form of 'passive consensus'.[123]

---

[121] Ibid., at para. 93: 'Several delegations expressed concern at the means by which the Commission had reached a conclusion on this matter and stressed that the Commission should try by all appropriate means to attempt to take such important decisions on the basis of consensus.' Also, as regards the adoption by vote of the Guidelines for the Design, Operation, Assessment and Accreditation of Food Import and Export Inspection and Certification Systems concerns were expressed, see Report of the 22nd Session of the Joint FAO/WHO Codex Alimentarius Commission, Geneva, 23-28 June 1997, ALINORM 97/37, para. 45: 'The United States, supported by Chile, India, Philippines and the United Arab Emirates, stated that it was highly inappropriate for such broad based and significant guidelines to be adopted before member countries had the opportunity to consider their legal impact on national legislation in the light of the WTO agreements. Concerns were also raised because the Executive Committee of the Codex Alimentarius Commission had just proposed that the Commission, through the Secretariat, consult with the WTO on the status of Codex guidelines and recommendations under the SPS Agreement, and that the Commission should await the outcome of these discussions before proceeding further.'

[122] See also, on the increased importance, G. Moore and A. Tavares, 'Mesures récentes prises par l'Organisation des Nations Unies pour l'alimentation et l'agriculture en conséquence de l'accord de l'Organisation Mondiale du Commerce sur l'application des mesures sanitaires et phytosanitaires', 43 *Annuaire Français de Droit International* (1997), p. 549.

[123] K. Zemanek, 'Majority Rule and Consensus Technique in Law-Making Diplomacy', in R. St. J. MacDonald and D.M. Johnston (eds.), *The Structure and Process of International Law* (Dordrecht, Martinus Nijhoff Publishers 1983), pp. 857-887.

As mentioned in Section 4.2 of Chapter II, the function of consensus-building (or active consensus) is to eliminate controversial points and to construct a proposal which does not raise objections.[124] One major difficulty with consensus-building is the slow progress of decision-making.[125] Although the structure of the Codex step procedure enhances the expression of interests by the different Members and observers,[126] it is held that the length and the repetitive steps tend to destroy the momentum established during the initiation of the elaboration of the measure.[127] Nevertheless, as expressed by B. Buzan:

'The criticism of slowness, however, only has substance if there is some faster way of proceeding towards general agreement.'[128]

For this purpose, this section will reconsider the structure of the Codex step-procedure. It will examine if the step procedure has been used, or could be used, as a tool to ensure the progress of consensus-building, while at the same time remaining an instrument to enhance the expression of interests.

In addition, at the moment of actual decision-making, a struggle can be perceived between the desire to progress and the need to take the interests expressed by the Codex Members into account. On the one hand, delaying a final decision in order to achieve agreement which is acceptable to all Codex Members comes at the expense of a fast decision-making process. On the other, adopting a measure to which some Codex Members still object may be detrimental to the legitimacy of the outcome.

3.1.1 *Final decision-making and the rules regulating consensus*

It has always been common practice for the Codex Alimentarius Commission to adopt Codex measures by consensus. Prior to the adoption of the WTO agreements, the final decision-making within the Codex Alimentarius Commission was based upon the understanding that all the parties involved should refrain from obstructing the adoption of decisions.[129] Several sources indicate the major role attributed to the Chairperson in this process to adopt Codex measures.[130] The Chairperson 'sensed' the majority of opinion, and, if no major opposition was raised, the measure was adopted.[131] This way of final decision-making was not subject to major criticism

---

[124] See Section 4.2 of Chapter II.
[125] B. Buzan, 'Negotiating by consensus: developments in technique at the United Nations Conference on the Law of the Sea', 75 *American Journal of International Law* (1981), p. 341.
[126] See also, Section 4.2 of Chapter II.
[127] G. Moore and A. Tavares, *supra* n. 122, p. 550.
[128] B. Buzan, *supra* n. 125, p. 341.
[129] F. Veggeland and S.O. Borgen, *supra* n. 117, p. 683. See also, Section 4.2 of Chapter II.
[130] L. Rosman, 'Public participation in international pesticide regulation: when the Codex Commission decides, who will listen?', 12 *Virginia Environmental Law Journal* (1993), p. 343.
[131] Ibid. As regards the meetings of the CCPR, L. Salter, *supra* n. 28, p. 75: 'Instead, the Chairman used a modified consensus procedure, in which both delegates and observers signaled their concerns

by the Codex Members, probably as the decision to adhere to the final text was taken by them individually through the acceptance procedure. Furthermore, the outcome resulted from extensive negotiations, enhanced by the structure of the standard-setting procedure.[132]

After the adoption of the WTO agreements and during the negotiations that preceded the adoption, this way of final decision-making by consensus, which favoured the progress of standard-setting above the acceptability of the measures to all Codex Members, no longer functioned. As already indicated above, consensus had become harder to achieve because expressing consent or objection during the standard-setting procedure has become more important for Codex Members. Several standards were adopted by vote during the Codex Alimentarius Commission sessions of 1991, 1995 and 1997, in order to regain the progress of standard-setting. These events triggered a renewed discussion on consensus as the method of final decision-making. One of the most controversial votes taken by the Codex Alimentarius Commission was the vote on the adoption of MRLs for five growth hormones, which were invoked by the United States and Canada in the case before the WTO Panel in *EC-Hormones*.[133] An earlier vote on the adoption of the MRLs in 1991 had failed.[134] During the 21st Session of the Codex Alimentarius Commission in 1995, the issue was presented again. After a vote on the adjournment of the debate,[135] a second vote on the adoption of the MRLs was taken.[136] On this occasion, the voting procedure resulted in a small majority in favour of the adoption of the MRLs.[137] Other standards and guidelines have also been adopted by vote.[138] For instance, a vote was taken during the 22nd Session of the Codex Alimentarius Commission on the revision of the Standard on Mineral Waters.[139] Likewise, and despite the comments made that the long-term consequences needed to be further clarified, the Codex Alimentarius Commission adopted the 'Draft Guidelines for

---

and agreements from the floor and the Chairman "read" the intent of the meeting and noted the decision.'

[132] See Section 4.2 of Chapter II.

[133] See also, G. Moore and A. Tavares, *supra* n. 122, p. 549.

[134] First vote was taken during the 19th Session of the Codex Alimentarius Commission in 1991, see Report of the 19th Session of the Joint FAO/WHO Codex Alimentarius Commission, Rome, 1-10 July 1991, ALINORM 91/40, para. 160.

[135] Report of the 21st Session of the Joint FAO/WHO Codex Alimentarius Commission, Rome, 3-8 July 1995, ALINORM 95/37, para. 44.

[136] Ibid., at para. 45.

[137] Ibid. 33 votes in favour of the adoption, 29 votes against and 7 abstentions.

[138] As regards the discussion on Bovine Somatotropins (BSt), voting occurred twice on the adjournment of the adoption. First time in 1995, 33 in favour, 31 against and 6 abstentions (Report of the 21st Session of the Joint FAO/WHO Codex Alimentarius Commission, Rome, 3-8 July 1995, ALINORM 95/37, para. 47). Second time in 1997, 38 in favour, 21 against and 13 abstentions (Report of the 22nd Session of the Joint FAO/WHO Codex Alimentarius Commission, Geneva, 23-28 June 1997, ALINORM 97/37, para. 69).

[139] Report of the 22nd Session of the Joint FAO/WHO Codex Alimentarius Commission, Geneva, 23-28 June 1997, ALINORM 97/37, paras. 89-90 (33 votes in favour of the adoption, 31 against and 10 abstentions). See also, D. McCrea, 'A View from Consumers', N. Rees and D. Watson (eds.), *International Standards for Food Safety* (Gaithersburg, Maryland, Aspen Publishers 2000), pp. 162-163.

the Design, Operation, Assessment and Accreditation of Food Import and Export Inspection and Certification Systems' by vote.[140]

After the adoption of several Codex standards and other related documents by vote, and the resentment expressed by numerous Codex Members regarding this method of final decision-making,[141] the Codex Alimentarius Commission proposed an amendment of the Rules of Procedures in order to emphasise the importance of taking decisions by consensus, and to discourage decision-making by voting.[142] The amended Rule XII of the Rules of Procedure states:

'The Commission shall make every effort to reach agreement on the adoption or amendment of standards by consensus. Decisions to adopt or amend standards may be taken by voting only if such efforts to reach consensus have failed.'[143]

However, the amendment of Rule XII only partly responds to the concerns expressed by Codex Members who emphasised the importance of taking important decisions, such as the adoption of standards and guidelines, by consensus. Rule XII still provides for the possibility of adopting standards by voting, and, although the threat of a voting procedure is not negative *per se*,[144] there is no clarity on when efforts to reach consensus have failed and when the Codex Alimentarius Commission should proceed with a vote. The amendment of Rule XII has not been accompanied by the inclusion of rules or guidelines that state precisely who should decide to proceed with a vote. In principle, the current rules allow any Codex Member to request a vote.[145] As the example of the voting procedure on the Draft Revised

---

[140] Report of the 22nd Session of the Joint FAO/WHO Codex Alimentarius Commission, Geneva, 23-28 June 1997, ALINORM 97/37, para. 44. The guidelines were adopted by 46 votes in favour, 16 against and 7 abstentions.

[141] Report of the 22nd Session of the Joint FAO/WHO Codex Alimentarius Commission, Geneva, 23-28 June 1997, ALINORM 97/37, para. 93: 'Several delegations expressed concern at the means by which the Commission had reached a conclusion on this matter and stressed that the Commission should try by all appropriate means to attempt to take such important decisions on the basis of consensus.' See also, Report of the 22nd Session of the Joint FAO/WHO Codex Alimentarius Commission, Geneva, 23-28 June 1997, ALINORM 97/37, para. 45: 'The United States, supported by Chile, India, Philippines and the United Arab Emirates, stated that it was highly inappropriate for such broad based and significant guidelines to be adopted before member countries had the opportunity to consider their legal impact on national legislation in the light of the WTO agreements. Concerns were also raised because the Executive Committee of the Codex Alimentarius Commission had just proposed that the Commission, through the Secretariat, consult with the WTO on the status of Codex guidelines and recommendations under the SPS Agreement, and that the Commission should await the outcome of these discussions before proceeding further.'

[142] Report of the 23rd Session of the Joint FAO/WHO Codex Alimentarius Commission, Rome 28 June-3 July 1999, ALINORM 99/37, paras. 61-62.

[143] Rules of Procedure, Procedural Manual, 16th edn., 2006, p. 16.

[144] C.D. Ehlermann and L. Ehring, 'Decision-Making in the World Trade Organization. Is the consensus practice of the World Trade Organization adequate for making, revising and implementing rules on international law?', 8 *Journal of International Economic Law* (2005), pp. 65-66. They explain that the practical impossibility of a vote within the WTO system creates a real danger of deadlock.

[145] Rule VIII.4 of the Rules of Procedure, Procedural Manual, 16th edn., 2006, p. 12.

Standard on Mineral Waters illustrates, it is the simple request of one of the Codex Members that triggered the voting procedure.[146] It is, however, controversial that, in a system based upon a consensual approach, the decision to proceed with a vote can be taken by one Member alone. One might at least expect the request of one Codex Member to be followed by a formal decision of the Codex Alimentarius Commission to proceed with the adoption of the measure by vote.[147]

There are other concerns with regard to the voting procedure itself. For example, in cases where voting on the adoption of standards takes place, decisions are taken by simple majority.[148] It is, however, questionable as to whether this way of voting is appropriate, given the fact that the consensual basis of decision-making is such an important element. The Codex Alimentarius Commission itself applies alternative ways of voting. For example, with regard to some internal matters, such as the amendment of its Agenda, it takes decisions by a majority vote of two-thirds.

Furthermore, the Codex Alimentarius Commission may also decide that voting should take place by secret ballot.[149] For instance, the adoption on the MRLs for growth hormones was decided by secret ballot.[150] The fact that the voting on this sensitive issue took place by secret ballot has been commented upon. For example, the EC held that this way of decision-making was incompatible with the Codex decision to increase transparency.[151]

Another serious concern regarding the final decision-making is the lack of a definition of the term 'consensus'.[152] Previous reports of the meetings of the Codex

---

[146] For instance, with regard to the draft standard on Mineral Waters, a roll-call vote was requested by one Codex Member, another Member requested a secret ballot vote. The first vote taken was not with regard whether or not to proceed with voting but on whether to proceed by a roll-call vote or by a secret ballot vote. The adoption of the draft guidelines were not preceded by any other decision, but directly adopted by voting.

[147] See, for discussion during the 15th Session of the Codex Committee on General Principles, a proposal was made to restrict the right of Codex Members to call for a roll-call vote. The amendment was, however, not adopted. See Report of the 15th Session of the Codex Committee on General Principles, ALINORM 01/33, 2000, paras. 71 and 72.

[148] Rule VIII.2 in conjunction with Rule XII.2 of the Rules of Procedure, Procedural Manual, 16th edn., 2006. The issue of a simple majority vote was discussed during the 14th Session of the Codex Committee on General Principles in 1999. Many delegations were in favour of having a larger majority of the vote cast seen the relevance of Codex texts as a reference in international trade. Other delegations held that in this way it would be more difficult to adopt standards. A proposal was made to require a two-third majority during the first two sessions at which the text was considered for adoption and if then the standard still had not been adopted to proceed to a simple majority of votes. See Report of the 14th Session of the Codex Committee on General Principles, ALINORM 99/33A, 1999, paras. 47 and 48.

[149] Rule VIII.5 of the Rules of Procedure, Procedural Manual, 16th edn., 2006, p. 12.

[150] One of the reasons of the request for a secret ballot vote was to ensure that the candidate EC Member countries could vote without pressure of the EC member states. The MRLs were adopted with 33 votes in favour, 29 votes against and 7 abstentions. While at the moment of deciding on the continuation of the discussion a total of 64 votes were expressed, this number had increased at the moment of adopting the MRLs to a total of 69 votes.

[151] Report of the 21st Session of the Joint FAO/WHO Codex Alimentarius Commission, Rome, 3-8 July 1995, ALINORM 95/37, para. 46. See also, D. McCrea, *supra* n. 139, p. 163.

[152] See also, Report of the evaluation of the Codex Alimentarius and other FAO and WHO food standards work, ALINORM 03/25/3, para. 132.

Alimentarius Commission illustrate that Codex Members and chairpersons have had a different understanding of the meaning of consensus. For instance, the lack of a clear understanding on the principle of consensus led to controversial decisions during the Codex Alimentarius Commission's session in 2001 that have since been subject to criticism.[153] For instance, with regard to the adoption of the position of the Codex Alimentarius Commission regarding situations with insufficient scientific information, it is stated that:

> '... the Chairperson indicated that there was no consensus but a majority of member countries had expressed themselves in favour of this proposal. On this basis the Commission adopted the above position and noted the reservations of the following countries: Austria, Belgium, Croatia, Finland, France, Germany, Greece, Hungary, Italy, Luxembourg, Malta, Netherlands, Portugal, Spain, Sudan, Sweden, Switzerland.'[154]

Although this decision by the Chairperson is in line with the practice of the Codex Alimentarius Commission that was observed prior to the adoption of the WTO agreements,[155] given the criticism received by Codex Members, this practice is no longer considered to be legitimate.

One final remark can be made with regard to the 'Statements of Principles', which aim to overcome the lack of consensus regarding other legitimate concerns. As explained in Section 3.3.2 of Chapter II, the 'Statements of Principles' urge the Codex Members not to prevent the adoption of Codex decisions by allowing them to abstain from acceptance. However, despite the abolition of the acceptance procedure, this statement is still included in the Procedural Manual.[156] Means of

---

[153] Ibid. See also, S. Suppan, *Governance in the Codex Alimentarius Commission* (Consumers International 2005), p. 18.

[154] See Report of the 24th Session of the Codex Alimentarius Commission, Geneva, 2-7 July 2001, ALINORM 01/41, para. 83. See also, para. 84: 'The Delegation of the United Kingdom expressed its disagreement with the manner in which the decision was made as it was essential to take decisions by consensus at the level of the Commission.' Although, the position that was adopted in this case constitutes an 'internal' position to be used within the context of the Codex Alimentarius Commission, a similar approach was followed with regard to the adoption of Codex measures: the MRLs for Aflatoxins in Milk. See Report of the 24th Session of the Codex Alimentarius Commission, Geneva, 2-7 July 2001, ALINORM 01/41, paras. 128 and 129: 'The Commission could not reach a consensus on this issue. (129) In view of the importance of establishing a level for the health protection of consumers, and in consideration that the higher level provided an adequate level of protection as determined by the Committee on Food Additives and Contaminants, the Commission adopted the maximum level of 05 mg/kg in milk. It was agreed that data supporting the lower level, if and when available, could be examined by the Committee on Food Additives and Contaminants at a future meeting if necessary. The Member States of the EU, as well as the delegations of Cyprus, Estonia, Ghana, Hungary, Nigeria, Norway, Poland, South Africa, Swaziland and Switzerland expressed their reservations on this decision ...'.

[155] L. Salter, *supra* n. 28, p. 75: 'Instead, the Chairman used a modified consensus procedure, in which both delegates and observers signaled their concerns and agreements from the floor and the Chairman "read" the intent of the meeting and noted the decision.'

[156] Statement 4 of the 'Statements of Principles Concerning the Role of Science in the Codex Decision-Making Process and the Extent to which Other Factors are Taken Into Account', Procedural Manual, 16th edn., 2006, p. 164.

responding to the Codex Members' objections which persist at the moment of final decision-making will have to be found in other ways.[157]

3.1.2 *Consensus-building: the role of Codex Committees*

As discussed in Section 4.2 of Chapter II, Codex Committees have an important task in consensus-building. Their responsibilities for consensus-building aim to facilitate the Codex Alimentarius Commission in establishing agreement by consensus. However, several remarks can be made with regard to their ability to serve as consensus-builders. Despite the fact that emphasis is put on the establishment of consensus before issues are sent to Commission level, the 'Guidelines for Committees' allow them to proceed to voting if no consensus can be established. The 'Guidelines for Committees' provide that:

> '… the chairpersons should always try to arrive at a consensus and should not ask the Committee to proceed to voting if agreement on the Committee's decision can be secured by consensus.' [158]

However, any possibility for the Codex Committees to proceed to a vote, even within a provision that strongly discourages voting, impedes the changes to Codex measures from being adopted by consensus at Commission level.

Furthermore, as mentioned in Section 4.2 of Chapter II, subsidiary committees may proceed to the establishment of working groups (both electronic as well as physical working groups) in order to facilitate consensus-building. However, the establishment of working groups is only useful to this end in cases where all interested parties participate.[159] If Codex Members that have not participated in the working group oppose the agreement reached by the working group, the discussion may start all over again at the Committee level. In this situation, it may very well be the case that the discussion is less open, as several Codex Members and other interested parties have already fully discussed the issue in question.

Similar to Commission level, the lack of a clear definition on consensus also impedes the Chairpersons of Codex Committees from carrying out their tasks in full. An example of this is illustrated by the decision of the Chairperson of the Codex Committee on Food Export and Import Inspection and Certification Systems to advance the 'guidelines on the judgement of equivalence of SPS measures' to the Codex Alimentarius Commission for adoption notwithstanding the lack of consensus.[160]

---

[157] See Section 4 of this chapter for further discussion.

[158] 'Guidelines for Codex Committees and ad hoc Intergovernmental Task Forces', Procedural Manual, 16th edn., 2006, p. 58.

[159] See, on further discussion on the issue of participation, Section 3.2 of this chapter.

[160] S. Suppan, *supra* n. 153, p. 18. Report of the 9th Session of the Codex Committee on Food Import and Export Inspection and Certification Systems, Perth, Australia, 11-15 December 2000, ALINORM 01/30A, para. 90.

3.1.3 Managing the procedure

As mentioned in Section 4.1 of Chapter II, the step procedure permits the Codex Alimentarius Commission to intervene twice during the development of the draft standard: at Step 5 (through the adoption or refusal of the proposed draft measure) and at Step 8 (through the adoption or refusal of the draft measure as Codex measure).[161] The flexible nature of the step procedure allows the Codex Alimentarius Commission to refer a (proposed) draft measure back to a previous step in the procedure for an unlimited number of times. When a failure to achieve consensus occurs at Commission level, the draft standards or related texts are sent back to Step 6 for reconsideration by the responsible Codex Committee.[162]

This mechanism could provide the Codex Alimentarius Commission with an instrument to ensure, on the one hand, that the considerations and concerns of all Codex Members have been taken into account during the procedure, and, on the other, that the preparation of draft standards progresses. One might expect that the more the standard in question is advanced in the step procedure, the more the discussion on fundamental and new issues is restricted. For instance, during the first phases of the procedure, discussions are open and the involved parties are invited to submit new issues, proposals and objections. In contrast, in more advanced steps, more efforts are put into finding a compromise text, and the inclusion of new issues, proposals and objections are less welcome. Managing consensus-building by the Codex Alimentarius Commission through the step procedure implies that the decision to advance a (proposed) draft measure to the next step is based upon certain criteria with regard to the agreement reached and the progress made. But has the Codex Alimentarius Commission applied certain criteria, and, if so, what have the criteria that it has used to decide upon whether to proceed with the preparation or adoption of a Codex measure or to refer it back to a previous step been?

The 'Procedures for the Elaboration of Codex Standards and Related Texts' used to contain several provisions that relate to the amendments which can be made during Step 8, prior to the adoption of the Codex measure. Only amendments which were of an editorial nature could be subject to immediate adoption by the Commission at Step 8. Substantive amendments agreed upon needed to be referred back to Committee level.[163] The submission of significant amendments was only allowed on condition that the amendments were submitted in writing prior to a date that

---

[161] As already indicated in Section 4 of Chapter II, if the Commission has decided at Step 1 to apply the accelerated procedure, the elaboration procedure is finalised at Step 5.

[162] For instance, the draft revised standards for Butter, Milk-fat Products, Evaporated Milk, Sweetened Condensed Milk, Milk and Cream Powders, Cheese, and Whey Cheese and the Draft Standard for Cheeses in Brine, Report of the 22nd Session of the Joint FAO/WHO Codex Alimentarius Commission, Geneva, 23-28 June 1997, ALINORM 97/37, para. 98. Or the MRLs on Aflatoxins in Milk, see Report of the 23rd Session of the Joint FAO/WHO Codex Alimentarius Commission, Rome 28 June-3 July 1999, ALINORM 99/37, para. 105.

[163] Provision 5 of the Guide to the Consideration of Standards at Step 8 of the Procedure for the Elaboration of Codex Standards including Consideration of any Statements relating to Economic Impact. Procedural Manual, 14th edn., 2004, p. 27.

allowed for all Members to have taken these amendments into consideration before the Commission meeting.[164] The idea behind this possibility of submitting substantive comments at Step 8 was to allow all Codex Members to participate fully in all matters, including those which had been dealt with by Codex Committees which they have not attended.[165] However, at the 2006 Session of the Codex Alimentarius Commission, the 'Guide to the Consideration of Standards at Step 8 of the Procedure for the Elaboration of Codex Standards including Consideration of any Statements relating to Economic Impact' has been deleted.

With regard to the considerations that determine a decision by the Codex Alimentarius Commission at Step 5, the 'Procedures' remain silent. Consequently, it is no longer clear in what way the first reading is to be differentiated from the second reading. However, Codex Alimentarius Commission reports do indicate that the Commission applies informal criteria when deciding whether to advance a draft measure beyond Step 5.[166] Generally, draft standards are only referred back at Step 5 if the lack of consensus concerns a fundamental element of the standard. For instance, the lack of consensus over the actual need for the standard or related text itself,[167] or over fundamental provisions of the proposed draft standard, have provided reasons for the Codex Alimentarius Commission to refer the standard or related text back to Step 3.[168] Likewise, disagreement on the scope and essential definitions has been another reason.[169] Generally, absence of complete scientific evaluation has also been a reason for the Commission either to hold the proposed draft standard at Step 5, or to refer it back to Step 3. For example, the draft maxi-

---

[164] Ibid., Provision 1, p. 26.

[165] Report of the 4th Session of the Codex Committee on General Principles, Paris, 4-8 March 1974, ALINORM 74/36, paras. 49-50.

[166] Documents illustrate that these criteria were already applied in an early stage. Report of the 4th Session of the Codex Committee on General Principles, Paris, 4-8 March 1974, ALINORM 74/36, para. 54. For the purpose of this section, the last seven reports of Commission sessions have been studied, the first one being the 21st Session of the Codex Alimentarius Commission in 1995 (the first Codex Alimentarius Commission meeting after the adoption of the WTO agreements).

[167] There was controversy over the actual development of the Guidelines for Vitamin and Mineral Supplements and the Commission sent it back to Step 3 for further comments and consideration, including the fundamental reconsideration of the need of the Guidelines. See the Report of the 22nd Session of the Joint FAO/WHO Codex Alimentarius Commission, Geneva, 23-28 June 1997, ALINORM 97/37, paras. 110-112.

[168] Several Codex Members raised their concern over the difficulty of implementing the mandatory labelling requirements of the proposed draft amendment to the guidelines on nutrition labelling and the guidelines were sent to Step 3. See Report of the 23rd Session of the Joint FAO/WHO Codex Alimentarius Commission, Rome 28 June-3 July 1999, ALINORM 99/37, para. 189 and para. 191.

[169] For instance, with regard to the scope and definitions of the standard on Spread fats, Report of the 21st Session of the Joint FAO/WHO Codex Alimentarius Commission, Rome, 3-8 July 1995, ALINORM 95/37, para. 66. Likewise, the inclusion of *Clupea bentincki* as new species on the list of sardine-type products in standard for Canned Sardines and Sardine-type Products. Report of the 23rd Session of the Joint FAO/WHO Codex Alimentarius Commission, Rome 28 June-3 July 1999, ALINORM 99/37, para. 118 and para. 120. Another example is the disagreement on the scope of the standard for ginseng products. See Report of the 28th Session of the Codex Alimentarius Commission, Rome 4-9 July 2005, ALINORM 05/28/41, paras. 73-74.

mum level for tin was held at Step 5, awaiting the results of a re-evaluation of the acute toxicity of tin by the JECFA.[170] Likewise, the submission of data by Japan to be considered by the JECFA with regard to the proposed draft of maximum levels for cadmium was a reason to refer the proposed draft standard back to Step 3.[171] Another example can be found in the return of the draft maximum levels of cadmium in rice, which are awaiting the evaluation of the JECFA.[172]

Notwithstanding the existence of these practices, the lack of clear provisions has frequently led to an inconsistent approach which has faced objections from the Codex Members. For instance, lack of consensus over more minor aspects has also been a reason for referring the draft standard in question back to Step 3, as was the case for the proposed draft standard for processed cereal-based foods for infants and young children.[173] The disagreement was about the deletion of the term 'starchy roots and stems' in the composition of cereal-based foods and about the age for the introduction of these foods not for children of 'four to six months' but for children of 'about six months'.[174] The proposed draft standard was sent back to Step 3, despite the opposition of several Codex Members.[175] Likewise, the proposed draft standard for a blend of sweetened condensed milk and vegetable fat was advanced to Step 6, despite the fact that some parts of the standards, such as the scope, needed further work.[176] Another example can be found in the inconsistent practice with regard to the scientific evaluations of the responsible expert bodies. On several other occasions, the Commission approved proposed draft maximum levels at Step 5, while JECFA evaluations were still pending.[177] This situation has raised questions about which different steps of risk analysis should be addressed during which stages of the elaboration procedure.[178]

---

[170] Report of the 23rd Session of the Joint FAO/WHO Codex Alimentarius Commission, Rome 28 June-3 July 1999, ALINORM 99/37, para. 185. The Delegation of India complained about the lack of inconsistency in approach of sending or adopted a standard where scientific evaluation was still pending (para. 186).

[171] Report of the 26th Session of the Codex Alimentarius Commission, Rome, 30 June-7 July, 2003, ALINORM 03/41, para. 125 and 126.

[172] Report of the 27th Session of the Codex Alimentarius Commission, Geneva, 28 June-3 July 2004, ALINORM 04/27/41, para. 68.

[173] Report of the 23rd Session of the Joint FAO/WHO Codex Alimentarius Commission, Rome 28 June-3 July 1999, ALINORM 99/37, paras. 179-182.

[174] Ibid., para. 179.

[175] Ibid., paras. 181-182.

[176] Report of the 27th Session of the Codex Alimentarius Commission, Geneva, 28 June-3 July 2004, ALINORM 04/27/41, para. 75.

[177] For instance, with regard to the guideline levels for aflatoxins. Report of the 21st Session of the Joint FAO/WHO Codex Alimentarius Commission, Rome, 3-8 July 1995, ALINORM 95/37, para. 79.

[178] Report of the 15th Session of the Codex Committee on General Principles, Paris 10-14 April 2000, ALINORM 01/33, para. 115: 'The Delegation of India recalled that the Commission at its 23rd Session had confirmed that the elaboration of Codex standards and related texts should be based on risk analysis. It requested the Committee to consider how the principles of risk analysis should be applied at various stages of the elaboration process. In particular, the Delegation drew attention to the development of certain Codes of Hygienic Practice under consideration by the Codex Committee on Food Hygiene at Step 3. The Delegation also drew attention to the consideration of Aflatoxin M1 in milk and

As regards amendments made at Step 8, which are not referred back to Committee level, most of them are of an editorial nature.[179] When comments have been submitted at a rather late stage during the standard-setting procedure, the standard in question, instead of being sent back to the responsible Codex Committee, is adopted at Step 8, with the decision to start a new elaboration procedure to revise the standard immediately. For instance, the Commission adopted the Draft Revised Standard for Bouillons and Consommés at Step 8, and proposed that a revision of the standard to take the late substantive comments into account be initiated immediately.[180] The diminishing use of the possibility of referring substantive changes back to Codex Committees at Step 6 provides a way of speeding up the developments of Codex measures.

The new rules introduced as a result of the Joint FAO/WHO Evaluation,[181] assign the Executive Committee with a new function, that of monitoring the progress of the development of standards.[182] The latter consists of a review of the status of development, which is considered against the time-frame that was agreed upon at the initiation of the procedure. The Executive Committee may propose corrective action, such as the extension of the time-frame, the cancellation of the work, or it may propose that another Codex Committee takes over the work. Furthermore, the Executive Committee has been assigned the task of examining the Codex measures 'for consistency with the mandate of Codex, the decisions of Codex and existing Codex texts, ... where appropriate, for format and presentation, and for linguistic consistency'.[183]

Several remarks can be made with regard to these new provisions. The monitoring function of the Executive Committee only addresses the problem of the progress

---

the provisions of Lead in various foods by the Codex Committee on Food Additives and Contaminants where, in the opinion of the Delegation, the measures proposed were not consistent with the current JECFA risk assessments and yet advanced to the further step. The Delegation proposed that the Committee in the future should consider how risk assessment would be applied to proposals for standards or related texts that were currently being considered by Committees or submitted to the Commission for adoption.'

[179] For instance, the draft revised standard for Cheese or the draft group standard for cheeses in brine, all amendments are not substantial, but merely editorial (Report of the 26th Session of the Codex Alimentarius Commission, Rome, 30 June-7 July, 2003, ALINORM 03/41, para. 94 and para. 98 and 99).

[180] Report of the 21st Session of the Joint FAO/WHO Codex Alimentarius Commission, Rome, 3-8 July 1995, ALINORM 95/37, para. 83.

[181] Although the recommendations of the Evaluation Team concerned the creation of a new Executive Board (the Codex Alimentarius Commission has not followed this proposal but decided to give new powers to the Executive Committee instead), the new function of the Executive Committee resemble those as initially proposed for the Executive Board. Report of the evaluation of the Codex Alimentarius and other FAO and WHO food standards work, ALINORM 03/25/3, Recommendations 9, 11 and 12. The new rules were adopted during the 27th Session of the Codex Alimentarius Commission in 2004, see Report of the 27th Session of the Codex Alimentarius Commission, Geneva, 28 June-3 July 2004, ALINORM 04/27/41, para. 13.

[182] Part II (Critical Review) of the 'Procedures for the elaboration of Codex standards and other related texts', Procedural Manual, 16th edn., 2006, pp. 22-23.

[183] Ibid.

of the development of standards on the basis of case-to-case decisions. If, in the future, it is not founded on a more structural basis, such as clear criteria for decision-making during the two readings of the step procedure, it may lead to a less transparent and arbitrary means of controlling the progress of consensus-building. This lack of clear criteria constitutes a shortcoming in the internal rules of the Codex Alimentarius Commission, particularly as the legitimacy of the decision-making is not evaluated by other institutions and largely depends on the knowledge of the Codex Members to ensure a system of checks and balances.

Furthermore, it is questionable as to whether the Executive Committee is the appropriate subsidiary body to examine the consistency of the proposed standards with the Procedural Manual, the format and presentation, and linguistic consistency. The Codex Committee on General Principles (CCGP) would seem to be a more appropriate place to which to assign such a task. Given its expertise in the matter and the fact that it is open to all Codex Members, the CCGP is more likely to be successful in ensuring that all proposed corrections are in line with the consensus achieved so far. As all Codex Members are able to participate in the discussion, the deliberations in the CCGP would prevent long discussions within the Codex Alimentarius Commission, on issues such as whether the proposed corrections constitute editorial or substantial corrections. To limit both the discussions within the CCGP and its workload, editorial amendments can be proposed by the Codex Secretariat.

3.2 Participation

Politicisation of the Codex standard-setting procedure has led to an increased number of both Codex Members and of state representatives attending the Codex meetings.[184] Furthermore, the participation of non-governmental organisations has also increased.[185] The authority accorded to Codex measures by the WTO agreements has increased the importance of participating in Codex meetings and of having a greater influence on the outcomes of the standard-setting procedure. This importance of fully participating in international standard-setting bodies has also been emphasised by the *SPS Agreement* and the *TBT Agreement* as mentioned in Section 5 of Chapter IV. Furthermore, the Codex Alimentarius Commission has also emphasised that the participation of all Members and interested parties has become more important than ever before.[186]

In the past, however, concern has often been expressed with regard to the dominant influence of the developed member countries and the industrial-interest INGOs

---

[184] F. Veggeland and S.O. Borgen, *supra* n. 117, pp. 687-688.

[185] G.E. Spencer, et al., 'Effects of Codex and GATT', 9 *Food Control* (1998), p. 179.

[186] Strategic Framework of the Codex Alimentarius Commission for the period 2003-2007. This framework states that 'full participation of all members and relevant intergovernmental and nongovernmental organizations in the work of the CAC and its subsidiary bodies is now more important than ever.' Report of the 24th Session of the Codex Alimentarius Commission, Geneva, 2-7 July 2001, ALINORM 01/41, Appendix II, para. 16.

(international non-governmental organisations).[187] The participants of these two groups have largely outnumbered the participants of developing member countries and public interest INGOs. As stipulated above, this section is first and foremost concerned with the *de jure* and *de facto* possibilities for Codex Members to participate in the standard-setting procedure. Nevertheless, the position of industrial-interest INGOs will be examined in order to define whether it may endanger the 'inter-state democracy' within the Codex. Furthermore, their position will be compared with the position of public interest groups in order to define the obstacles to the equal participation of these two types of international non-governmental organisations.

3.2.1 Participation possibilities of developing countries

Being an intergovernmental body, participation in the decision-making process of the Codex Alimentarius Commission and its subsidiary bodies is open to all states. Any state that is either a member of the FAO or the WHO can become a Member of the Codex Alimentarius Commission. All Codex Members *de jure* have equal rights at every stage of the decision procedure. However, when it comes to the equal opportunities of all Codex Members with regard to *de facto* participation, the situation is different.

The Codex Alimentarius Commission has often been characterised as an 'industrialised-country' club. It is true that at the beginning of its existence, its membership was mainly dominated by industrialised countries. However, since the 1970s, developing countries have constituted the majority of the Codex Members. Despite the increase in the number of developing countries, the voice of the industrialised Codex Members has still remained strong. This is partly due to the expertise that these Codex Members have with the technical and regulatory issues relating to food safety and quality. The developing country Codex Members, on the other hand, often have neither the technical, nor the regulatory, expertise. Nevertheless, it is mainly the developing country Codex Members which have used the Codex standards as a basis for their national regulations, as it reflects the current scientific knowledge and is believed to facilitate international trade,[188] whereas the devel-

---

[187] N. Avery, et al., *Cracking the Codex* (London, National Food Alliance 1993), p. 1. G.E. Spencer, et al., *supra* n. 185, p. 179. L. Salter, *supra* n. 28, pp. 70-71 and p. 74. L. Rosman, *supra* n. 130, p. 346. D.L. Leive, *supra* n. 21, pp. 435-436. D.G. Victor, *Effective Multilateral Regulation of Industrial Activity: Institutions for Policing and Adjusting Binding and Nonbinding Legal Commitments*, Ph.D. Thesis (Massachusetts, Harvard University, Institute of Technology 1997), pp. 198-201. D. McCrea, *supra* n. 139, p. 155. S. Suppan, *supra* n. 153.

[188] S. Henson, et al., *supra* n. 119: 'Developing countries regard international standards as a resource-efficient approach to establishing technical regulations at the national level, which reflect current scientific knowledge and facilitate international trade. This supports previous research findings that developing countries have a tendency to adopt standards established elsewhere, whether by international standards-setting organisations or in developing countries, rather than play an active role in the setting of standards themselves. This raises serious issues regarding the appropriateness of such standards in the context of developing countries, even after adaptation to local circumstances, as well

oped country Codex Members have been more hesitant to modify their regulations accordingly.[189]

3.2.1.1   Obstacles to a *de facto* effective participation of developing countries

Effective participation of developing Codex Members can be impeded by three types of obstacles:

1. obstacles that are inherent to the domestic situation of the developing Codex Members;
2. obstacles that may be classified as affecting the equal possibilities of participating in the standard-setting procedure (also referred to as the principle of impartial participation); and:
3. obstacles inherent to the current standard-setting procedure that, although not in conflict with the principle of impartial participation as such, may add to the disadvantaged position that the developing country Codex Members already possess.

Obstacles inherent to the domestic situation of developing country Codex Members result from the fact that they often lack financial, technical and human resources.[190] This deficit can put them in a disadvantaged position during the elaboration procedure, and can, as a result, lead to the creation of international standards that are inappropriate for them.[191] First, the lack of financial resources obstructs the representatives of the developing member countries from actually attending the Codex meetings. Nevertheless, even in cases where developing Codex Members attend the meetings, they may still lack the capacity to ensure their input into the standards-setting procedure adequately.[192] Some developing member coun-

---

as sovereignty and/or international agencies on domestic regulation in developing countries. These issues have particular poignancy when developing countries are unable to participate effectively in international standards-setting organisations.'

[189] See also, Section 4 of Chapter III with regard to the role of Codex standards in the EC.

[190] L.T. Marovatsanga, 'The need for Developing Countries to Improve National Infrastructure to Contribute to International Standards', in N. Rees and D. Watson (eds.), *International Standards for Food Safety* (Gaithersburg, Maryland, Aspen Publishers 2000), p. 140. See also, D. McCrea, *supra* n. 139, p. 153 and p. 155. D.L. Leive, *supra* n. 21, pp. 435-436.

[191] Department for International Development, 'Standards as Barriers to Trade: Issues for Development', Background Briefing, September 2001, p. 2. S. Henson, et al., *supra* n. 119.

[192] L.T. Marovatsanga, *supra* n. 190, p. 138: 'Unfortunately, many developing countries with economies in transition lack the necessary food control infrastructure to implement both strategy and programs to protect their consumers and to overcome their disadvantaged status in the international trade in food. These problems include: 1. inadequate or outdated legislation and regulations, 2. inadequate resources and/or failure to maximize available resources, 3. failure to develop a national food control strategy and poorly implemented or managed program and activities, 4. inadequately equipped laboratories and inspectorate, 5. inadequately trained and technically deficient personnel, 6. poor coordination and cooperation among food control agencies, other concerned government agencies, academia, industry, and consumers in order to suit international standard setting, 7. lack of political will and commitment on food safety and standards by developing countries.'

tries lack a domestic institutional and legal infrastructure that is updated and do not have the legal framework in which the standard in question can be implemented.[193] The speedy process of standard-setting means that their legal and institutional infrastructure cannot keep pace and they face a situation of always lagging behind the international standard-setting, which only seems to worsen. Furthermore, they often lack experience in enforcing food standards, which means that the major problems of enforcement that they may encounter are not always clear. Their laboratories and inspection services are often not adequately equipped to verify compliance. This incapacity to verify compliance with the regulatory instruments in place leads to difficulties in determining the appropriateness and effectiveness of these instruments. In addition, they often lack the expertise both to conduct risk assessments at domestic level and to ensure the competence of their representatives who attend the Codex meetings. Additionally, as already mentioned in Section 2.3 of this chapter, there are difficulties collecting scientific data from developing countries. This may lead to the adoption of Codex MRLs and other food safety standards that do not correspond with their national situations and their national priorities.[194] In practice, these food standards may be difficult for the food producers and distributors to adhere to in developing member countries. Lastly, adequate input may also prove to be insufficient due to political constraints at domestic level. Co-ordination between the different governmental agencies and other stakeholders is not always adequately addressed at domestic level, and this may obstruct a clear standpoint (in particular, in cases where a Codex standard is dealt with by several different Codex Committees).[195]

Second, the principle of impartial participation aims to prevent the privileged treatment of certain interested parties with regard to all steps in the standard-setting procedure, such as access to participation in the work, submission of comments on drafts, consideration of both the views expressed and the comments made, the obtaining of information and documents, the dissemination of the international standard, the fees charged for documents, the right to transpose the international standard into a regional standard, and the revision of the international standard. In particular, the preparation of the position, or discussion, papers or draft proposed standards is of interest in this context, which is a task that is increasingly assigned to Codex Members in the initial phase of the elaboration procedure. The proposed draft standard can have an important influence on the final result, as fundamental and structural changes to the initial texts are more complicated to make (for both political as well as technical reasons).[196] The allocation of this task is, therefore, a delicate

---

[193] L.T. Marovatsanga, *supra* n. 190, p. 138.

[194] One of the complaints often heard from the developing countries is that standards or MRLs which reflected their priorities have hardly been developed. One good example is the commodity of spices. This case is quite complicated as data on spices are not available. For example, in India the high number of farmers cultivating spices makes it very hard to collect data on Good Agricultural Practices and residue trials.

[195] This concern is not only present in developing countries. It is a major concern equally present in most other Codex Members.

[196] H.G. Schermers and N.M. Blokker, *supra* n. 13, p. 502.

matter. It frequently occurs that the Codex Member that has submitted the proposal to elaborate a Codex standard is also the Member that has drafted the proposed draft standard, which only contributes to the influence that one Codex Member may have. The drafting of standards is increasingly undertaken by a drafting group which consists of several Codex Members and is open to all interested Members. However, an examination of the assignment of this task, which has taken place over the last three years, indicates that the majority of the proposed draft standards and the discussion papers are prepared under the authority of developed member countries, despite the recommendation of the Codex Alimentarius Commission to appoint a co-author from a developing country Codex Member, for position papers, where the main author is from a developed country Codex Member.[197]

Another aspect that puts developed Codex Members in a privileged position is the hosting of Codex committees. Most Codex committees are hosted by developed member countries and take place in Europe and North America. This means that travel costs are generally higher for developing member countries. This adds to their difficulties in attending Codex meetings due to financial resources. Consequently, developed member countries often outnumber the delegations from developing member countries.[198]

Other factors that may constitute obstructions to the *de facto* participation of developing member countries relate, for instance, to the complex institutional system and procedure, the various working groups which stimulate consensus-building, and the flexibility of the standard-setting procedure. Lack of insight and knowledge of these procedural issues complicates the situation for developing countries to participate effectively in the decision-making process even further.[199]

3.2.1.2 Ways to stimulate the participation of developing countries

Concerns raised with regard to the participation of developing member countries in the elaboration procedure are not new. Approximately ten years after the establishment of the Joint FAO/WHO Food Standards Programme, and due to the impact of the 1972 Stockholm Declaration on Human Environment, the problematical position of developing countries was emphasised.[200] The establishment of a trust fund

---

[197] Out of the 20 approved new work items by the Executive Committee in 2000, 4 discussion papers or drafts were developed under the authority of a developing country (of which three working items were very closely related and all prepared by Malaysia) and 10 by industrialised countries. Similarly, out of the 14 approved new work items by the Executive Committee in 2002, only one discussion paper was prepared by a developing country (Iran as regards the proposed draft code of practice for the Prevention and reduction of Aflatoxins in Tree Nuts) and 8 by industrialised countries. Recommendation of the Codex Alimentarius Commission see Report of the 23rd Session of the Joint FAO/WHO Codex Alimentarius Commission, Rome, 28 June-3 July 1999, ALINORM 99/37, para. 56(f).

[198] G. Sander, 'Gesundheitsschutz in der WTO - eine neue Bedeutung des Codex Alimentarius im Lebensmittelrecht?', 3 *Zeitschrift für Europarechtliche Studien* (2000). N. Avery, et al., *supra* n. 187, p. 34.

[199] See also, Section 3.3 of this chapter on transparency.

[200] Report of the 9th Session of the Codex Alimentarius Commission, Rome, 6-17 November 1972, ALINORM 72/35, paras. 5-8 and 10.

to ensure adequate participation for the developing countries was proposed in 1978.[201] Although the possibility of hosting Codex Sessions in developing countries was mentioned during the 14th Session of the Codex Alimentarius Commission in 1981,[202] it was not until the 1990s that the discussion to assist the developing member countries took a more serious dimension. As a result of the 1991 FAO/WHO Conference on Food Standards, Chemicals in Food and Food Trade, the Codex Alimentarius Commission emphasised the necessity of establishing a financing framework to enhance the participation of developing member countries.[203] The Trust Fund was launched in 2003 and started operating in 2004. The first initiatives of the co-chairing and hosting of Codex Committees in developing countries took place in 2000.

3.2.1.2.1 Co-chairing and hosting of Codex Committee meetings in developing countries

The participation of developing countries is, in some cases, enhanced by the concept of co-chairing of the Codex Committees. This has been undertaken by the host countries of numerous Codex Committees, such as the Codex Committee on Food Additives and Contaminants (CCFAC) in China (2000) and in Tanzania (2003), the Codex Committee on Food Hygiene (CCFH) in Thailand (2001), the Codex Committee on Residues of Veterinary Drugs in Foods (CCRVDF) in Mexico (2006) and the Codex Committee on Pesticide Residues (CCPR) in India (2004) and Brazil (2006). This means that, while a developed member countries finances the meeting and organisation, the meeting (which normally takes place in its country) does, in fact, take place in a developing member country. Such projects undertaken by the Netherlands as the former formal host country of the CCFAC and the CCPR in the past have demonstrated that this encourages the participation of developing country members of the same region in which the meeting is taking place. Furthermore, it also motivates the active participation of the host developing member country to become involved in the Codex decision-making procedure in the future, as in the case of China, which is now formal host country of both the Codex Committee on Food Additives (CCFA) and the Codex Committee on Pesticides Residues (CCPR). On the other hand, these projects have the disadvantage that they often do not reduce the travel costs for developing member countries coming from other regions than that of the host country. Due to the global infrastructure, the participation of developing countries from other regions is, in these cases, often more complicated than when meetings take place in Europe or North America.

---

[201] See Report of the 12th Session of the Codex Alimentarius Commission, Rome, 17-28 April 1978, ALINORM 78/41, para. 133.
[202] Report of the 14th Session of the Codex Alimentarius Commission, Geneva, 29 June-10 July, 1981, ALINORM 81/39, paras. 135-147.
[203] Report of the 19th Session of the Joint FAO/WHO Codex Alimentarius Commission, Rome, 1-10 July 1991, ALINORM 91/40, paras. 64-66.

3.2.1.2.2 The Trust Fund for the participation of developing countries and countries in transition in the work of the Codex Alimentarius Commission

As explained in Chapter I, the Trust Fund for the participation of developing countries and countries in transition in the work of the Codex Alimentarius Commission' (hereinafter referred to as the Trust Fund) was launched during the 25th Session of the Codex Alimentarius Commission in February 2003, and has been operational since 1 March 2004.[204] The minimum threshold that was defined to trigger the operation of the Trust Fund was an amount of $500,000 US and the overall target for fund-raising over 12 years was set at $40 million US.[205] During the period of 2003-2004, a total of $1,543,090 US was contributed to the fund.[206] The Trust Fund aims, amongst other things, to promote the participation of developing countries by financially supporting their visits to the meetings of the Codex Alimentarius Commission and its subsidiary committees.[207] Other objectives of the Trust Fund, besides the promotion of the participation of developing countries, include the promotion of effective participation through adequate preparation and active participation of different Codex Committees/Task Forces.[208] Likewise, capacity-building to enhance the scientific and technical capacities of the participants is included in the Trust Fund objectives.[209]

The member countries which are eligible to the donations of the Trust Fund are divided in three categories, based on the data of the World Bank: 1. Low-income countries (LIC), 2. Lower middle-income countries (LMIC), 3. Upper middle-income countries (UMIC). When distributing the donations to the eligible member countries, priority has been given to the participation of the low-income countries, and on focusing on the strengthening of participation of these countries.[210] In order to be eligible for funding from the Trust Fund, the Members have to fulfil several criteria:

---

[204] See Section 3.6 of Chapter I.

[205] FAO/WHO Co-operative Programme, FAO/WHO Project and Fund for Enhanced Participation in Codex, Project Document, 17 June 2003, p. 9.

[206] FAO/WHO Project and Fund for Enhanced Participation in Codex, 5th Progress Report (January-June 2005), CAC/28 INF/12, p. 2.

[207] FAO/WHO Co-operative Programme, FAO/WHO Project and Fund for Enhanced Participation in Codex, Project Document, 17 June 2003, p. 4. Participation is limited to Codex Commission and its subsidiary bodies. It does not include participation in the expert bodies.

[208] Ibid.

[209] Ibid.

[210] FAO/WHO Co-operative Programme, FAO/WHO Project and Fund for Enhanced Participation in Codex, Project Document, 17 June 2003, p. 7. Priorities in the programme are expressed in percentages. 60% will be attributed to the participation of developing countries, of which 40% to low-income countries and 20% to lower middle-income countries. 30% will be dedicated to capacity-building concentrating on the preparation and active participation of the meetings (15% to low-income countries and 15% to lower middle-income countries) and 10% will finance the enhancement of scientific and technical expertise of the participants (3% to low-income countries, 3% to lower middle-income countries and 4% to upper middle-income countries). This scheme is a flexible scheme and is expected to be adjusted according to the experiences.

1. the country has to be a Member of the Codex Alimentarius Commission;
2. the country has to have an identified Codex contact point;
3. the country has to present a workplan that relates to one of the three objectives of the Trust Fund; and
4. the country has to demonstrate that co-ordination between the relevant governmental entities is taking place.[211]

Although the initiative is a step forward in enhancing the participation of the developing member countries, the Trust Fund will only be able to respond to part of the problem, given the number of Codex Members in need of financial resources, and given the numerous meetings of the Codex Committees. Furthermore, the sustainability of the impact which results from the efforts is not guaranteed if the efforts are not undertaken in combination with other activities which ensure a more transparent and impartial standard-setting procedure.[212]

3.2.2 Analysing the intergovernmental character of participation rights

The intergovernmental character of the Codex Alimentarius Commission implies that its decision-making powers are exercised by the representatives of Member countries.[213] The fact that only countries can become Members of the Codex Alimentarius Commission is a clear reflection of this intergovernmental nature. However, a lack of a clear distinction between the participation rights of Members *vis-à-vis* the rights of international non-governmental organisations (INGOs) that have observer-status, can affect the intergovernmental character of the organisation and its decision-making procedure. Privileges, either *de jure* or *de facto*, accorded to INGOs as observers may have consequences for the *de facto* participation of Codex Members. In particular, industrial INGOs are strongly represented at Codex meetings. In the past, it was not uncommon for the representatives of these groups to outnumber the government representatives of developing member countries.[214]

---

[211] FAO/WHO Co-operative Programme, FAO/WHO Project and Fund for Enhanced Participation in Codex, Project Document, 17 June 2003, p. 8.

[212] See Sections 3.2.2 and 3.3 of this chapter for further discussion on these characteristics. Furthermore, it needs to be noted that coherence of the trust fund with other capacity-building initiatives becomes important. Numerous capacity-building projects are undertaken by various international organisations that indirectly may influence and promote the effective participation of developing member countries in the elaboration procedure of the Codex. At the Doha WTO Ministerial meeting in Qatar in 2001, heads of several international organisations agreed to co-operate and strengthen the capacity in implementing SPS issues (Ministerial Conference, Implementation-related issues and concerns, Decision of 14 November 2001, Doha, WT/MIN(01)/17, para. 3.5 (with regard to SPS) and para. 5.3 (TBT). This resulted in the launch of the Standards and Trade Development Facility which aims to coordinate the capacity-building projects undertaken by the relevant international organisations and to undertake joint capacity-building projects.

[213] H.G. Schermers and N.M. Blokker, *supra* n. 13, para. 46.

[214] N. Avery, et al., *supra* n. 187, p. 34.

According them with privileges may come at the expense of developing member countries, particularly in situations where their interests are in conflict.[215]

The right to vote is a clear example of a right that has only been accorded to Codex Members and which reflects the intergovernmental character. However, in a decision-making process that adopts decisions by consensus, this right is negligible, and other rights become more important, such as the right to take an active part in the deliberations preceding the adoption of the Codex measures. In smaller groups of a technical nature, such as Codex Committees, the input of INGO observers can be especially influential. As an informal rule, INGOs are invited to speak only after Codex Members have spoken on the issue.[216] This customary rule is not always applied very strictly.[217] Furthermore, if a discussion is deadlocked, the Chairperson may call upon INGOs to offer new proposals to advance the discussions.[218]

The right to request the initiation of the standard-setting procedure is not clearly laid down in the Procedural Manual. It is neither amongst the rights of Codex Members, nor amongst the privileges as enumerated in the 'Principles of NGO Participation'.[219] Reports indicate that INGOs have also launched a request for the initiation of new work, such as the request of the International Dairy Federation during the 6th Session of the Codex Committee on Milk and Milk products.[220]

This situation is complicated by the fact that there is an additional way that may result in the initiation of the Codex standards-setting procedure: through a direct request for the re-evaluation of a substance to the Joint Secretariat of the Expert Committees. Industry, as the promoter of data, may directly address its request to the Joint Secretariat of the expert body, in the case of a *re-evaluation* of a substance.[221]

---

[215] N. Avery, M. Drake and T. Lang demonstrate that the assumption that developing Codex Members do not have higher food safety standards in place than the standards discussed in Codex is false. They give examples that illustrate this such as the Codex acceptable residue limits for the pesticide Lindane, a pesticide not permitted in Bolivia, Ecuador, Egypt or Guatemala, or the ban of Indonesia on 57 pesticides in rice production. N. Avery, et al., *supra* n. 187, p. 36.

[216] S. Suppan, *supra* n. 153, p. 18. D. McCrea, *supra* n. 139, pp. 163-164.

[217] For example, during the 24th Session of the Codex Alimentarius Commission, which the author attended, Consumers International was invited to speak before other Codex Members had the word.

[218] S. Suppan, *supra* n. 153, p. 18.

[219] Principles concerning the participation of international non-governmental organisations in the work of the Codex Alimentarius Commission (the right to initiate is not mentioned among the privileges), Procedural Manual, 16th edn., 2006, pp. 34-40.

[220] Also the report of the 23rd Session of the Codex Alimentarius Commission illustrates that the recommendation of the International Association of Consumer Food Organisations (IACFO) has triggered the decision of the Commission to request the Codex Committee examine the recommendation. See Report of the 23rd Session of the Joint FAO/WHO Codex Alimentarius Commission, Rome 28 June-3 July 1999, ALINORM 99/37, para. 202.

[221] FAO procedural guidelines for the Joint FAO/WHO Expert Committee on Food Additives, Rome, September 2002, Annex 1. WHO procedural guidelines for the Joint FAO/WHO Expert Committee on Food Additives, Geneva, January 2001, Annex 1. WHO Procedural guidelines for the Joint FAO/WHO Meeting on Pesticides Residues, Geneva, January 2001, Annex 1. The Joint Secretariat or the expert body itself hold a similar right.

Codex observers have the right to express their points of view and to submit memoranda or written views. Reports of Codex Commission and Committees' meetings demonstrate that INGOs frequently also prepare discussion papers for proposed draft standards and thus are, in this manner, actively involved in the initial phase of the elaboration procedure.[222] Likewise, as mentioned in Section 2.4 of this chapter, the IDF is accorded a special role in the standard-setting procedure, which is related to the development of standards on milk or milk products as its recommendations have to be taken into account during Step 2. Given the importance of a first draft of a standard throughout the rest of the procedure, it can be concluded that, despite the fact that, formally, INGOs only have observer-status, in reality, their influence on the decision-making procedure is, in some situations, not outdone by that of Codex Members.

3.2.3 The participation of industry INGOs vis-à-vis public interest INGOs

INGOs can have an influence on the decision-making process in three ways:

1. through attendance to the Codex Committees and Commission meetings as observer;
2. as member to the national delegations; or
3. indirectly by expressing their concerns in an earlier stage of the process at a national or regional level before a national or regional standpoint has been adopted.

At a Codex level, both industrial INGOs and public interest INGOs have similar rights to participate in the standards-setting procedure. However, it cannot be denied that *de facto* industrial-interest INGOs are in a stronger position to influence the decision-making procedure than public interest INGOs are. This can partly be ascribed to their different backgrounds and circumstances. Industrial-interest INGOs, as promoters of food production technology, possess a high degree of expertise in the area and a large source of information. As such, they are regarded as useful advisors and contributors to the standards-setting procedure. In a previous edition of a formal explanatory guide on the Codex Alimentarius written by the FAO and the WHO, this role of the food industry was explicitly mentioned.[223] Furthermore,

---

[222] The International Dairy Federation prepared a discussion paper on the proposed draft standard for processed cheese. Similarly, the International Consultative Group on Food Irradiation prepared a discussion paper on the proposed draft revision of the recommended International Code of Practice for the Operation of Irradiation Facilities Used for the Treatment of Foods and the draft for the revision of the Code of Practice for the Processing and Handling of Quick Frozen Foods was prepared by the International Institute of Refrigeration.

[223] L. Rosman, *supra* n. 130, p. 345, citing Joint FAO/WHO Food Standards Programme, *Introducing Codex Alimentarius*, 1990: 'Industry and trade ... have an important role in Codex. They can and do make valuable contribution of terms of scientific and economic information. Acting both as advisers to government representatives in national delegations and through international industry associations, they bring a great wealth of information and advice to the Codex discussions.'

the fact that they are the promoters of food production technologies and therefore possess the necessary data for scientific assessment automatically grants the industrial INGOs a particular role in the Codex Alimentarius Commission as data sponsor. In this role, as already mentioned in Section 3.5.2 of Chapter I, they are heard by expert committees. This task should not be considered as according them a privileged right to participate *vis-à-vis* public interest INGOs, as it is restricted to the scientific interrogation of the data that the scientists of the sponsors have made available. A cautious approach is taken in order to avoid the evaluations of expert bodies being influenced or inhibited by actions of the sponsors, either in fact or in appearance.[224] However, there is no doubt that their task as data promoters puts them in an advantageous position. Furthermore, industrial INGOs often have access to sufficient financial resources that allow them to participate in the relevant Codex meetings and to lobby to defend their interests.

On the other hand, public interest INGOs may lack the expertise to participate effectively in the Codex standards-setting procedure.[225] Their position is definitely weakened by their lack of funding to participate and lobby at the relevant Codex meetings.[226] Several studies and observations conducted over the years indicate an unbalanced number of representatives in favour of industry,[227] both as part of the national delegations,[228] as well as in their capacity as observer INGOs.[229] A study conducted in 1985 indicated the absence of public-interest groups during the CCPR meeting, while there were 36 representatives from industry, including 34 individuals from GIFAP, the international organisation representing the pesticide manufacturers.[230] A publication by Rosman indicates that, during the following years, only 2 out of the 73 INGO participants present at the CCPR meetings were public interest INGOs.[231] A profound study undertaken by Avery, Drake and Lang at the beginning of the 1990s illustrates that similar patterns were observed at meetings of other Codex Committees and at the Codex Alimentarius Commission.[232]

At the 1991 FAO/WHO Conference on Food Standards, Chemicals in Food and Food Trade, attention was paid to the position of public-interest INGOs.[233] As a result, at its session in 1991, the Codex Alimentarius Commission emphasised the responsibility of Codex Members to achieve greater participation of consumer

---

[224] WHO Procedural guidelines for the Joint FAO/WHO Expert Committee on Food Additives, Geneva January 2001, Section 6. FAO procedural guidelines for the Joint FAO/WHO Expert Committee on Food Additives, Rome, September 2002, Section 6. WHO Procedural guidelines for the Joint FAO/WHO Meeting on Pesticides Residues, Geneva, January 2001, Section 6.
[225] C.C. van der Tweel, *supra* n. 117, p. 39.
[226] N. Avery, et al., *supra* n. 187, p. 5. D. McCrea, *supra* n. 139, p. 153.
[227] D. McCrea, *supra* n. 139, p. 155. D.G. Victor, *supra* n. 187, pp. 198-200.
[228] G. Sander, *supra* n. 198.
[229] N. Avery, et al., *supra* n. 187. L. Rosman, *supra* n. 130, pp. 345-346. S. Suppan, *supra* n. 153, p. 17. L. Salter, *supra* n. 28, p. 74.
[230] L. Salter, *supra* n. 28, p. 74.
[231] L. Rosman, *supra* n. 130, p. 346.
[232] N. Avery, et al., *supra* n. 187.
[233] Ibid., at p. 38.

organisations in the Codex.[234] Since the beginning of the 1990s, the position of public interest INGOs has slightly improved.[235] For example, the Codex Alimentarius Commission agreed that the Codex Committee on Food Labelling should take the recommendations of the International Association of Consumer Food Organisations (IACFO) into account in order to revise the 'General Standard for the Labelling of Pre-packaged Foods: Quantitative Declaration of Ingredients'.[236] The recommendations triggered the initiation of new work and a discussion paper prepared by the IACFO served as a basis for the discussions.[237] Likewise, Consumers International, a public-interest INGO, has been able to submit proposals to advance the debate in situations where a discussion between Codex Members was deadlocked.[238] Furthermore, a Member with very close connections to Consumers International was included in the Expert Panel that assisted in the evaluation of the Joint FAO/WHO Food Standards Programme undertaken by the FAO and the WHO.[239] Although on several occasions the Codex Alimentarius Commission and its parent organisations have stimulated a more active participation of public interests INGOs (in the past, funds have even been provided to enable public interest INGOs to participate in Codex meetings),[240] the Codex Alimentarius Commission has clearly emphasised its priority to ensure the participation of all its Members. When, at its session in 1997, the Codex Alimentarius Commission agreed to examine the possibilities of establishing a Trust Fund to enhance participation, the initial idea was to promote the participation of both developing Codex Members and public interest INGOs.[241] However, during its session in 1999, there was overall agreement that financing the participation of developing member countries was a first priority, and the idea of establishing a Trust Fund for consumer interest INGOs was abandoned.[242] Another example that illustrates the Codex Alimentarius Commission's priority to ensure the participation of its Members above the participation of public interest INGOs is the debate on whether to open the Executive Committee to INGOs as observers. Objections were raised by numerous Codex Members that thought it unfair to ex-

---

[234] Report of the 19th Session of the Joint FAO/WHO Codex Alimentarius Commission, Rome, 1-10 July 1991, ALINORM 91/40, para. 57.

[235] See, for instance, C.C. van der Tweel, *supra* n. 117, pp. 38-39. D.G. Victor, *supra* n. 187, p. 201.

[236] Report of the 23rd Session of the Joint FAO/WHO Codex Alimentarius Commission, Rome, 28 June-3 July 1999, ALINORM 99/37, para. 202.

[237] Report of the 28th Session of the Codex Committee on Food Labelling, Ottawa, Canada, 5-9 May 2000, ALINORM 01/22E, paras. 78-81. Report of the 29th Session of the Codex Committee on Food Labelling, Ottawa, Canada, 1-4 May 2001, ALINORM 01/22A, para. 112.

[238] S. Suppan, *supra* n. 153, p. 18.

[239] Ibid., p. 10.

[240] N. Avery, et al., *supra* n. 187, p. 39.

[241] Report of the 22nd Session of the Joint FAO/WHO Codex Alimentarius Commission, Geneva, 23-28 June 1997, ALINORM 97/37, para. 159.

[242] Report of the 23rd Session of the Joint FAO/WHO Codex Alimentarius Commission, Rome, 28 June-3 July 1999, ALINORM 99/37, para. 43. See also, D. McCrea, *supra* n. 139, p. 153.

tend observer status to some INGOs, while the Executive Committee remained closed to most Codex Members.[243] At a later session, the idea was rejected.[244]

The Codex Alimentarius Commission puts the responsibility of ensuring that public interest groups are heard largely back to its Members.[245] It has recommended its Members to establish National Codex consultative committees (NCCCs) for this purpose.[246] These NCCCs have the task of hearing views from the interested parties in order to include them in their national position. Furthermore, it is the responsibility of Codex Members to compose their national delegations. They may choose to include a representative from a public interest INGO as a member of their delegation.

The extent to which Codex Members promote the participation of public interest INGOs varies widely.[247] For instance, not all Codex Members have established an NCCC.[248] Among the Codex Members that have established NCCCs, the extent to which these NCCCs are open to public-interest NGOs also differs.[249] Furthermore, although they are invited to participate as members of the national delegation, public interest INGOs may not always be able to finance this participation. In some cases, Codex Members provide special funds in order to contribute to their travel costs.[250]

The attribution of the responsibility for enhancing the inclusion of consumer interests to its Members is again a clear reflection of the intergovernmental character of the Codex Alimentarius Commission. However, the Codex Co-ordinating Committees play an active role in assessing consumer representative participation at national levels.[251] Consumer participation is a standing item on their agenda and Codex Members exchange experiences.[252] In 1999, the Codex Alimentarius Commission agreed to develop objectives to measure the state of consumer participation in national Codex Committee meetings.[253] During its session in 2000, the Codex

---

[243] Report of the 23rd Session of the Joint FAO/WHO Codex Alimentarius Commission, Rome, 28 June-3 July 1999, ALINORM 99/37, para. 44.

[244] Report of the 21st (Extraordinary) Session of the Codex Committee on General Principles, Paris, 8-12 November 2004, ALINORM 05/28/33, paras. 63-69.

[245] See, for instance, Report of the 20th Session of the Joint FAO/WHO Codex Alimentarius Commission, Geneva, 28 June-7 July 1993 ALINORM 93/40, para. 52.

[246] Report of the 21st Session of the Joint FAO/WHO Codex Alimentarius Commission, Rome, 3-8 July 1995, ALINORM 95/37, para. 90. Also during the 1991 FAO/WHO Conference on Food Standards, Chemicals in Food and Food Trade, this was recommended. See N. Avery, et al., *supra* n. 187, p. 38.

[247] D. McCrea, *supra* n. 139, p. 154. S. Suppan, *supra* n. 153, p. 12.

[248] D. McCrea, *supra* n. 139, p. 153.

[249] Ibid., at p. 154.

[250] Ibid., at p. 152.

[251] See, on Codex Co-ordinating Committees (or Regional Co-ordinating Committees), Section 3.4.1.3 of Chapter I.

[252] Report of the 15th Session of the Codex Committee on General Principles, Paris 10-14 April 2000, ALINORM 01/33, para. 110.

[253] Report of the 23rd Session of the Joint FAO/WHO Codex Alimentarius Commission, Rome 28 June-3 July 1999, ALINORM 99/37, para. 43. S. Suppan, *supra* n. 153, p. 12.

Committee on General Principles developed these measurable objectives which serve as a basis for assessment to be undertaken by Codex Members and discussed within the context of Codex Co-ordinating Committees.[254] However, despite the recommendation that Codex Co-ordinating Committees should report to the Codex Alimentarius Commission every two years, consumer participation is not a standing agenda item of the latter body.[255]

3.3 Transparency

When examining the legitimacy of the Codex standard-setting procedure, transparency is an important element. It constitutes an instrument for optimising participation and control.[256] It also contributes to access to the information that enables Codex Members to make informed choices, influence the decision-making process and defend their interests. Moreover, it allows Codex Members to control the process, for example, by responding to any incorrect application of procedural rules. It can further help Codex Members to develop arguments when challenging national regulations before WTO Panels and to enable the Panels to examine whether these arguments are well-founded.[257] Transparency is also an important tool for enhancing confidence.[258] In the state-oriented procedure of the Codex, which is characterised by the fact that national delegations represent their countries, transparency permits interest groups and people who have not been present during the decision-making process to be informed.[259]

During the 1990s, the Secretariat of the Codex Alimentarius Commission took important steps to increase public access to Codex documents and procedures, for example, by publishing them on the Internet.[260] However, concerns are still expressed with regard to the comprehensibility of the procedures and the documents.[261]

The concept of transparency discussed in this section is wider than the concept of openness, as transparency also includes issues such as simplicity and comprehensibility.[262] It includes both the transparency of the standard-setting procedure in general and the transparency of the specific decision-making process in hand. The

---

[254] Report of the 15th Session of the Codex Committee on General Principles, Paris 10-14 April 2000, ALINORM 01/33, para. 110.

[255] Ibid.

[256] T. Larsson, 'How open can a government be? The Swedish experience', in V. Deckmyn and I. Thomson (eds.), *Openness and Transparency in the European Union* (Maastricht, European Institute of Public Administration 1998), p. 41.

[257] See, for an analogical reasoning as regards the European Union, B. Vesterdorf, 'Transparency-not just a vogue word', 22 *Fordham International Law Journal* (1999), p. 906.

[258] A. Cosbey, *supra* n. 117, p. 18.

[259] See also, M.C.W. Pinto, 'Democratization of international relations and its implications for development and application of international law', 5 *Asian Yearbook of International Law* (1997), p. 121.

[260] D. McCrea, *supra* n. 139, p. 151.

[261] Ibid.

[262] See, for this distinction, T. Larsson, *supra* n. 256, p. 40.

first dimension of transparency refers to the comprehensibility and clarity of the standard-setting procedure itself and the procedural rules that regulate it. The second dimension aims to ensure access to information of importance to decisions that have been taken or will be taken.[263] This dimension includes aspects as the timely access to documents and reports of meetings, the completeness of reports and the use of translations to ensure comprehensibility.

3.3.1 Transparency and the standard-setting procedure

As already outlined in Section 4.2 of Chapter II and Section 3.1 of this chapter, the Codex elaboration procedure is a flexible and complex procedure which consists of two rounds in which several bodies are involved. This structure of the procedure was designed to optimise consensus-building.[264] However, the procedure is perceived as complex, partly due to the many subsidiary bodies involved, and partly due to the various rounds, and also due to the possibility of sending draft proposed standards back to previous steps.[265] Furthermore, a consensus decision-making procedure can be far from transparent. This is not only due to the private character of the negotiations that precede the formal consensus which may result in an incompleteness of the arguments in the reports of the meetings.[266] This lack of transparency also results from the fact that the actual consensus can be achieved at any time during the preceding negotiations.[267] As discussed in Section 3.1.3, in contrast to what the structure of the Codex step procedure seems to suggest (the structure of two readings), there is hardly any guidance on the status of the agreement to be achieved at a given 'step' of the procedure. The Rules of Procedure do not provide clear criteria when, for example, a (proposed) draft standard passes to the next step, or when the elaboration procedure is discontinued. Given the already complex nature of the Codex standard-setting procedure, clear criteria could play a significant role in enhancing the transparency of the standard-setting procedure and in streamlining the input as wished for by the Codex Members at a given moment of the procedure. It allows Codex Members to control the process, for example, the decisions of the Codex Alimentarius Commission to advance (proposed) draft texts to

---

[263] This definition is based on the definition of Vesterdorf, see B. Vesterdorf, *supra* n. 257, p. 902.

[264] Section 4.2 of Chapter II.

[265] The Codex Committees may even establish *ad hoc* working groups to further consensus-building, which amounts to the number of discussion forums, see Section 3.1.2.

[266] H.G. Schermers and N.M. Blokker, *supra* n. 13, pp. 533-534. M. Footer, 'The Role of Consensus in GATT/WTO Decision-making', 17 *Northwestern Journal of International Law and Business* (1996-1997), pp. 653-680.

[267] As held by Allott (Ph. Allott, 'Making the new international law: law of the sea as law of the future', 40 *International Journal* (1985), pp. 443-444) the elimination of significant opposition to a text through the consensus procedure may result from different ways. He mentions: '1. by a change of attitude on the part of a negotiating party which decides to assent to what it has theretofore opposed, 2. by the modification of the text leading to a termination of significant opposition to it, 3. by a holding-off in relation to a given text, neither giving assent nor maintaining active opposition to it.' Clearly, there is no specific moment in the decision-making process when this is achieved.

next steps (see Section 3.1.3), and it gives them a better insight in order to determine which issues will be raised at what moment of the Codex standard-setting procedure.

3.3.2 Transparency and the Procedural Manual

The opening statement of the introduction to the Procedural Manual indicates:

> 'The Procedural Manual of the Codex Alimentarius Commission is intended to help Member governments participate effectively in the work of the Joint FAO/WHO Food Standards Programme ...'.

In order to participate effectively, knowledge of the internal rules is essential. Comprehensibility of the rules enhances this. However, the Procedural Manual is not an example of clarity.[268] Internal working procedures of the Codex Committees are not always written down and may vary according to the Committee concerned.[269] Furthermore, the Procedural Manual is clearly the result of an *ad hoc* codification of different practices and agreements relating to the internal procedures and definitions that have been established over the years. Consequently, several parts have become incoherent, obsolete or incomplete. The impact, for instance, on the elaboration procedure resulting from the decision to maintain a 'horizontal approach' has not been consistently laid down in the Procedural Manual. For instance, as concluded in Section 2.2, the relations between Codex Committees regulated in 'Relations between Commodity Committees and General Committees' no longer reflect the whole structure of inter-Committee relations.

With the increased need for transparency in the elaboration procedure of Codex documents, several fundamental changes are required and it may even be submitted that a thorough analysis of the Procedural Manual with regard to whether it responds to its task of providing transparency and security for its participants is needed. Recently, a discussion on the structure and presentation of the Procedural Manual is taking place in the Codex Committee on General Principles.[270]

---

[268] See also, request of the Codex Committee on General Principles to the Secretariat to prepare a discussion paper with suggestions how to make the Procedural Manual 'more user-friendly', Report of the 20th Session of the Codex Committee on General Principles, Paris, 3-7 May 2004, ALINORM 04/27/33A, paras. 133-134.

[269] For instance, Codex Members and other interested Codex observers may submit written comments to the Codex Committee in question, which can then amend the (proposed) draft standard accordingly during its meeting at Steps 4 or 7. However, there is no deadline for submission of comments regulated in the Procedural Manual. The deadline for submission of comments is set by the Secretariat of the Codex Committee itself at an *ad hoc* basis and is announced in the report of the preceding session. See also, the proposal of the Codex Secretariat to include internal working procedures of subsidiary bodies in the Procedural Manual, which suggests the presence of these informal documents. Report of the 22nd Session of the Codex Committee on General Principles, Paris, 11-15 April 2005, ALINORM 05/28/33A, para. 98.

[270] Report of the 22nd Session of the Codex Committee on General Principles, Paris, 11-15 April 2005, ALINORM 05/28/33A, paras. 98-105. Report of the 24th Session of the Codex Committee on General Principles, Paris, 2-6 April 2007, ALINORM 07/30/33, paras. 156-165.

Another concern is with the regular amendments made to the Procedural Manual and their publication. For instance, during the 2005 session of the Codex Alimentarius Commission (which took place in July), important amendments were made to the Procedural Manual, such as the abolition of the acceptance procedure (including recommendations to amend the Statutes), and amendments were made to the Rules of Procedure in order to enlarge the membership of the Executive Committee. The entry into force occurs at a different moment. Amendments to the Statutes of the Codex Alimentarius Commission have to be approved by the World Health Assembly and the FAO Conference, and amendments to the Rules of Procedure have to be approved by the Director-Generals of both the FAO and the WHO.[271] Amendments to other documents contained in the Procedural Manual or the inclusion of new documents enter into force upon adoption by the Codex Alimentarius Commission. However, a new version of the Procedural Manual is published only once during the period between two sessions of the Codex Alimentarius Commission. This had the consequence that, until January 2006, approximately six months after the entry into force of most of the amendments made, the preceding edition of the Procedural Manual was still announced as the 'current version' on the official website of the Codex Alimentarius Commission. Although the amendments are included in the Annexes of the report of the Codex Alimentarius Commission in question, this situation is not an example of clarity.

3.3.3 Openness of the meetings of the involved bodies

The fact that the Codex Commission outsources tasks to external bodies results in work being developed by bodies which are not open to (all) Codex Members. In the past, transparency relating to the work of expert bodies has often been an issue.[272] Meetings of expert bodies are conducted behind closed doors. In Section 2.3 of this chapter, it has been argued that the responsibilities of the risk assessors and the risk managers is not always clearly distinct. This problem has been addressed by the adoption of the 'Working Principles' and the 'Working Procedures'. However, the allocation of tasks also requires a clear reporting of the scientific decisions which form the basis of the conclusions of the expert bodies and the political decisions that have led to scientific assumptions and decisions. The 'Working Principles for Risk Analysis for Application in the Framework of the Codex Alimentarius' recommends that the three components of risk analysis (risk assessment, risk management and risk communication) be fully and systematically documented in a transparent manner. Furthermore, it states that reports of the risk assessment 'should

---

[271] See Section 3.1.1.3 of Chapter I.

[272] Report of the 11th Session of the Codex Committee on General Principles, ALINORM 95/33, 1994, para. 35: 'The Committee proposed that the Commission should advise FAO and WHO of the desire of this Committee that a greater transparency should be incorporated into the working procedures of expert groups …'. Report of the evaluation of the Codex Alimentarius and other FAO and WHO food standards work, ALINORM 03/25/3, paras. 193-194.

indicate any constraints, uncertainties, assumptions and their impact on the risk assessment. Minority opinions should also be recorded'.

With regard to other international organisations which are given the task of preparing documents, the need for transparency is no different. Although their meetings do not necessarily take place behind closed doors, not all Codex Members may have access to the meetings. For instance, the International Dairy Federation is only open to representatives of the dairy industry or semi-governmental bodies. The UNECE has a restricted membership and non-European Codex Members do not have the right to participate. Besides the question of whether these organisations should actually be able to prepare documents which are based upon the political preferences of these organisations, the principle of transparency requires the clear reporting of both the recommendations and all scientific or political decisions upon which these recommendations are based.

Another body that is not open to all Codex Members is the Executive Committee.[273] When the Executive Committee still had the authority to decide upon the initiation of new work and to advance standards beyond Step 5, the issue of the restricted membership and participation was frequently discussed.[274] Observer status in the Executive Committee for both Codex Members, who are not members of the Executive Committee, and INGOs, was repeatedly requested and remained a controversial issue.[275] In 2004, as a result of the Evaluation of the Functioning of the FAO/WHO Food Standards Programme in particular, the Codex Alimentarius Commission attributed a new task to the Executive Committee. With the adoption of the amendments to the 'Procedures for the Elaboration of Codex Standards and Related Texts', the Executive Committee is no longer entitled to approve new work and adopt proposed standards at Step 5.[276] As its role is reduced to that of an advisory body, its legitimacy with regard to participation is less at stake. Consequently, the Codex Committee on General Principles has decided not to open up the meetings of the Executive Committee to all interested Codex Members and observers.[277] Instead, it has chosen to consider other methods in order to ensure transparency, such as webcasting of the meetings, or the establishment of listening rooms.[278]

---

[273] See Section 3.2 of Chapter I.

[274] The Executive Committee consists of the Chairperson and the Vice-Chairpersons of the Codex Alimentarius Commission, the Co-ordinators of Co-ordinating Codex Committees and seven other Members elected by the Commission, each from the geographical locations as defined by the Codex Alimentarius Commission.

[275] Report of the 11th Session of the Codex Committee on General Principles, ALINORM 95/33, 1994, para. 54. Report of the 16th Session of the Codex Committee on General Principles, ALINORM 01/33A, 2001, para. 103 and further. Report of the 19th Session of the Codex Committee on General Principles, ALINORM 04/27/33, 2003, para. 13 and further, and 42 and 43.

[276] Part 2 of the 'Procedures for the Elaboration of Codex Standards and Related Texts', Procedural Manual, 16th edn., 2006, pp. 21-22.

[277] Report of the 19th (Extraordinary) Session of the Codex Committee on General Principles, Paris, 17-21 November 2003, ALINORM 04/27/33, paras. 40-45. S. Suppan, *supra* n. 153, p. 17

[278] Ibid., paras. 40-45.

3.3.4 *Access to documents*

The Procedural Manual provides that Codex Committees should send all documents ((proposed) draft standards, as well as working and discussion papers) to the Codex Members and other interested Codex observers at least two months before the opening of the session.[279] However, on occasions preparatory documents have not been circulated or have been circulated too late due to time constraints.[280] The numerous Codex committee meetings during the year, and, consequently, the tight time-schedule between the sessions of Codex committees and the Codex Commission renders the distribution of materials two months before the Commission session problematical.[281] This can result in the written comments submitted by Codex Members or observers often not be circulated to other Members and observers prior to the Codex Committee meeting. Instead, they have to be distributed as a room paper leaving the other parties with little time to reflect on the issues which are being brought before to the Codex Committee. Furthermore, the late distribution of important documents makes it almost impossible for some delegations to consult with interest groups at national level.[282]

3.3.5 *Problems related to translations*

Particular problems are encountered with regard to access to translated documents. This should be considered as an important cause for concern as numerous developing member countries do not have English as their first language. There have been several complaints with regard to the late distribution of translated documents and the correctness of translations.[283] There have also been requests to provide documents in all the official languages of the FAO (Arabic, Chinese, English, French and Spanish). Subject to the availability of resources, the Codex Alimentarius Commission commits itself to provide the working papers, the reports of its sessions and the sessions of the Executive Committee, the Procedural Manual, information docu-

---

[279] Guidelines for Committees, Procedural Manual, 16th edn., 2006, p. 51.

[280] See for example, the explanations with regard to the proposed draft guidelines for evaluating acceptable methods of analysis as contained in para. 20 of the Report of the 24th Session of CCMAS, ALINORM 03/23, 2002 or Report of the 25th Session of the CCMAS, ALINORM 04/27/23, 2004, para. 8. See also D. McCrea, *supra* n. 139, p. 165.

[281] Report of the 22nd Session of the Joint FAO/WHO Codex Alimentarius Commission, Geneva, 23-28 June 1997, ALINORM 97/37, para. 191, 'The Secretariat informed the Commission that the deadline for distribution of working documents was normally two months, however some Committees had been held shortly before the Commission and the planning might need to be reconsidered in this respect …'.

[282] See also, D. McCrea, *supra* n. 139, p. 165.

[283] Report of the 22nd Session of the Joint FAO/WHO Codex Alimentarius Commission, Geneva, 23-28 June 1997, ALINORM 97/37, para. 193. As regards to the correctness of translations during the sessions, see, for instance, D. McCrea, *supra* n. 139, p. 165: 'Even providing adequate translations in Spanish, an official Codex language, was a problem at the 1998 session [of the CCHFSDU], and thereby excluded effective participation from Spanish-speaking delegations.'

ments and final Codex texts in all five languages.[284] However, the 13th edition of the Procedural Manual, which was the result of the amendments made by the Codex Alimentarius Commission in 2003, was only provided in English due to financial constraints. It remains a problem that, despite the fact that the reports of the Codex Alimentarius Commission sessions and the sessions of the Executive Committee are in all five languages, the reports of the other the sessions of the other Codex committees are still only available in the three languages.

4. Substantive legitimacy

As indicated in the introduction to this research, the acceptance of the legitimacy of international rules depends on whether these rules are in conformity with shared values and interests.[285] The shared values and interests incorporated in the objectives of the Codex Alimentarius include the protection of consumer health and the ensurance of fair practices in the food trade.[286] Consequently, the substantive legitimacy of the Codex Alimentarius will largely be measured on the basis of its aptness to address these objectives. Furthermore, the discussion on the substantive legitimacy of the Codex Alimentarius also addresses the justification of the existence of Codex measures in addition to national measures. Is normative intervention at international level necessary, or are issues better addressed at national level? As explained in Chapter II, the fundamental reason for the adoption of Codex measures as international standards is the harmonisation of national food requirements in order to facilitate international trade in food products.[287] However, since the Uruguay Round that preceded the adoption of the WTO agreements, concerns have been expressed, in particular, by interest groups and institutions of the developed countries that are Codex Members, that the threat of a WTO enforcement of Codex measures would lower national food standards.[288] Consequently, the search for harmonisation could come at the cost of having to address the protection of consumer health and the ensurance of fair practices at national level. These concerns identify an important element that may endanger the substantive legitimacy of Codex measures: the scope of harmonisation. The challenge for the Codex Alimentarius Commission is to find the right balance.

In Chapter II, it was demonstrated that the content of Codex measures, in particular, the content of the standards, is fairly detailed and does not include many

---

[284] Report of the 23rd Session of the Joint FAO/WHO Codex Alimentarius Commission, Rome, 28 June-3 July 1999, ALINORM 99/37, para. 232.

[285] J.H.H. Weiler, *The Constitution of Europe. 'Do the new clothes have an emperor?' and other essays on European integration* (Cambridge, Cambridge University Press 1999), p. 80. R. Howse, 'The legitimacy of the World Trade Organization', in J.M. Coicaud and V. Heiskanen (eds.), *The Legitimacy of international organizations* (Tokyo, United Nations University Press 2001), p. 363.

[286] Principle 1 of the General Principles of the Codex Alimentarius, Procedural Manual, 16th edn., 2006.

[287] See Principle 1 of the General Principles of the Codex Alimentarius, Procedural Manual, 15th edn., 2005.

[288] L. Rosman, *supra* n. 130, pp. 329-365. J. Braithwaite and P. Drahos, *supra* n. 2, p. 403.

options from which Codex Members can choose to address national situations and concerns.[289] It was explained that the flexibility was mainly laid down in the acceptance procedure.[290] With the abolition of the acceptance procedure, the question that the Codex Alimentarius Commission faces is whether it should insert more flexibility in the Codex standards in order to respond to the diversity of national circumstances, or whether this flexibility should mainly be part of the WTO provisions and subject to judicial decisions.

Although the topic was not explicitly addressed in the Codex Alimentarius Commission or in its subsidiary bodies in the context of the FAO/WHO evaluation, a tendency can be detected that the Codex Alimentarius Commission is increasingly using other ways to respond to the diversity of the legitimate concerns of Codex Members in a specific standard. On several occasions, a final text is adopted, while the footnotes of the reports or of the final documents themselves contain statements that one or several Codex member countries have expressed reservations *vis-à-vis* their position.[291] At Committee level, this possibility of including reservations on the content of Codex standards is laid down in the 'Guidelines for Committees'. The statements in the footnotes of the reports:

'… should not merely use a phrase such as: "The delegation of X reserved its position" but should make clear the extent of the delegation's opposition to a particular decision of the Committee, and state whether they were simply opposed to the decision or wished for a further opportunity to consider the question.'[292]

In contrast, in the sections of the reports of Codex Alimentarius Commission relating to the adoption of draft standards at Step 8, little is mentioned on the reasons for the reservations made by Codex Members. However, this technique of including reservations in the reports does not touch upon the detailed character of Codex standards and does not provide them with more flexibility. It seems simply to serve as a tool to facilitate consensus and offer an alternative way for Codex Members to express their objections.[293]

---

[289] Section 5.1 of Chapter II.

[290] Section 5 of Chapter II.

[291] See, for instance, Report of the 28th Session of the Codex Alimentarius Commission, Rome 4-9 July 2005, ALINORM 05/28/4, paras. 56 (reservation by Cuba to the adoption of MRL for carbofuran in maize and the MRL for clorpyrifos for potatoes), 58 (reservation by EC to the adoption of Interim MRLs), 59 (reservation by Tunesia and Cuba to the adoption of the Code of Practice to minimize and contain antimicrobial resistance).

[292] Guidelines for Committees, section on 'Conduct of Meetings', Procedural Manual, 16th edn., 2006, pp. 53-54.

[293] Besides, this technique may result in the false expectations to have more flexibility under the WTO agreements. Footnotes and reports may serve as a tool to defend arguments justifying deviations from international standards in the context of the WTO agreements, as they give insights of the political and scientific considerations that have been taken into account during the elaboration procedure and the objections and difficulties that Codex Members may have with the adoption of a certain Codex document. However, it will not be accepted as substitute the actual justifications themselves for such deviations.

This is different when a Codex Committee or the Codex Alimentarius Commission as a whole recognises both the particular circumstances of the Codex Member(s) in question and the fact that the international standard may not be appropriate for this Codex Member. Such observation would result in the inclusion of a provision in the Codex standard itself. Interesting, in this regard, is the adoption of the Revised Standard for Olive Oils and Olive Pomace Oils at the 26th Session of the Codex Alimentarius Commission,[294] in which it agreed to adopt the revised standard on the condition that a footnote was included that stated that 'pending the results of the International Olive Oil Council survey and further consideration by the Committee on Fats and Oils, *national limits may remain in place*'.[295] The inclusion of this footnote was a response to the fact that more information on the natural and geographical variations was expected, which would assist in determining an appropriate level of linolenic acid in oil for the definition of certain olive oils (those from New Zealand in particular) as high quality olive oils.

This solution is of particular interest as it may indicate the commencement of the explicit inclusion of 'safeguard clauses' in the Codex documents, something that, to date, has hardly been used. Safeguard clauses, or explicit options to deviate from the Codex standard, could be a useful tool to respond to the concern of substantive legitimacy. In this way, the Codex Alimentarius Commission will take the initiative in order to define the flexibility to allow Codex Members to address values and interests at national level. How 'safeguard clauses' inserted in Codex standards would be interpreted by WTO panels would probably depend on the formulation of the 'safeguard clauses'. The question in this context is whether the 'safeguard clause' will be interpreted in the light of the obligation to base the national technical measure on the Codex measure, or whether it will serve as a reference point for determining whether the deviation from the Codex measure is justified. If 'safeguard clauses' define additional options that can be used by Codex Members to ensure their higher level of protection, such as the safeguard clause inserted in the Codex standard on butter,[296] the implementation of this option would – most probably – not have to be justified. If, however, 'safeguard clauses' are formulated in more general terms, such as the explicit recognition that Codex Members may adopt stricter measures in specific circumstances, then they are likely to be taken into account in the examination of the justifications provided by the Codex Member in question.

Furthermore, the language used and the commitment expressed in different Codex measures becomes increasingly important. For instance, contrary to provisions on hygienic practices contained in commodity standards, recommended codes of practice which often contain general provisions on hygienic practices are only ad-

---

[294] Report of the 26th Session of the Codex Alimentarius Commission, Rome, July 2003, ALINORM 03/41, para. 83.

[295] Ibid.

[296] Codex Standard for Butter, Codex Stan A-1-1971, Rev. 1-1999, 'Butter may be labelled to indicate whether it is salted or unsalted according to national legislation.' See also, section 5.2 of Chapter II.

visory of nature.[297] This is reflected by the use of the words 'should' instead of 'shall'. Likewise, specific provisions on methods of analysis and sampling contained in Codex Commodity standards are of an obligatory nature, whereas the general provisions on methods of analysis and sampling as laid down in Guidelines are formulated in voluntary terms. As most of the recommended codes of practice and guidelines serve as instrument to assist Codex Members in implementing and enforcing Codex standards at national level, it would be logical that these measures should continue to reflect their voluntary nature. It is the responsibility of the Codex Alimentarius Commission to ensure that the precision and clarity of their measures augments, particularly now that WTO Panels interpret these measures.[298] This may be part of the new task of examining proposed standards on their consistency with existing Codex texts, format and presentation, and linguistic consistency that is assigned to the Executive Committee.

5. Conclusions

Many concerns relating to the legitimacy of the Codex Alimentarius Commission have been expressed during the last years. During the Joint FAO/WHO Evaluation of the Codex Alimentarius Commission, several of these concerns were addressed and the Codex Alimentarius Commission has adjusted, or is in the process of adjusting, its internal rules accordingly. However, the adjustments do not always fully respond to the concerns of legitimacy. The following can be concluded.

When it comes to institutional legitimacy, there are several issues relating to the attribution of powers to the Codex Alimentarius Commission and the delegation of its powers to other bodies that require special attention. Neither the Codex Alimentarius Commission, nor its parent organisations have explicit competence to adopt measures that become binding without the consecutive consent of their members. This lack of formal competence within the 'constitutions' of the Codex Alimentarius Commission and its parent organisations has not raised any concerns so far. However, the indirect attribution of powers through the reference contained in the WTO agreements to international standard-setting bodies may, in the long-term, be too weak a basis to legitimise firmly this new normative power of the Codex Alimentarius Commission.

The delegation of tasks to Codex Committees, which has allowed the Codex Alimentarius Commission to build an institutional framework without depending on its parent organisations, tends, however, to a decentralisation of the standard-

---

[297] Report of the 5th Session of the Codex Alimentarius Commission, Rome, 20 February-1 March 1968, ALINORM 68/35, paras. 47 and 49: 'The Commission agreed that codes of hygienic practice were advisory. Parts of these codes, especially those dealing with end product specifications could, however, be included in Codex standards and could then become mandatory. The Commission thought that codes of practice should not be published in the standards section of the Codex Alimentarius. The Codex Alimentarius should contain appropriate cross-references to these codes.'

[298] See Section 5.1 of Chapter IV.

setting procedure. As Codex Committees are financial independent and each possesses a high level of expertise in a particular field, some subsidiary committees have acquired a strong position within the standard-setting procedure. This, in itself, is not a problem, provided that the delegation of powers to the Codex Committee is carefully supervised by the Codex Alimentarius Commission and that their work is well co-ordinated. Two concerns, in particular, have drawn attention: the power of Codex Committees to propose new work, and the definition of inter-Committee relationships. In the past, the Codex Alimentarius Commission or the Executive Committee was not assured in a consistent manner that the initiatives of new work undertaken by Codex Committees corresponded with the overall 'Criteria for Establishing Work Priorities'. Likewise, the regulation of inter-Committee relationships is incomplete and co-ordination efforts by the Codex Alimentarius Commission can hardly be considered as being systematic. Several amendments resulting from the Joint FAO/WHO Evaluation report have, in part, responded to these concerns. The Executive Committee in its new capacity has the responsibility of advising the Codex Alimentarius Commission on the proposals for new work and on the co-ordination of any work that falls within the mandate of several committees. However, the capacity of the Executive Committee to test new work initiatives against the Criteria for Establishing Work Priorities does not seem to cover proposals for new work of individual MRLs. This would, as a result, leave an important component of Codex standards outside of the scope of this supervision. Furthermore, the amendments to the Procedural Manual do not include a clear provision that allows the Executive Committee to advise the Codex Alimentarius Commission to undertake corrective actions to augment co-ordination between the Codex Committees.

As risk managers, Codex Committees do not always clearly define the mandate, or risk assessment policy, that accompanies the request of consultations to be executed by expert bodies. In the past, the lack of a clear mandate made expert bodies take political decisions and resulted in the blurring of scientific and political considerations. The 'Working Principles for Risk Analysis for Application in the Framework of the Codex Alimentarius' adopted in 2003 respond to these concerns, as, for the first time, they define the separation of responsibilities between the risk assessors and the risk managers, and specify that the definition of risk assessment policy is a component of risk management. However, in practice, this separation of responsibilities has not always been applied, and a more active and more direct interaction between the Codex Committees and the expert bodies has been suggested. The independence of the consulted scientific experts was also not adequately ensured, as has been demonstrated by the infiltration of the tobacco industry in the proceedings of the JMPR. As a result of the Report of the Expert Committee evaluating the Tobacco Industry documents, internal rules to ensure the independence of experts have been adopted, such as the requirement to sign a declaration of interest form, and rules which regulate the interaction of the temporary advisor and the industry as the sponsor of scientific data. However, doubts have been expressed with regard to the adequate application of these rules. Furthermore, concerns have been raised that there is no indication as to what the FAO and the WHO do with the declared interests.

In the past, the allocation of certain tasks to external international organisations served as a tool to promote co-operation with these organisations. However, delegation of tasks to other international organisations that are not bound by the internal rules of the Codex Alimentarius Commission complicates the supervision of the executed tasks. Furthermore, the international organisations (both governmental and non-governmental) have a different membership, different working procedures and different priorities. The 'Guidelines on the Co-operation with Other Intergovernmental Organisations' adopted in 2005 do not address all these concerns, as they do not regulate the co-operation of the Codex Alimentarius Commission with non-governmental organisation. In addition, it is not clear how the Guidelines relate to the role of some international organisations, such as the UNECE and the IDF, which have been explicitly laid down in the Procedural Manual.

The fact that the binding force of some elements of Codex measures results from their mere adoption and no longer requires the explicit consent of governments has led to a 'politicisation' of the standard-setting procedure, and has resulted in the need to reconsider the legitimacy of the Codex standard-setting procedure. Codex Members have become keener on ensuring that their interests are taken into account during the procedure. The establishment of consensus as a means of final decision-making had become harder to achieve and, on several occasions, decisions were taken by vote. This renewed the discussion of consensus, which resulted in the adoption of an amendment of the Codex Rules of Procedures, and emphasised the importance of adopting measures by consensus. However, the rules on the final adoption of Codex measures are not clear and are even partly in contrast with the consensual approach. Voting is still possible if efforts to achieve consensus have failed. However, there is still no definition of consensus. Furthermore, the decisions to proceed with a vote can be taken by one Codex Member alone and are taken on the basis of a simple majority, and may even occur by secret ballot. With regard to consensus-building during the standard-setting procedure, the flexible structure of the step-procedure is an important instrument in the hands of the Codex Alimentarius Commission in order to ensure that the consideration of all interests and the progress of standard-setting go hand in hand. However, the internal rules do not specify the progress of consensus-building that needs to be achieved in order to advance the (proposed) draft measures beyond the first (Step 5) or the second reading (Step 8), and the Codex Alimentarius Commission practice of sending (proposed) draft standards back to a previous step in the procedure is inconsistent. With the amendments of the Rules of Procedure, the Executive Committee has been assigned the task of monitoring the progress of standards development and of advising the Codex Alimentarius Commission accordingly. Nevertheless, the new rules still do not specify on the basis of which criteria (proposed) draft standards are to be sent back to previous steps.

With regard to the principle of the equal and impartial participation of all stakeholders, attention has been drawn to a dominant influence of developed member countries and the industrial-interest INGOs on the standard-setting procedure *vis-à-vis* developing member countries and public-interests INGOs. Developing member countries often do not have the financial, technical and human resources to participate actively in the standard-setting procedure. Furthermore, the assignment of tasks during the procedure, such as the preparation of a proposed draft measure and the hosting and chairing of Codex Committee meetings, is more frequently undertaken by developed member countries. Likewise, the complex institutional and procedural framework complicates active participation even further. Important projects have been launched, such as the co-chairing and hosting of Codex Committees by developing member countries and the establishment of the Trust Fund to promote the participation of developing member countries. These efforts, without doubt, constitute an important step in enhancing the participation of developing member countries. However, in order to ensure their active participation on the long-term, these efforts need to be accompanied by other activities which ensure a more transparent and impartial standard-setting procedure. In addition, public-interests INGOs are in a less favourable position than the industrial-interest INGO, as the latter group of INGOs often have a higher level of expertise and better access to more financial resources. Although the Codex Alimentarius Commission has emphasised the importance of promoting more active participation by public-interests INGOs, it places the responsibility for ensuring that their interests are taken into account with the Members. In clearly defining the priority of ensuring the participation of all its Members and leaving the participation of INGOs to the responsibility of the Members, the Codex Alimentarius Commission confirms that it is, first and foremost, an intergovernmental institution. However, this intergovernmental character is not always ensured. In some cases, observers from INGOs have similar rights as the Codex Members, which can result in a more influential position of particularly industrial-interest INGOs *vis-à-vis* some developing member countries.

The aim of the Procedural Manual is to enhance effective participation. Thus, the Procedural Manual is, indeed, an important instrument to the promotion of the transparency of the complex institutional and procedural structure of the standard-setting procedure. However, the Procedural Manual is not an example of clarity. The Procedural Manual is the result of an *ad hoc* codification over time, which is reflected in the fact that some parts have become incoherent, obsolete or incomplete. In order to ensure that the Procedural Manual may be used as an instrument to enhance the transparency of the standard-setting procedure, it will have to be thoroughly revised and restructured. Furthermore, the delegation of tasks to 'outside bodies', such as expert bodies and other international organisations, to which not all Codex Members will have access, requires comprehensive reporting in order to ensure the transparency of the decisions taken as well as the interests taken into consideration. Other concerns of transparency relate to the fact that preparatory documents are not always circulated in sufficient time to allow for adequate preparation of the meetings, and relate to the lack, or incorrect translation of documents and reports that sometimes occurs.

As already mentioned in Chapter IV, Codex Members can no longer deviate without constraint from the, sometimes, strict and detailed provisions of the Codex measures. This means that the content and the language used in the text of the Codex measures have become more important. It also raises the question as to whether Codex standards rightly address certain concerns or whether these concerns are more appropriately addressed at national level. Although no general discussion has taken place in the Codex Alimentarius Commission on this issue, some actions of the Codex Alimentarius Commission reveal that it is looking for ways to respond to this concern. Footnotes to the reports of meetings in which Codex measures have been adopted contain the reservations expressed by Codex Members and observers to certain provisions of the Codex measures. Moreover, the explicit inclusion of 'safeguard clauses' within the Codex measures themselves can also be found.

CONCLUSIONS

1. General conclusions

The increased recognition of the need for international co-operation in the field of food standards and for the co-ordination of already ongoing initiatives resulted in the launching of the Joint FAO/WHO Food Standards Programme in 1962. The operation of this programme is mainly executed by the Codex Alimentarius Commission, a subsidiary body of both the Food and Agriculture Organisation (FAO) and the World Health Organisation (WHO). In addition to the Codex Alimentarius Commission, the FAO and the WHO established other bodies to operate within the scope of the Joint FAO/WHO Food Standards Programme: the Joint FAO/WHO expert bodies, responsible for the execution of risk assessments, and the Joint FAO/WHO Consultative Group for the Trust Fund, responsible for the management of the Trust Fund set up to assist the participation of developing member countries. The latter bodies fall under the direct responsibility of the FAO and the WHO, and function independently from the Codex Alimentarius Commission. The authority of the Codex Alimentarius Commission is restricted due to its subsidiary position: for financial resources and for the management of its Secretariat, it depends on its parent organisations. Yet, in spite of its subsidiary position, the Codex Alimentarius Commission has been able to establish an important institutional framework, as a result of some important delegated powers, such as the establishment of subsidiary bodies and the co-ordination of international food standard activities. This framework consists of subsidiary bodies that are financially supported by Codex Members, and which, as a result, are financially independent of the FAO and the WHO. Furthermore, the Codex Alimentarius Commission delegates some tasks relating to the preparation of draft standards, such as the preparation of draft proposed standards, to other international organisations. This institutional framework has enabled the Codex Alimentarius Commission to give content, *ratione materiae*, to its mandate.

The core element of the mandate of the Codex Alimentarius Commission is the creation of the Codex Alimentarius: a collection of uniformly-defined food standards which aim to provide assistance to the harmonisation of food requirements. In its standard-setting activities, the Codex Alimentarius Commission operates in an autonomous way, and adopted standards do not need approval from its parent organisations. Over the years, the Codex Alimentarius has become a complex system of food requirements, consisting of approximately 216 standards, 47 recommended codes of practice, 40 guidelines, maximum residue levels (MRLs) of pesticide residues for over 360 commodities and MRLs for 47 different veterinary drugs, and 7 other measures. It consists of measures which aim to protect human health and fair practices in the food trade. These measures are often detailed in nature, and cover the regulation of a large range of issues, including scientific and political considerations. Furthermore, Codex standards are formulated in a way that reflects the intention of their being mandatory in nature. 'Optional clauses' and 'safeguard clauses' that provide Codex Members with some discretion are not common practice. Reaching agreement on strict and detailed provisions that need to respond to a large variety of national circumstances is not an easy task. The Codex Alimentarius Commission has applied several instruments to facilitate agreement. It has adopted a standard-setting procedure that is composed of two readings, both at a Codex Committee level as well as at a Codex Alimentarius Commission level. The standard-setting procedure functions in a flexible way: a standard or other related text in the process of elaboration can be sent back to previous stages of the procedure if the desired level of agreement has not yet been reached. Furthermore, the presence of a flexible and voluntary acceptance procedure has greatly facilitated agreement on detailed food requirements. Although the acceptance procedure – which had never really worked – was abolished in 2005, the consequences of its existence are still in place. In order to become binding upon Codex Members, standards had to be accepted through their explicit consent. Furthermore, Codex Members were permitted to accept standards with specific deviations, which, in a way, meant that Codex Members could determine their own optional clauses and safeguard clauses.

Despite the lack of binding force of Codex standards within the context of the Joint FAO/WHO Food Standards Programme, the EC has attached great importance to them. During the first phase of the existence of the EC up to around 1987, in particular, reference to Codex measures and their scientific basis can be frequently found in both the legislative process of EC food regulation and in the jurisprudence of the European Court of Justice on Articles 28 and 30 EC, thus forcing member states to take account of international standards or their scientific basis when preparing and adopting their national rules. The frequent use of Codex measures during this period is most probably linked to the ability of the Codex Alimentarius Commission to adopt food measures more quickly than the EC institutions during this initial phase. This explains why one can observe a diminishing influence of Codex measures on the content of both secondary legislation and decisions of the European Court of Justice from the moment that new institutional and procedural settings – which allowed secondary food regulation to be adopted more easily and faster by the EC institutions – were put into place in 1987. If it were not for the adoption of the WTO agreements in 1994, this tendency would most likely have persisted. However, the adoption of the WTO agreements and the resulting obligations for the EC institutions as Members of the WTO meant that the Community institutions could no longer neglect the importance of the Codex measures. It was, in particular, the *EC-Hormones* decisions of the WTO Panel and the Appellate Body that made the Community institutions realise that the status of Codex measures had definitively changed. Regulation 178/2002 on the general principles and requirements of food law and the establishment of the European Food Safety Authority, better known as the 'General Food Law' thus adopted in 2002, in Article 5(3), a general obligation to take the Codex standards, directed at both member states as well as the Community institutions, into consideration when adopting and preparing food law. This means that Codex measures have 'again' acquired the status of a 'reference point' in the justification of measures. Recently adopted EC secondary food legislation demonstrates that the Community institutions have frequently taken Codex measures or their scientific basis into consideration, which has led to adjustments of several EC measures or to an explicit explanation of the reasons for a deviation from Codex measures. However, the question of whether Codex measures as a reference point will be a decisive element in the Court's examination of the justifications of food measures remains to be seen. With these changes over time, the position of the EC Commission *vis-à-vis* the Codex Alimentarius Commission has also changed considerably from a mere promoter of the acceptance of Codex standards by the EC member states to an active actor in the Codex standard-setting procedure, which, being the 'defender' of the interests of the Community, tries to set the Codex standard in accordance with the Community position. This position was re-inforced in 2003 when the EC became a full Member of the Codex Alimentarius Commission.

Without doubt, the increased attention by the EC to Codex standards can largely be attributed to the fact that, within the context of the WTO agreements, Codex measures have acquired an increased status in comparison to that which they had under the former Codex acceptance procedure.[1] Through the obligation to use international standards as a basis for national measures as laid down in the *SPS Agreement* and the *TBT Agreement*, and through the explicit recognition of Codex standards as international standards (Annex A.3(a) of the *SPS Agreement* and the Panel decision in *EC-Sardines* with regard to the *TBT Agreement*), it is submitted that some elements of the standards have obtained a *de facto* binding force. However, it has to be noted that this binding force is not direct. In order words, the binding force does not result from the Codex measures themselves, but only applies through the binding provisions of the *SPS Agreement* and the *TBT Agreement*. Consequently, outside of the WTO agreements, Codex measures remain voluntary; for example, Codex measures remain voluntary for those Codex Members that have not adhered to the WTO agreements.

Furthermore, the binding force does not concern Codex standards as a whole, but is restricted to some elements of the standards. This can be deduced from the application by panels and the Appellate Body of the relevant SPS and TBT provisions, which comes down to a *de facto* binding force of the Codex standards to a certain extent. Reference can also be made to the clear distinction made the Appellate Body in *EC-Hormones* between the requirement to conform national measures with international standards and the obligation to base national measures on international standards. It stated that the obligation to base measures on international standards means that only some, and not all, elements of the standards (in contrast

---

[1] See Sections 3 and 4 of Chapter IV.

to a requirement to conform the measures to the international standards) need to be incorporated in the national measure. Unfortunately, the Appellate Body did not define the elements which need to be incorporated. However, closer examination of the *EC-Sardines* decisions reveals the possibility that this restriction in scope may very probably be linked with the restricted scope of the objective of the WTO *vis-à-vis* that of the Codex Alimentarius Commission. This would mean that only the elements of Codex standards that aim to further trade liberalisation have become binding, and that the elements that are directed to protect human health and fair practices in the food trade remain of a voluntary nature. A good example to illustrate this is the provision of the Codex Stan 94, which regulates that sardine-types other than Sardina pilchardus, 'shall be named "X Sardines" of a country, a geographic area, the species, or the common name of the species in accordance with the law and custom of the country in which the product is sold'. The application of this Codex provision by the Panel and the Appellate Body in *EC-Sardines* illustrates that a national measure needs to allow the circulation of food products that are named under one of the enumerated designations. However, the national measure does not have to require that all sardine-type products are named in accordance with the Codex provision in order to circulate. This element will not be enforced by WTO panels.

The obligation to base national measures on Codex standards is not absolute. Under both the *SPS Agreement* and the *TBT Agreement*, WTO Members have an autonomous right to deviate from Codex standards. However, this right is restricted to the cases in which the Codex standards are not sufficient or are not appropriate to achieve the legitimate objectives pursued. Furthermore, the notification procedure under both agreements requires WTO Members to notify deviations from international standards and to justify these deviations upon request to other Members. Deviations must be justifiable, and, in the examination of the justifications, both the panels and the Appellate Body use Codex standards as reference point. As a result of the automatic presumption (also referred to as the presumption of consistency) that Codex standards are necessary trade-restrictive measures under both agreements, their position as a reference point has a central place in the examination. Consequently, deviations from Codex standards require a solid justification, which only confirms the *de facto* binding force of the relevant elements.

This *de facto* binding nature of some of the elements of Codex standards under the *SPS Agreement* and the *TBT Agreement* is a considerable change compared to the status that was accorded to them under the Codex acceptance procedure. Under the Codex acceptance procedure, binding force would only result from the explicit acceptance by countries and was only applicable to standards and MRLs. In contrast, under the *SPS Agreement* and the *TBT Agreement*, the application of the obligation to use Codex measures as a basis results from the final adoption of the measures and concerns all measures, including recommended codes of practice and guidelines. This also means that the Codex acceptance procedure was considered by the Panel in *EC-Sardines* not to be relevant under the WTO Agreement.

The reference to international standards under the *TBT Agreement* and the *SPS*

Agreement and the explicit recognition of Codex standards as such also has institutional consequences. As the WTO is not a standardising body itself and relies on 'outside' international standard-setting bodies, such as the Codex Alimentarius Commission, for the adoption of harmonisation measures, the relationship between the WTO and the Codex Alimentarius Commission can be characterised by a separation of powers: the Codex Alimentarius Commission being responsible for the 'legislative' acts on the one hand, and the WTO being responsible for the promotion and enforcement of the application of harmonisation measures on the other. This separation of powers between the WTO and the Codex Alimentarius Commission was strengthened by the fact that the Codex Alimentarius Commission abolished the acceptance procedure and has consequently done away with the mechanism through which it promoted the application of its measures. Although, as mentioned above, the acceptance procedure has become irrelevant within the context of the WTO agreements, it could still have been used as a tool to promote the application of Codex measures, especially in their capacity as minimum platform measures to ensure a minimum level of protection.

Two WTO bodies are of particular interest in the relationship between the WTO and the Codex Alimentarius Commission, and their actions may easily have consequences for the operation of the Codex Alimentarius Commission:

1. WTO panels and the Appellate Body as adjudicating bodies responsible for settling conflicts on the use of Codex measures; and
2. The SPS Committee as 'regular forum for consultation' on the process of harmonisation and the use of international standards.

As the adjudicating body responsible for settling conflicts on the use of Codex measures, the panels and the Appellate Body have to interpret and apply these measures. However, the rules regulating the WTO dispute settlement mechanism, laid down in the Dispute Settlement Understanding (DSU), and the current practice of the Appellate Body reveal several shortcomings to this end. No judicial review on whether Codex measures have been adopted in accordance with the internal rules of the Codex Alimentarius Commission has been undertaken. Furthermore, the application of the Vienna rules of interpretation when interpreting Codex measures has not been ascertained and neither has the competence to consult 'outside sources'. In cases where panels apply neither the Vienna rules of interpretation nor consult with the Codex Alimentarius Commission or its Secretariat, their interpretation of Codex measures may run counter to the initial meaning, purpose and function of the measures.

In addition to the dispute settlement mechanism, another WTO body promotes the application of Codex standards and is thus involved in the relationship between the WTO and the Codex Alimentarius Commission: the SPS Committee as 'regular forum for consultation' on the process of harmonisation and the use of international standards. For the moment, the relationship with the Codex Alimentarius Commission, which partially consists of an overlap of mandate, is no reason for concern.

The *SPS Agreement* provides for inter-institutional co-operation and there is no hierarchal structure between the SPS Committee and the Codex Alimentarius Commission.

The reference to Codex measures as a result of the adoption of the WTO agreements has had many consequences for the Codex Alimentarius Commission. With the increased status of Codex measures, the authority of the Codex Alimentarius Commission has also increased and has led many international organisations and scholars to question the legitimacy of its procedures. Several concerns have been addressed during the Joint FAO/WHO Evaluation of the Codex Alimentarius Commission, which has led to important adjustments of the Procedural Manual, which include a new task for the Executive Committee to monitor the standard-setting progress. However, some concerns of legitimacy still remain. In the context of this research, these concerns have been distinguished in three categories for reasons of clarity:

1. Institutional legitimacy relating to the attribution and delegation of powers within the framework of the Codex standard-setting procedure;
2. Procedural legitimacy related to the organisation and regulation of the preparation and adoption of Codex measures and whether equal participation rights are ensured; and
3. Substantive legitimacy related to the content of Codex measures themselves, in particular the harmonisation techniques incorporated in the Codex measures.

a. **Institutional legitimacy: the Codex Alimentarius Commission and the attribution of powers**

With regard to institutional legitimacy, the following can be observed. The mandate of the Codex Alimentarius Commission does not foresee a formal power to adopt binding provisions, which may raise concerns of legitimacy in the long run. Furthermore, the delegation of tasks to Codex Committees, which are financially independent and possess a high level of expertise, tends towards a de-centralisation of the standard-setting procedure. This means that, in order to ensure a coherent functioning of its subsidiary bodies, the Codex Alimentarius Commission needs to supervise and co-ordinate their activities carefully. Important adjustments to the Procedural Manual have been made to allow for better supervision, in particular, with regard to new work initiatives of Codex Committees. It is, however, unfortunate that the new task of the Codex Executive Committee does not clearly include an important component of new work initiatives, namely, new work proposals for individual MRLs for pesticides, veterinary drugs, food additives and MLs (maximum levels) for contaminants. When it comes to the inclusion of instruments that enable the Codex Alimentarius Commission to co-ordinate the activities of its subsidiary bodies, few structural adjustments have been proposed. With regard to the operation of the Joint FAO/WHO expert bodies, some important regulatory rules have been introduced by the FAO and the WHO. These rules are, indeed, an important improvement to ensure the independence of experts and to ensure a relationship between the scientific expert bodies and the Codex Alimentarius Commission that is based upon the separation between risk assessment and risk management. However, doubts have been expressed with regard to the adequate application of these rules by the expert bodies, the Codex Committees and both the FAO and the WHO as the parent organisations of the expert bodies. In addition, the delegation of tasks to other international organisations is a serious threat to the legitimacy of the standard-setting procedure, as these organisations often have very different priorities, membership and working procedures. The 'Guidelines on the co-operation with other intergovernmental organisations' adopted in 2005 hardly contribute to an improvement of this concern, as they fail to regulate the co-operation with non-governmental organisations and seem to be in contrast with earlier established provisions laid down in the Procedural Manual which explicitly attribute tasks to some of these organisations.

b. **Procedural legitimacy: the Codex standard-setting procedure**

The increased status of Codex standards has also raised several concerns with regard to the procedural legitimacy of the standard-setting procedure. These concerns are:

- The lack of transparency of the standard-setting procedure which largely results from unclear and sometimes incoherent internal rules laid down in the Procedural Manual;
- A dominant position of developed member countries and industrial-interests INGOs in the standard-setting procedure; and
- Decision-making based upon a consensual approach, which is generally perceived as the legitimate way of decision-making by Codex Members, is not adequately ensured by the internal rules laid down in the Procedural Manual.

Although the Codex Secretariat has greatly improved the access to documents and the reports of the meetings, one fundamental concern with regard to the transparency of the Codex standard-setting procedure remains. As the standard-setting procedure is complex, the function of the Procedural Manual of the Codex Alimentarius Commission to ensure transparency acquires increasing importance. However, the Procedural Manual is far from being an example of clarity. As it is the result of an *ad hoc* codification over time, some parts have become incoherent, obsolete or incomplete. Other concerns regarding transparency relate to the fact that preparatory documents are not always circulated in due time in order to allow for adequate preparation of the meetings and to the fact that there is sometimes a lack or incorrectness of the translation of documents and reports.

Attention has often been drawn to a dominant influence of the developed member countries and the industrial-interest INGOs on the standard-setting procedure *vis-à-vis* the developing member countries and the public-interest INGOs. This concern has clearly been addressed by the Codex Alimentarius Commission, whose first priority is the promotion of the active participation of the developing member countries. Its initiatives, such as the establishment of the Trust Fund for the participation of both the developing countries and the countries in transition in the work of the Codex Alimentarius Commission, as well as the co-chairing and hosting of Codex meetings in the developing member countries, undoubtedly constitute an important step in enhancing the participation of developing member countries. However, in order to ensure their active participation in the long-term, these efforts need to be accompanied by other activities which ensure a more transparent and impartial standard-setting procedure. The attribution of tasks to the developing member countries is still underdeveloped if one compares it to the work undertaken by the developed member countries. In this situation, the delegation of tasks to industrial-interest INGOs seems to be out of place. Specific mention can be made to the explicit attribution of tasks which is still laid down in the Procedural Manual, to the International Dairy Federation, as a result of the Federation's important contribution to international standard-setting prior to the establishment of the Codex Alimentarius Commission. With regard to the position of public interest INGOs, the Codex Alimentarius Commission leaves most of the responsibility to improve participation to the Codex Members.

The fact that the *de facto* bindingness, which follows from the reference to Codex standards in the WTO agreements, results from the mere adoption of Codex measures and no longer requires the explicit consent of governments has led to a 'politicisation' of the standard-setting procedure. Today, consensus has become harder to achieve. In order to respond to the many concerns raised by Codex Members with regard to the final adoption of standards that have occurred by vote, the Codex Alimentarius Commission introduced an amendment of the Codex Rules of Procedures in 1999, which stated that the Codex Alimentarius Commission 'shall make every effort' to take decisions to adopt or amend standards by consensus. However, it has been argued in this book that important shortcomings with regard to the clarity of the internal rules relating to final decision-making still remain. For instance, there is no definition of consensus, and final adoption of Codex measures can still occur by vote which can, in principle, be triggered by any Codex Member. In situations where voting does occur, this is done by a simple-majority, which seems to be in sharp contrast with the consensual approach. Furthermore, a consistent application of the flexible structure of the step-procedure, which is an important instrument in the hands of the Codex Alimentarius Commission in order to ensure that the consideration of all interests and the progress of standard-setting go hand in hand, is not ensured. This is caused by the lack of clear internal rules, which leads to a failure to distinguish the progress of consensus-building to be achieved during the first reading from the progress to be achieved during the second reading of the Codex standard-setting procedure.

c. **Substantive legitimacy: Codex measures as instruments for harmonisation**

When it comes to the question of substantive legitimacy, which, in the context of this research, concentrates on the ways and techniques incorporated in Codex measures to advance the harmonisation of food requirements, the following can be concluded. In the past, the Codex acceptance procedure was an important instrument to allow the acceptability of rather strict harmonisation techniques in the Codex standards, as the acceptance procedure allowed Codex Members to deviate or withhold from the application of Codex standards. However, the irrelevance of the acceptance procedure under the WTO agreements means that the majority of options to deviate from, or to withhold the application of the adopted Codex standards have disappeared. The Codex standards themselves consists of many strict and detailed provisions and do not leave Codex Members with a high degree of discretion to respond to domestic concerns. This raises the question as to whether Codex standards rightly address certain concerns or whether these concerns are more appropriately addressed at national level. The incorporation of different harmonisation techniques which would allow Codex Members to have more discretionary powers has not been considered by the Codex Alimentarius Commission or its parent organisation on a more structural basis than that of the negotiations which preceded the adoption of some individual standards.

2. MAIN CONCERNS, RECOMMENDATIONS AND SUGGESTIONS

The above makes it clear that the increased status of Codex measures within the international and European context has led to some important consequences. These consequences are related to the issue of the fragmentation of international law on the one hand and to the fact that legitimacy of the Codex Alimentarius Commission, its standard-setting procedure and its measures have not yet been sufficiently ensured on the other.

2.1 **Consequences resulting from the fragmentation of law**

The fact that harmonisation of food measures is regulated in the different *fora* of the WTO, the EC and the Codex Alimentarius Commission has had some important consequences, some of which can be solved relatively easily, others which require more thorough consideration.

a. *Lack of legal certainty because of the lack of recognition of increased status of Codex measures by WTO bodies*

Probably one of the most important concerns relating to the fragmentation of law between the WTO and the Codex Alimentarius Commission is the lack of recognition of the increased status by WTO bodies themselves, such as the Appellate Body and the SPS Committee. While rejecting a binding character of the Codex standards, the Appellate Body has not defined their status under the *SPS Agreement* and the *TBT Agreement*. On the other hand, we have observed that the application of the relevant provisions has been strict, and justification for the necessity of the measures at dispute has been measured against the Codex standard and its scientific basis at hand. In this, we concluded that, in practice, this leads to a *de facto* bindingness of some elements of Codex standards. However, the fact that the WTO bodies do not explicitly recognise this status remains problematical, in that it causes legal uncertainty. Furthermore, the fact that panels and the Appellate Body do not apply the Vienna rules of interpretation to Codex standards, and that, as a consequence, there is always the risk that their interpretation of Codex standards may run counter to the initial meaning of these standards only adds to the legal uncertainty. Finding a way out of this opposing approach will not be an easy task. The best solution would be the adoption of an authoritative interpretation of the relevant WTO provisions by the Ministerial Conference. However, this is not the quickest option, as such an adoption would take time, due to the fact that the Ministerial Conference does not meet frequently and due to the delicateness of the matter. There are other ways, which do not have the same legal authority, but which may contribute to finding ways of diminishing this legal uncertainty: for example, by raising these issues in the Dispute Settlement Body when relevant, and turning them into a point of enduring discussion within the SPS Committee and the TBT Committee.

b. *The abolition of the acceptance procedure*

Another issue that is related to the fragmentation of the secondary rules between the WTO and the Codex Alimentarius Commission is the abolition of the Codex acceptance procedure. The mandate of the WTO is restricted to the enforcement of the application of Codex standards to ensure trade liberalisation; or, in other words, the application of Codex standards as trade standards. For this reason, the WTO dispute settlement mechanism will not examine whether national measures are based on Codex standards in order to ensure a minimum level of protection. This means that Codex standards only function as a ceiling, not as a floor. As a consequence, the abolition of the acceptance procedure by the Codex Alimentarius Commission in 2005, means that there is no longer a mechanism that promotes the application of Codex measures as minimum platform standards, in order to ensure a minimum level of protection. Corrective action to adjust the fragmentation of secondary rules has reduced the promotion of an important dimension of the function of the Codex measures. The fact that their status as trade standards is a stronger status than their status as minimum platform standards is, in itself, not a problem, as long as the quality of the content of these provisions is ensured, and that it will not result in a 'race to the bottom'. Furthermore, the use of Codex measures as minimum platform standards can be promoted through other ways, based upon the recognition that these provisions reflect a high level of expertise and quality. An important way, which has already been undertaken by the FAO and the WHO, is through capacity-building activities, such as regional and national projects undertaken in Africa, Asia, Europe, Latin America, the Near East and the South West Pacific, in order to assist numerous countries to strengthen national food control systems.

c. *EC and the 'implementation' of the WTO obligations*

The fragmentation of law, which mainly consists in a fragmentation of primary rules, has, in the past, led to a conflict of rules, between the EC on the one hand, and the WTO and its reference to Codex measures on the other, has been largely resolved. The conflict of rules was clearly demonstrated by two WTO decisions (*EC-Hormones* and *EC-Sardines*) in which the EC legislation was found to be in conflict with the WTO obligations. The adoption of Article 5(3) of Regulation 178/2002, which represents a type of 'implementation' of the relevant SPS and TBT provisions, is a step forward to the promotion of the use of Codex standards by EC institutions and brings the obligations within the EC largely into accordance with the WTO obligations. However, it has to be acknowledged that the formulation of Article 5(3) is less strict than the WTO obligations, which may result in the fact that Codex standards are outweighed by other considerations, such as the findings of the EC scientific committees or agencies. This, in itself, is not in contrast with the WTO obligations, as long as there is a solid justification for deviating from Codex measures. This means that, although justifying deviations from Codex measures is not an explicit obligation under Regulation 178/2002, Community institutions should, in our view, indicate the specific reasons for their decisions, as several recently adopted or amended secondary food measures already exemplify. This goes hand in hand with the general obligation of Community institutions to state reasons under Article 253 EC.

2.2 Towards a more structural approach to ensure the legitimacy of the Codex Alimentarius Commission, its standard-setting procedure and its standards

When it comes to the legitimacy of the Codex Alimentarius Commission, it can be concluded that serious efforts have been made to improve the legitimacy of the Codex Alimentarius Commission. However, some structural shortcomings which are largely linked to the inadequate exploitation of the most fundamental instruments in the hands of the Codex Alimentarius Commission, such as the character of the Codex standard-setting procedure and the Procedural Manual, remain.

a. *Use of the structure of two readings*

One strong instrument of the Codex Alimentarius Commission designed to ensure legitimacy is its 8-step procedure, which consists of two readings. It is an excellent instrument in the hands of the Codex Alimentarius Commission, which enables it to supervise and co-ordinate the delegated tasks to other bodies (subsidiary Codex bodies, expert bodies and other international organisations). Furthermore, it can greatly contribute to the progress of consensus-building. However, the structure of two readings has not been fully explored. The major shortcoming of the current regulation of the Codex standard-setting procedure is that the internal rules of the Codex Alimentarius Commission hardly distinguish the first reading from the second reading. To make better use of the structure of two readings, the following can be considered. In order to facilitate consensus-building, the two readings should be more clearly differentiated. In other words, the negotiations during the first reading (Steps 3-5) should concentrate on different elements than those undertaken during the second reading (Steps 6-8). For instance, the goal to attain at the end of the first reading is to achieve consensus on the fundamental elements of the standards, such as the necessity of the standard, the title and scope of the standard, the definitions and the completion of scientific opinions of the expert bodies and the risk management options proposed. This means that, if no consensus has been reached on these issues, the proposed draft standard will not pass beyond Step 5. On the other hand, if the proposed draft standard has been forwarded to Step 6, these issues should no longer be a topic of the negotiations. In this way, the negotiations during the second reading can concentrate on less fundamental issues. In order to ensure a structural approach and a more transparent procedure, it is highly recommended that criteria are developed and included in the Procedural Manual. The co-ordination of the work undertaken by different Codex Committees can follow this same line. If a fundamental element of a proposed draft standard needs to be considered by another Codex Committee, it should be submitted to this Committee at Step 3 during the first reading, whereas, for more minor amendments, it can pass the relevant Committee during Step 6 of the second reading. To this end, the meeting of Codex chairs can be formalised and can give advice to the Codex Alimentarius Commission.

b. *Revision of the Procedural Manual*

Another important instrument in the hands of the Codex Alimentarius Commission, which has not received the attention it should have had, is the Procedural Manual. The latter aims to ensure transparency within the institutional and procedural operation of the standard-setting procedure in general. As we have concluded, the Procedural Manual is far from being an example of clarity, as some parts have become obsolete, incomplete, and incoherent. However, its quality is essential in ensuring the transparency of an institution and a procedure as complex as that of the Codex Alimentarius Commission and in assisting the developing member countries, in particular, in participating more effectively. In fact, it is the knowledge – on the part of Codex Members – of the internal rules of the Codex Alimentarius Commission that provides for a type of 'checks and balances' system in order to avoid the unlawful adoption of Codex standards, which is necessary in the absence of the judicial review elsewhere. For these reasons, a thorough revision of the Procedural

Manual should be considered as being a top priority. The following examples can be taken into account.

First, several inconsistencies should be removed from the Procedural Manual. For example, both the explicit reference to the IDF in the 'Procedures for the elaboration of Codex standards and related texts' and the Working party on Agricultural Quality Standards of the UNECE in the terms of reference of the Codex Committee on Fresh Fruits and Vegetables are in contrast with the requirement of same principles of membership (meaning that membership is open to all members and associate members of the FAO and WHO) laid down in the recently adopted guidelines on the co-operation with other international intergovernmental organisations. Another example is Statement 4 of the 'Statements of Principles Concerning the Role of Science in the Codex Decision-Making Process and the Extent to which Other Factors are Taken Into Account'. The statement aims to overcome a lack of consensus by making Codex Members aware of the possibility of abstaining from acceptance. However, this possibility no longer exists now that the acceptance procedure has been abolished.

Second, several provisions relating to decision-making, laid down in the Rules of Procedure, could be more clearly defined in order to avoid illegitimate decision-making, or, in other words, to avoid the adoption of standards in a manner that is in conflict with the common way of decision-making that is perceived as legitimate by Codex Members. The inclusion of the definition of consensus is essential to this end, as, at present, there is no agreement as to what consensus really consists of. As explained in Chapter V, this has led to the controversial situation in which the Chair of the Codex Alimentarius Commission concluded that the adoption of decisions which were considered to be illegitimate by several Codex Members. A definition of consensus will serve as an instrument in the hands of Codex Members to object more successfully to these controversial situations. Other issues can be reconsidered. For instance, it would be more appropriate for the Codex Alimentarius Commission as a whole to decide to proceed with a vote in cases where there is a failure to obtain a consensus (at present, any Codex Member can, in principle, decide to proceed to a vote). This can, for instance, be decided upon a simple-majority vote. Another example is that, in cases of voting, the voting on the final adoption of standards occurs by simple-majority. In order to ensure a more consensual approach, a two-thirds majority vote could be considered. A last point is whether voting on the final adoption of standards by secret ballot is acceptable.

Third, the codification of clear criteria on the status of consensus over a (draft) standard to be achieved in order to adopt it at Step 5 or Step 8, as discussed above under the heading 'use of the structure of two readings', is recommended.

Fourth, the formalisation and codification of some current informal practices, such as the above-mentioned Codex chairs meeting, and the relationships between general Codex Committees can also be considered.

c. *Reporting*

Reporting is an important tool in order to ensure transparency on the progress of the preparation of a particular standard during the standard-setting procedure. This is particularly true when it concerns the reporting of meetings to which not all Codex Members have access. The submission of the preliminary reports of the expert bodies to the relevant Codex Committee could be a way of establishing a more active and a more direct interaction. This could be combined with discussing risk assessment policy issues, for instance, during a joint meeting of the Codex Committee and the expert body in question. Reports of meetings of other international organisations may also be helpful to ensure the transparency of their discussions on the preparation of a proposed draft standard.

d. *The inclusion of safeguard clauses in Codex standards*

There is an increasing use within the Codex Alimentarius Commission to include footnotes containing the reservations of Codex Members relating to an adopted standard in the report of the relevant meeting. This indicates the increasing need to adopt more flexible standards that leave Codex Members with some discretion or with some options. It has, however, to be remembered that these reservations have no status under the WTO agreements and will, most probably, not be taken into due consideration by WTO panels and the Appellate Body. In this context, it is interesting to note that the Codex Alimentarius Commission has started to include 'safeguard clauses' in Codex measures themselves. This development should be encouraged. It assists in achieving consensus more easily and recognises that some issues may be better addressed at national level. Furthermore, the fact that the Codex Alimentarius Commission as a whole agrees on the issues that may be better addressed at national level, will most probably be accepted more easily as basis for justifying deviations from the Codex standards by WTO panels.

e. *The new role of the Executive Committee*

The new role of the Executive Committee as the consultative body to the Codex Alimentarius Commission can be considered to be an improvement as it assists the Codex Alimentarius Commission in its role as the supervisor of the delegated powers to other bodies and as the manager of the progress made during the standard-setting procedure. However, it has to be remembered that the Executive Committee consists of a limited number of Codex Members, and, for this reason, cannot replace the Codex Alimentarius Commission in its function as final decision-maker. Thus, the delegation of certain tasks to the Executive Committee may not necessarily lead to a more effective functioning standard-setting procedure. For instance, its new role to ensure linguistic consistency is one task that may not be appropriate to the Executive Committee. Clearly, the language of Codex measures has become more important after the adoption of the WTO agreements. Given the fact that linguistic amendments during the standard-setting procedure run the risk of having an impact on the meaning of the provision in question, advice given by a restricted group of Codex Members may easily reopen the discussion and jeopardise the consensus that has already been established. A more appropriate body would be the CCGP, given its current specialisation and given the fact that all Codex Members may have access to the discussions. This would mean that the CCGP would need more time to meet and to interact more closely with the other subsidiary Codex bodies. This could, for example, be organised in the following way: during the first reading, questions relating to linguistic and legal consistency could be discussed in the CCGP, while, during the second reading, all Codex measures that had been submitted for final approval by the Codex Alimentarius Commission would have to pass the CCGP. An impressive task, in which the CCGP would have to be assisted by the Codex Secretariat.

f. *The Codex Secretariat*

It is clear from the augmented workload that rests upon the shoulders of the Codex Alimentarius Commission that its Secretariat is simply too small to assist it adequately on a day-to-day basis. In order to respond to the responsibility that comes with the increased authority of the Codex Alimentarius Commission, the Secretariat should expand. For the same reason, it would be advisable for the Secretariat to have its own legal consultants who could work on the legal matters which arise in the context of the Codex Alimentarius Commission, and who can propose amendments to (proposed) draft standards in order to ensure consistency and conformity. In this way, the Codex Alimentarius Commission can also carry out the revision of Procedural Manual and ensure its transparency and coherence in the long-term. The major dilemma to such an enlargement has always been the financial restraints of its parent organisations, as has again been reflected in the report of the session of the Codex Alimentarius Commission of 2006. If such financial restraints persist, the Codex Alimentarius Commission and its parent organisations may want to consider establishing a Codex Secretariat, which is financially and administratively independent from its parent organisations.

g. *The Codex Alimentarius Commission as an independent international organisation?*

This leads to the question of whether it is time to consider the subsidiary position of the Codex Alimentarius Commission, and whether it should be 'upgraded' as being an independent international organisation. It could be held that this would take away some complications that the Codex Alimentarius Commission is faced with, such as the financial and administrative dependency of the Codex Secretariat on its parent organisations, the lack of power to enter into agreements with international organisations, the need for approval of the competent bodies of its parent organisations for amendments proposed to its Statutes and Rules of Procedure.

However, it is believed that, apart from the concern expressed above as regards the Codex Secretariat, the operation of the Codex Alimentarius Commission suffers little from its subsidiary position. Over the years, it has been able to construct an institutional and procedural framework that allow the Codex Alimentarius Commission to operate in a rather autonomous way. It is even believed that the subsidiary position of the Codex Alimentarius Commission comes with some important advantages, one of them being the relation with two important international organisations (the FAO and the WHO) that are renown for their expertise in the area. Its subsidiary position allows the Codex Alimentarius Commission to have an important link with the expert bodies of the FAO and the WHO, while at the same time their relationship remains one of 'separation of powers' between the risk assessors and risk managers. At this moment, the advantages of its subsidiary position outweigh the disadvantages, and transforming the Codex Alimentarius Commission into an independent international organisation may unnecessarily raise questions with regard to aspects of its institutional and procedural structure that currently function well.

The suggestions mentioned above need further examination. What is emphasised here is that there is an urgent need to provide a more explicit definition and recognition of both the functions of the WTO bodies and the Codex Alimentarius Commission in the light of the new relationship so as to promote legal certainty. Furthermore, it is believed that the recent adjustments to the institutional and procedural framework of the Codex Alimentarius Commission do not sufficiently reconsider the core instruments of the Codex Alimentarius Commission. What most of the suggestions above have in common is that they seek to make adjustments of the core instruments, such as the Procedural Manual, the two reading structure standard-setting procedure and the Codex Secretariat, in the light of their original objective in order to optimise their operation. It is believed that a better use of these already existing instruments will greatly contribute to ensuring the legitimacy of the Codex Alimentarius Commission, its procedure and its measures upon a more structural basis.

ANNEX I

CODEX INTERNATIONAL INDIVIDUAL STANDARD FOR GOUDA
CODEX STAN C-5-1966

1 DESIGNATION OF CHEESE

Gouda[1]

2 DEPOSITING COUNTRY

The Netherlands (country of origin)

3 RAW MATERIALS

3.1 KIND OF MILK: cow's milk

3.2 AUTHORIZED ADDITIONS:

- cultures of harmless lactic acid producing bacteria (starter)
- rennet or other suitable coagulating enzymes
- sodium chloride
- water
- calcium chloride, max. 200 mg/kg of the milk used
- annatto[2] and beta carotene, singly or in combination max. 600 mg/kg of cheese
- sodium and potassium nitrate, max. 50 mg/kg of cheese

4 PRINCIPAL CHARACTERISTICS OF THE CHEESE READY FOR CONSUMPTION

4.1 TYPE (CONSISTENCY): semi-hard

4.2 SHAPE:

a) cylindrical, with convex sides, curving smoothly into the flat top and bottom; the rate height/diameter varying from 1/4 to 1/3

b) flat block with square and/or rectangular sides (not being a loaf) and with or without rind

c) loaf, the length of the long side more than twice that of the shortest

4.3 DIMENSIONS AND WEIGHTS

4.3.1 Dimensions

a) cylindrical, with convex sides (as under 4.2a) fixed by prescribed shape (4.2.a) and weights (4.3.2a)

b) flat block (as under 4.2.b) fixed by prescribed shape (4.2.b) and weights (4.3.2.b)

c) loaf (as under 4.2.c) fixed by prescribed shape (4.2.c) and weights (4.3.2.c)

4.3.2 Weights

a) cylindrical, with convex sides (as under 4.2.a): from 2.5 to 30 kg

b) flat block (as under 4.2.b): not less than 5 kg

c) loaf (as under 4.2.c): from 2.5 to 5 kg

4.4 RIND

4.4.1 Consistency: hard

---

[1] Or such other synonym (e.g. Goudycki) derived from the name Gouda as will clearly indicate this variety
[2] temporarily endorsed

4.4.2 Appearance: dry or coated with either wax, a suspension of plastic or a film of vegetable oil

4.4.3 Colour: yellowish

4.5 BODY

4.5.1 Texture: firm, suitable for cutting

4.5.2 Colour: straw coloured

4.6 HOLES

4.6.1 Distribution: from few to plentiful, all over the interior of the cheese, distributed regularly as well as irregularly

4.6.2 Shape: more or less round

4.6.3 Size: varying from a pin's head to a pea

4.6.4 Appearance: not defined

4.7 MINIMUM FAT CONTENT IN THE DRY MATTER: 48%

4.8 MAXIMUM MOISTURE CONTENT: 43%

MINIMUM DRY MATTER CONTENT: 57%

4.9 OTHER PRINCIPAL CHARACTERISTICS: Gouda cheese is not normally consumed before it is five weeks old

5 BABY GOUDA

Small cheeses complying with the requirements for Gouda cheeses - except those under 4.2, 4.3, 4.8 and 4.9 - may be designated as "Baby Gouda", provided they comply with the following:

5.1 SHAPE: cylindrical with convex sides, curving smoothly into the flat top and bottom; the rate height/diameter is about 1/2

5.2 DIMENSIONS AND WEIGHTS

5.2.1 Dimensions: filed by prescribed shape (5.1) and weights (5.2.2)

5.2.2 Weights: from 0.180 to 1.500 kg

5.3 MAXIMUM MOISTURE CONTENT: 45%

MINIMUM DRY MATTER CONTENT: 55%

5.4 OTHER PRINCIPAL CHARACTERISTICS: Baby Gouda is not normally consumed before it is three weeks old

6 METHOD OF MANUFACTURE

6.1 METHOD OF COAGULATION: rennet or other suitable coagulating enzymes

6.2 HEAT TREATMENT: the curd is heated with or without the aid of warm water

6.3 FERMENTATION PROCEDURE: chiefly lactic acid

6.4 MATURATION PROCEDURE: maturation during storage at a temperature preferably between 10° and 20°C

6.5 OTHER PRINCIPAL CHARACTERISTICS: salted in brine after manufacture

7 SAMPLING AND ANALYSIS

See Volume 13 of the *Codex Alimentarius*.

8   MARKING AND LABELLING

Only cheese conforming with this standard may be designated "Gouda". It shall be labelled in conformity with the appropriate sections of Article 4 of FAO/WHO Standard A.6 "General Standard for Cheese"[3]. The cheese mentioned under 5 may be designated "Gouda" provided that the designation is accompanied by the prefix "Baby".

---

[3]   Currently Section 7 of the Codex General Standard for Cheese (CODEX STAN A-6-1978, Rev.1-1999)

ANNEX II

CODEX STANDARD FOR CHOCOLATE AND CHOCOLATE PRODUCTS
(CODEX STAN 87-1981, Rev. 1 - 2003)

1 SCOPE

The standard applies to chocolate and chocolate products intended for human consumption and listed in Section 2. Chocolate and chocolate products shall be prepared from cocoa and cocoa materials with sugars and may contain sweeteners, milk products, flavouring substances and other food ingredients.

2 DESCRIPTION AND ESSENTIAL COMPOSITION FACTORS

Chocolate is the generic name for the homogenous products complying with the descriptions below and summarized in Table 1. It is obtained by an adequate manufacturing process from cocoa materials which may be combined with milk products, sugars and/or sweeteners, and other additives listed in section 3 of the present standard. Other edible foodstuffs, excluding added flour and starch (except for products in sections 2.1.1.1 and 2.1.2.1 of this Standard) and animal fats other than milk fat, may be added to form various chocolate products. These combined additions shall be limited to 40 % of the total weight of the finished product, subject to the labelling provisions under Section 5.

The addition of vegetable fats other than cocoa butter shall not exceed 5% of the finished product, after deduction of the total weight of any other added edible foodstuffs, without reducing the minimum contents of cocoa materials. Where required by the authorities having jurisdiction, the nature of the vegetable fats permitted for this purpose may be prescribed in applicable legislation.

2.1 CHOCOLATE TYPES (COMPOSITION)

2.1.1 Chocolate

Chocolate (in some regions also named *bittersweet chocolate, semi-sweet chocolate, dark chocolate* or *"chocolat fondant"*) shall contain, on a dry matter basis, not less than 35% total cocoa solids, of which not less than 18% shall be cocoa butter and not less than 14% fat-free cocoa solids.

2.1.1.1 *Chocolate a la taza* is the product described under Section 2.1.1 of this Standard and containing a maximum of 8% m/m flour and/or starch from wheat, maize or rice.

2.1.2 Sweet Chocolate

Sweet Chocolate shall contain, on a dry matter basis, not less than 30% total cocoa solids, of which at least 18% shall be cocoa butter and at least 12% fat-free cocoa solids.

2.1.2.1 *Chocolate familiar a la taza* is the product described under Section 2.1.2 of this Standard and containing a maximum of 18% m/m flour and/or starch from wheat, maize or rice.

2.1.3 Couverture Chocolate

Couverture Chocolate shall contain, on a dry matter basis, not less than 35% total cocoa solids of which not less than 31% shall be cocoa butter and not less than 2.5% of fat-free cocoa solids.

2.1.4 Milk Chocolate

Milk Chocolate shall contain, on a dry matter basis, not less than 25% cocoa solids (including a minimum of 2.5% fat-free cocoa solids) and a specified minimum of milk solids between 12% and 14% (including a minimum of milk fat between 2.5% and 3.5%). The minimum content for milk solids and milk fat shall be applied by the authority having jurisdiction in accordance with applicable legislation. "Milk solids" refers to the addition of milk ingredients in their natural proportions, except that milk fat may be added, or removed.

Where required by the competent authority, a minimum content of cocoa butter plus milk fat may also be set.

2.1.5 Family Milk Chocolate

Family Milk Chocolate shall contain, on a dry matter basis, not less than 20% cocoa solids (including a minimum of 2.5% fat-free cocoa solids) and not less than 20% milk solids (including a minimum of 5% milk fat). "Milk solids" refers to the addition of milk ingredients in their natural proportions, except that milk fat may be added, or removed.

Where required by the competent authority, a minimum content of cocoa butter plus milk fat may also be set.

2.1.6 Milk Chocolate Couverture

Milk Chocolate Couverture shall contain, on a dry matter basis, not less than 25% cocoa solids (including a minimum of 2.5% non-fat cocoa solids) and not less than 14% milk solids (including a minimum of 3.5% milk fat) and a total fat of not less than 31%. "Milk solids" refers to the addition of milk ingredients in their natural proportions, except that milk fat may be added, or removed.

2.1.7 Other chocolate products

2.1.7.1 White Chocolate

White Chocolate shall contain, on a dry matter basis, not less than 20% cocoa butter and not less than 14% milk solids (including a minimum milk fat in a range of 2.5% to 3.5% as applied by the authority having jurisdiction in accordance with applicable legislation). "Milk solids" refers to the addition of milk ingredients in their natural proportions, except that milk fat may be added, or removed.

Where required by the competent authority, a minimum content of cocoa butter plus milk fat may also be set.

2.1.7.2 Gianduja Chocolate

"*Gianduja*" (or one of the derivatives of the word "*Gianduja*") *Chocolate* is the product obtained, firstly, from chocolate having a minimum total dry cocoa solids content of 32%, including a minimum dry non-fat cocoa solids content of 8%, and, secondly, from finely ground hazelnuts such that the product contains not less than 20 % and not more than 40% of hazelnuts.

The following may be added:

- (a) milk and/or dry milk solids obtained by evaporation, in such proportion that the finished product does not contain more than 5% dry milk solids ;
- (b) almonds, hazelnuts and other nut varieties, either whole or broken, in such quantities that, together with the ground hazelnuts, they do not exceed 60% of the total weight of the product.

2.1.7.3 Gianduja Milk Chocolate

"*Gianduja*" (or one of the derivatives of the word "*Gianduja*") *Milk Chocolate* is the product obtained, firstly, from milk chocolate having a minimum dry milk solids content of 10% and, secondly, from finely ground hazelnuts such that the product contains not less than 15 % and not more than 40% of hazelnuts. "Milk solids" refers to the addition of milk ingredients in their natural proportions, except that milk fat may be added or removed.

The following may be added: Almonds, hazelnuts and other nut varieties, either whole or broken, in such quantities that, together with the ground hazelnuts, they do not exceed 60% of the total weight of the product.

Where required by the competent authority, a minimum content of cocoa butter plus milk fat may also be set.

2.1.7.4 Chocolate para mesa

Chocolate para mesa is unrefined chocolate in which the grain size of sugars is larger than 70 microns.

2.1.7.4.1 *Chocolate para mesa*

Chocolate para mesa shall contain, on a dry matter basis, not less than 20% total cocoa solids (including a minimum of 11% cocoa butter and a minimum of 9% fat-free cocoa solids).

2.1.7.4.2 *Semi-bitter chocolate para mesa*

Semi-bitter Chocolate para mesa shall contain, on a dry matter basis, not less than 30% total cocoa solids (including a minimum of 15% cocoa butter and a minimum of 14% fat-free cocoa solids).

2.1.7.4.3 *Bitter chocolate para mesa*

Bitter Chocolate para mesa shall contain, on a dry matter basis, not less than 40% total cocoa solids (including a minimum of 22% cocoa butter and a minimum of 18% fat-free cocoa solids).

2.2 CHOCOLAT TYPES (FORMS)

2.2.1 *Chocolate Vermicelli and Chocolate Flakes*

Chocolate Vermicelli and Chocolate Flakes are cocoa products obtained by a mixing, extrusion and hardening technique which gives unique, crisp textural properties to the products. Vermicelli are presented in the form of short, cylindrical grains and flakes in the form of small flat pieces.

2.2.1.1 **Chocolate Vermicelli / Chocolate Flakes**

Chocolate Vermicelli / Chocolate Flakes shall contain, on a dry matter basis, not less than 32% total cocoa solids, of which at least 12% shall be cocoa butter and 14% fat-free cocoa solids.

2.2.1.2 **Milk Chocolate Vermicelli / Milk Chocolate Flakes**

Milk Chocolate Vermicelli / Milk Chocolate Flakes shall contain, on a dry matter basis, not less than 20% cocoa solids (including a minimum of 2.5% fat-free cocoa solids) and not less than 12% milk solids (including a minimum of 3% milk fat). "Milk solids" refers to the addition of milk ingredients in their natural proportions, except that milk fat may be added, or removed.

Where required by the competent authority, a minimum content of cocoa butter plus milk fat may also be set.

2.2.2 *Filled Chocolate*

Filled Chocolate is a product covered by a coating of one or more of the Chocolates defined in Section 2.1, with exception of *chocolate a la taza, chocolate familiar a la taza* and products defined in section 2.1.7.4 (*chocolate para mesa*), the centre of which is clearly distinct, through its composition, from the external coating. Filled Chocolate does not include Flour Confectionery, Pastry, Biscuit or Ice Cream products. The chocolate part of the coating must make up at least 25% of the total weight of the product concerned.

If the centre part of the product is made up of a component or components for which a separate Codex Standard exists, the component(s) must comply with the applicable standard.

2.2.3 *A Chocolate or Praline*

A Chocolate or *Praline* designates the product in a single mouthful size, where the amount of the chocolate component shall not be less than 25% of the total weight of the product. The product shall consist of either filled chocolate or a single or combination of the chocolates as defined under Section 2.1, with exception of *chocolate a la taza, chocolate familiar a la taza* and products defined in section 2.1.7.4 (*chocolate para mesa*).

TABLE 1. SUMMARY TABLE OF COMPOSITIONAL REQUIREMENTS OF SECTION 2[1]

(% calculated on the dry matter in the product and after deduction of the weight of the other edible foodstuffs authorized under Section 2)

PRODUCTS	CONSTITUENTS (%)						
2. Chocolate Types	Cocoa Butter	Fat-free Cocoa Solids	Total Cocoa Solids	Milk Fat	Total Milk Solids	Starch / Flour	Ground Hazelnuts
2.1 CHOCOLATE TYPES (COMPOSITION)							
2.1.1 Chocolate	≥18	≥14	≥35				
2.1.1.1 Chocolate a la taza	≥18	≥14	≥35			< 8	
2.1.2 Sweet Chocolate	≥18	≥12	≥30				
2.1.2.1 Chocolate familiar a la taza	≥18	≥12	≥30			< 18	
2.1.3 Couverture Chocolate	≥31	≥2.5	≥35				
2.1.4 Milk Chocolate		≥2.5	≥25	≥2.5-3.5	≥12-14		
2.1.5 Family Milk Chocolate		≥2.5	≥20	≥5	≥20		
2.1.6 Milk Chocolate Couverture		≥2.5	≥25	≥3.5	≥14		
2.1.7 Other chocolate products							
2.1.7.1. White Chocolate	≥20			≥2.5-3.5	≥14		
2.1.7.2 Gianduja Chocolate		≥8	≥32				≥20 and ≤40
2.1.7.3 Gianduja Milk Chocolate		≥2.5	≥25	≥2.5-3.5	≥10		≥15 and ≤40
2.1.7.4 Chocolate para mesa							
2.1.7.4.1 Chocolate para mesa	≥ 11	≥ 9	≥ 20				
2.1.7.4.2 Semi-bitter chocolate para mesa	≥15	≥14	≥ 30				
2.1.7.4.3 Bitter chocolate para mesa	≥ 22	≥18	≥ 40				
2.2 CHOCOLATE TYPES (forms)							
2.2.1 Chocolate Vermicelli / Chocolate Flakes							
2.2.1.1 Chocolate Vermicelli / Chocolate Flakes	≥12	≥14	≥32				
2.2.1.2 Milk Chocolate Vermicelli / Milk Chocolate Flakes		≥2.5	≥ 20	≥3	≥12		
2.2.2 Filled Chocolate (see section 2.2.2)							
2.2.3 A Chocolate or Praline (see section 2.2.3)							

---

[1] "***Milk solids***" refers to the addition of milk ingredients in their natural proportions except that milk fat may be added or removed.

3 FOOD ADDITIVES

The food additives listed below may be used and only within the limits specified.

Other additives from the General Standard for Food Additives (GSFA) approved list may be used, subject to the authority having jurisdiction in accordance with applicable legislation.

3.1 Alkalizing and neutralizing agents carried over as a result of processing cocoa materials in proportion to the maximum quantity as provided for.

3.2 ACIDITY REGULATORS   **Maximum Level**

503(i)	Ammonium carbonate	
527	Ammonium hydroxide	
503(ii)	Ammonium hydrogen carbonate	
170(i)	Calcium carbonate	
330	Citric acid	
504(i)	Magnesium carbonate	
528	Magnesium hydroxide	
530	Magnesium oxide	Limited by GMP
501(i)	Potassium carbonate	
525	Potassium hydroxide	
501(ii)	Potassium hydrogen carbonate	
500(i)	Sodium carbonate	
524	Sodium hydroxide	
500(ii)	Sodium hydrogen carbonate	
526	Calcium hydroxide	
338	Orthophosphoric acid	2.5 g/kg expressed as P_2O_5 in finished cocoa and chocolate products
334	L-Tartaric acid	5 g/kg in finished cocoa and chocolate products

3.3 EMULSIFIERS   **Maximum Level**   **Products**

471	Mono- and di-glycerides of fatty acids		Products described under 2.1 and 2.2
322	Lecithins	GMP	" "

422	Glycerol			"	"
442	Ammonium salts of phosphatidic acids	10 g/kg		"	"
476	Polyglycerol esters interesterified recinoleic acid	5 g/kg	15 g/kg	"	"
491	Sorbitan monostearate	10 g/kg	in combination	"	"
492	Sorbitan tristearate	10 g/kg		"	"
435	Polyoxyethylene (20) sorbitan monostearate	10 g/kg			

3.4 FLAVOURING AGENTS

3.4.1	Natural flavours as defined in the Codex Alimentarius, and their synthetic equivalents, except those which would imitate natural chocolate or milk flavours[2]	GMP	Products described under 2.1 and 2.2
3.4.2	Vanillin	1 g/kg	Products described under 2.1 and 2.2
3.4.3	Ethyl vanillin	in combination	Products described under 2.1 and 2.2

3.5 SWEETENERS

950	Acesulfame K	500 mg/kg	Products described under 2.1 and 2.2	
951	Aspartame	2 000 mg/kg	"	"
952	Cyclamic acid and its Na and Ca salts	500 mg/kg	"	"
954	Saccharin and its Na and Ca salts	500 mg/kg	"	"
957	Thaumatin		"	"

---

[2] Temporarily endorsed

420	Sorbitol		" "
421	Mannitol		" "
953	Isomalt	GMP	" "
965	Maltitol		" "
966	Lactitol		" "
967	Xylitol		" "

3.6 GLAZING AGENTS

414	Gum Arabic (Acacia gum)		Products described under 2.1 and 2.2
440	Pectin		" "
901	Beeswax, white and yellow	GMP	" "
902	Candelilla wax		" "
904	Shellac		" "

3.7 ANTIOXIDANTS

304	Ascorbyl palmitate	200 mg/kg	Products described under 2.1.7.1 calculated on a fat content basis
319	Tertiary butylhydroquine		"
320	Butylated hydroxyanisole	200 mg/kg singly or in combination	"
321	Butylated hydroxytoluene		"
310	Propylgallate		"
307	α-Tocopherol	750 mg/kg	"

3.8 COLOURS
(FOR DECORATION PURPOSE ONLY)

175	Gold	GMP	Products described under 2.1 and 2.2
174	Silver	GMP	

3.9 BULKING AGENTS

1200	Polydextrose A and N	GMP	Products described under 2.1 and 2.2

3.10 PROCESSING AID

	Maximum Level	
Hexane (62°C - 82°C)	1 mg/kg	calculated on a fat content basis

4 HYGIENE

4.1 It is recommended that the products covered by the provisions of this standard be prepared and handled in accordance with the appropriate sections of the Recommended International Code of Practice – General Principles of Food Hygiene (CAC/RCP 1-1969, Rev 3-1997), and other relevant Codex texts such as Codes of Hygienic Practice and Codes of Practice.

4.2 The products should comply with any microbiological criteria established in accordance with the Principles for the Establishment and Application of Microbiological Criteria for Foods (CAC/GL 21-1997).

5 LABELLING

In addition to the requirements of the Codex General Standard for the Labelling of Prepackaged Foods (CODEX STAN 1-1985 Rev. 1-1991), the following declarations shall be made:

5.1 NAME OF THE FOOD

5.1.1 Products described under Sections 2.1 and 2.2 of this Standard and complying with the appropriate requirements of the relevant section shall be designated according to the name listed in Section 2 under subsequent section and subject to the provisions under Section 5 of this Standard. The products defined in section 2.1.1 may be described as "*Bittersweet chocolate*", "*Semi-sweet chocolate*", "*Dark chocolate*" or "*Chocolat fondant*".

5.1.1.1 When sugars are fully or partly replaced by sweeteners, an appropriate declaration should be included in proximity of the sales designation of the chocolate, mentioning the presence of sweeteners. *Example*: "X Chocolate with sweeteners".

5.1.1.2 The use of vegetable fats in addition to Cocoa butter in accordance with the provisions of Section 2 shall be indicated on the label in association with the name and/or the representation of the product.

The authorities having jurisdiction may prescribe the specific manner in which this declaration shall be made.

5.1.2   Filled Chocolate

5.1.2.1   Products described under Section 2.2.2. shall be designated "Filled Chocolate", "X Filled Chocolate", "Chocolate with X Filling" or "Chocolate with X Centre", where "X" is descriptive of the nature of the filling.

5.1.2.2   The type of chocolate used in the external coating may be specified, whereby the designations used shall be the same as stated under Section 5.1.1 of this Standard.

5.1.2.3   An appropriate statement shall inform the consumer about the nature of the centre.

5.1.3   A Chocolate or Praline

Products in a single mouthful size described under Section 2.2.3 of this Standard shall be designated "*A Chocolate*" or "*Praline*".

5.1.4   Assorted Chocolates

Where the products described under Section 2.1 or 2.2 with exception of *chocolate a la taza*, *chocolate familiar a la taza* and *chocolate para mesa* are sold in assortments, the product name may be replaced by the words "*Assorted Chocolates*" or "*Assorted filled Chocolates*", "*Assorted Chocolate Vermicelli*", etc. In that case, there shall be a single list of ingredients for all the products in the assortment or alternatively lists of ingredients by products.

5.1.5   Other Information Required

5.1.5.1   Any characterizing flavour, other than chocolate flavour shall be in the designation of the product.

5.1.5.2   Ingredients, which are especially aromatic and characterize the product shall form part of the name of the product (e.g. Mocca Chocolate).

5.1.6   Use of the Term Chocolate

Products not defined under this Standard, and where the chocolate taste is solely derived from non-fat cocoa solids, can carry the term "chocolate" in their designations in accordance with the provisions or customs applicable in the country in which the product is sold to the final consumer and this to designate other products which cannot be confused with those defined in this Standard.

5.2   DECLARATION OF MINIMUM COCOA CONTENT

When required by the authority having jurisdiction, products described under Section 2.1 of this Standard, except for *white chocolat*, shall carry a declaration of cocoa solids. For the purpose of this declaration, the percentages declared shall be made on the chocolate part of the product after the deduction of the other permitted edible foodstuffs.

5.3   LABELLING OF NON-RETAIL CONTAINERS

5.3.1   Information required in Section 5.1 and 5.2 of this Standard and Section 4 of the *Codex General Standard for the Labelling of Prepackaged Foods* shall be given either on the container or in accompanying documents, except that the name of the product, lot identification, and the name and address of the manufacturer, packer, distributor and/or importer shall appear on the container.

5.3.2   However, lot identification, and the name and address of the manufacturer, packer, distributor and/or importer may be replaced by an identification mark provided that such a mark is clearly identifiable with the accompanying documents.

6 METHODS OF ANALYSIS AND SAMPLING

6.1 DETERMINATION OF CENTRE AND COATING OF FILLED CHOCOLATE

All methods approved for the chocolate type used for the coating and those approved for the type of centre concerned.

6.2 DETERMINATION OF COCOA BUTTER

According to AOAC 963.15 or IOCCC 14-1972.

6.3 DETERMINATION OF FAT-FREE COCOA SOLID

According to AOAC 931.05.

6.4 DETERMINATION OF FAT-FREE MILK SOLIDS

According to IOCCC 17-1973 or AOAC 939.02.

6.5 DETERMINATION OF MILK FAT

According to IOCCC 5-1962 or AOAC 945.34, 925.41B, 920.80.

6.6 DETERMINATION OF MOISTURE

According to IOCCC 26-1988 or AOAC 977.10 (Karl Fischer method); or AOAC 931.04 or IOCCC 1-1952 (gravimetry).

6.7 DETERMINATION OF TOTAL FAT

According to AOAC 963.15.

6.8 DETERMINATION OF NON-COCOA BUTTER VEGETABLE FAT IN CHOCOLATE AND CHOCOLATE PRODUCTS

The following methods of analysis are the best available at the present time. Further systematic improvement is required. Documentation identifying the type of commercial blends of non-cocoa butter vegetable fats used must be made available upon request by competent authorities.

6.8.1 Detection of Non-Cocoa Butter Vegetable Fats in Chocolate

Detecting sterol breakdown products in refined vegetable fats added to chocolate by method AOCS Ce 10/02 (02).

*6.8.2 Quantitative Determination of Non-Cocoa Butter Vegetable Fats**

Determination of the triacyglycerols (C50, C52, C54) present in cocoa butters and non-cocoa butter vegetable fats by GC-FID in *J. Amer. Oil Chem. Soc.* (1980), **57**, 286-293. In milk chocolate, there is a need to correct for the milk fat

- **Interpretation:**

---

* This method is intended to measure vegetable fats which are cocoa butter equivalents (CBE) i.e. SOS type triglycerides. Other vegetable fats can only be added in very limited amounts before they affect the physical properties of chocolate in a detrimental way. These can be determined by conventional methods i.e. fatty acid and triacyglycerol analyses.

When type of non-cocoa butter vegetable fat is known, the amount of non-cocoa butter vegetable fat is calculated according to *J. Amer. Oil Chem. Soc.* (1980), **57**, 286-293.

When type of non-cocoa butter vegetable fat is not known, the calculation is made according to *J. Amer. Oil Chem. Soc.* (1982), **61 (3)**, 576-581.

ANNEX III

GENERAL STANDARD FOR THE LABELLING OF PREPACKAGED FOODS

CODEX STAN 1-1985 (Rev. 1-1991)[1]

1. SCOPE

This standard applies to the labelling of all prepackaged foods to be offered as such to the consumer or for catering purposes and to certain aspects relating to the presentation thereof.

2. DEFINITION OF TERMS

For the purpose of this standard:

"Claim" means any representation which states, suggests or implies that a food has particular qualities relating to its origin, nutritional properties, nature, processing, composition or any other quality.

"Consumer" means persons and families purchasing and receiving food in order to meet their personal needs.

"Container" means any packaging of food for delivery as a single item, whether by completely or partially enclosing the food and includes wrappers. A container may enclose several units or types of packages when such is offered to the consumer.

For use in **Date Marking** of prepackaged food:

"Date of Manufacture" means the date on which the food becomes the product as described.

"Date of Packaging" means the date on which the food is placed in the immediate container in which it will be ultimately sold.

"Sell-by-Date" means the last date of offer for sale to the consumer after which there remains a reasonable storage period in the home.

"Date of Minimum Durability" ("best before") means the date which signifies the end of the period under any stated storage conditions during which the product will remain fully marketable and will retain any specific qualities for which tacit or express claims have been made. However, beyond the date the food may still be perfectly satisfactory.

"Use-by Date" (Recommended Last Consumption Date, Expiration Date) means the date which signifies the end of the estimated period under any stated storage conditions, after which the product probably will not have the quality attributes normally expected by the consumers. After this date, the food should not be regarded as marketable.

"Food" means any substance, whether processed, semi-processed or raw, which is intended for human consumption, and includes drinks, chewing gum and any substance which has been used in the manufacture, preparation or treatment of "food" but does not include cosmetics or tobacco or substances used only as drugs.

"Food Additive" means any substance not normally consumed as a food by itself and not normally used as a typical ingredient of the food, whether or not it has nutritive value, the intentional addition of which to food for a technological (including organoleptic) purpose in the manufacture, processing, preparation, treatment, packing, packaging, transport or holding of such food results, or may be reasonably expected to result, (directly or indirectly) in it or its by-products becoming a component of or otherwise affecting the characteristics of such foods. The term does not include "contaminants" or substances added to food for maintaining or improving nutritional qualities.

"Ingredient" means any substance, including a food additive, used in the manufacture or preparation of a food and present in the final product although possibly in a modified form.

"Label" means any tag, brand, mark, pictorial or other descriptive matter, written, printed, stencilled, marked, embossed or impressed on, or attached to, a container of food.

"Labelling" includes any written, printed or graphic matter that is present on the label, accompanies the food, or is displayed near the food, including that for the purpose of promoting its sale or disposal.

"Lot" means a definitive quantity of a commodity produced essentially under the same conditions.

"Prepackaged" means packaged or made up in advance in a container, ready for offer to the consumer, or for catering purposes.

---

[1] The Codex General Standard for the Labelling of Prepackaged Foods was adopted by the Codex Alimentarius Commission at its 14th Session, 1981 and subsequently revised in 1985 and 1991 by the 16th and 19th Sessions. It was amended by the 23rd, 24th, 26th and 28th Sessions in 1999, 2001, 2003 and 2005.

"**Processing Aid**" means a substance or material, not including apparatus or utensils, and not consumed as a food ingredient by itself, intentionally used in the processing of raw materials, foods or its ingredients, to fulfil a certain technological purpose during treatment or processing and which may result in the non-intentional but unavoidable presence of residues or derivatives in the final product.

"**Foods for Catering Purposes**" means those foods for use in restaurants, canteens, schools, hospitals and similar institutions where food is offered for immediate consumption.

3. GENERAL PRINCIPLES

3.1 Prepackaged food shall not be described or presented on any label or in any labelling in a manner that is false, misleading or deceptive or is likely to create an erroneous impression regarding its character in any respect.[2]

3.2 Prepackaged food shall not be described or presented on any label or in any labelling by words, pictorial or other devices which refer to or are suggestive either directly or indirectly, of any other product with which such food might be confused, or in such a manner as to lead the purchaser or consumer to suppose that the food is connected with such other product.

4. MANDATORY LABELLING OF PREPACKAGED FOODS

The following information shall appear on the label of prepackaged foods as applicable to the food being labelled, except to the extent otherwise expressly provided in an individual Codex standard:

4.1 THE NAME OF THE FOOD

4.1.1 The name shall indicate the true nature of the food and normally be specific and not generic:

4.1.1.1 Where a name or names have been established for a food in a Codex standard, at least one of these names shall be used.

4.1.1.2 In other cases, the name prescribed by national legislation shall be used.

4.1.1.3 In the absence of any such name, either a common or usual name existing by common usage as an appropriate descriptive term which was not misleading or confusing to the consumer shall be used.

4.1.1.4 A "coined", "fanciful", "brand" name, or "trade mark" may be used provided it accompanies one of the names provided in Subsections 4.1.1.1 to 4.1.1.3.

4.1.2 There shall appear on the label either in conjunction with, or in close proximity to, the name of the food, such additional words or phrases as necessary to avoid misleading or confusing the consumer in regard to the true nature and physical condition of the food including but not limited to the type of packing medium, style, and the condition or type of treatment it has undergone; for example: dried, concentrated, reconstituted, smoked.

4.2 LIST OF INGREDIENTS

4.2.1 Except for single ingredient foods, a list of ingredients shall be declared on the label.

4.2.1.1 The list of ingredients shall be headed or preceded by an appropriate title which consists of or includes the term ' ingredient' .

4.2.1.2 All ingredients shall be listed in descending order of ingoing weight (m/m) at the time of the manufacture of the food.

4.2.1.3 Where an ingredient is itself the product of two or more ingredients, such a compound ingredient may be declared, as such, in the list of ingredients, provided that it is immediately accompanied by a list, in brackets, of its ingredients in descending order of proportion (m/m). Where a compound ingredient (for which a name has been established in a Codex standard or in national legislation) constitutes less than 5% of the food, the ingredients, other than food additives which serve a technological function in the finished product, need not be declared.

4.2.1.4 The following foods and ingredients are known to cause hypersensitivity and shall always be declared:[3]

- Cereals containing gluten; i.e., wheat, rye, barley, oats, spelt or their hybridized strains and products of these;
- Crustacea and products of these;

---

[2] Examples of descriptions or presentations to which these General Principles refer are given in the Codex General Guidelines on Claims.

[3] Future additions to and/or deletions from this list will be considered by the Codex Committee on Food Labelling taking into account the advice provided by the Joint FAO/WHO Expert Committee on Food Additives (J ECFA).

- Eggs and egg products;
- Fish and fish products;
- Peanuts, soybeans and products of these;
- Milk and milk products (lactose included);
- Tree nuts and nut products; and
- Sulphite in concentrations of 10 mg/kg or more.

4.2.1.5 Added water shall be declared in the list of ingredients except when the water forms part of an ingredient such as brine, syrup or broth used in a compound food and declared as such in the list of ingredients. Water or other volatile ingredients evaporated in the course of manufacture need not be declared.

4.2.1.6 As an alternative to the general provisions of this section, dehydrated or condensed foods which are intended to be reconstituted by the addition of water only, the ingredients may be listed in order of proportion (m/m) in the reconstituted product provided that a statement such as "ingredients of the product when prepared in accordance with the directions on the label" is included.

4.2.2 The presence in any food or food ingredients obtained through biotechnology of an allergen transferred from any of the products listed in Section 4.2.1.4 shall be declared.

When it is not possible to provide adequate information on the presence of an allergen through labelling, the food containing the allergen should not be marketed.

4.2.3 A specific name shall be used for ingredients in the list of ingredients in accordance with the provisions set out in Section 4.1 (Name of the Food) except that:

4.2.3.1 Except for those ingredients listed in section 4.2.1.4, and unless a general class name would be more informative, the following class names may be used:

NAME OF CLASSES	CLASS NAMES
Refined oils other than olive	' Oil' together with either the term ' vegetable' or ' animal' , qualified by the term ' hydrogenated' or' partially-hydrogenated' , as appropriate
Refined fats	' Fat' together with either, the term ' vegetable' or ' animal' , as appropriate
Starches, other than chemically modified starches	' Starch'
All species of fish where the fish constitutes an ingredient of another food and provided that the labelling and presentation of such food does not refer to a specific species of fish	' Fish'
All types of poultrymeat where such meat constitutes an ingredient of another food and provided that the labelling and presentation of such a food does not refer to a specific type of poultrymeat	' Poultrymeat'
All types of cheese where the cheese or mixture of cheeses constitutes an ingredient of another food and provided that the labelling and presentation of such food does not refer to a specific type of cheese	' Cheese'

Name of Classes	Class Names
All spices and spice extracts not exceeding 2% by weight either singly or in combination in the food	'Spice', 'spices', or 'mixed spices', as appropriate
All herbs or parts of herbs not exceeding 2% by weight either singly or in combination in the food	'Herbs' or 'mixed herbs', as appropriate
All types of gum preparations used in the manufacture of gum base for chewing gum	'Gum base'
All types of sucrose	'Sugar'
Anhydrous dextrose and dextrose monohydrate	'Dextrose' or 'glucose'
All types of caseinates	'Caseinates'
Milk Protein	Milk products containing a minimum of 50% of milk protein (m/m) in dry matter *
Press, expeller or refined cocoa butter	'Cocoa butter'
All crystallized fruit not exceeding 10% of the weight of the food	'Crystallized fruit'

* Calculation of milk protein content: Kjeldahl nitrogen x 6.38

4.2.3.2   Notwithstanding the provision set out in Section 4.2.3.1, pork fat, lard and beef fat shall always be declared by their specific names.

4.2.3.3   For food additives falling in the respective classes and appearing in lists of food additives permitted for use in foods generally, the following class titles shall be used together with the specific name or recognized numerical identification as required by national legislation.[4]

- Acidity Regulator
- Acids
- Anticaking Agent
- Antifoaming Agent
- Antioxidant
- Bulking Agent
- Colour
- Colour Retention Agent
- Emulsifier
- Emulsifying Salt
- Firming Agent
- Flour Treatment Agent
- Flavour Enhancer
- Foaming Agent
- Gelling Agent
- Glazing Agent
- Humectant
- Preservative
- Propellant
- Raising Agent
- Stabilizer
- Sweetener
- Thickener

---

[4]   Governments accepting the standard should indicate the requirements in force in their countries.

4.2.3.4 The following class titles may be used for food additives falling in the respective classes and appearing in lists of food additives permitted generally for use in foods:

- Flavour(s) and Flavouring(s)

Modified Starch(es) The expression "flavours" may be qualified by "natural", "nature identical", "artificial" or a combination of these words as appropriate.

4.2.4 Processing Aids and Carry-Over of Food Additives

4.2.4.1 A food additive carried over into a food in a significant quantity or in an amount sufficient to perform a technological function in that food as a result of the use of raw materials or other ingredients in which the additive was used shall be included in the list of ingredients. The exemption does not apply to food additive and processing aids listed in section 4.2.1.4.

4.2.4.2 A food additive carried over into foods at a level less than that required to achieve a technological function, and processing aids, are exempted from declaration in the list of ingredients. The exemption does not apply to food additives and processing aids listed in section 4.2.1.4.

4.3 NET CONTENTS AND DRAINED WEIGHT

4.3.1 The net contents shall be declared in the metric system ("Système International" units).[5]

4.3.2 The net contents shall be declared in the following manner:

  (i) for liquid foods, by volume;

  (ii) for solid foods, by weight;

  (iii) for semi-solid or viscous foods, either by weight or volume.

4.3.3 In addition to the declaration of net contents, a food packed in a liquid medium shall carry a declaration in the metric system of the drained weight of the food. For the purposes of this requirement, liquid medium means water, aqueous solutions of sugar and salt, fruit and vegetable juices in canned fruits and vegetables only, or vinegar, either singly or in combination.[6]

4.4 NAME AND ADDRESS

The name and address of the manufacturer, packer, distributor, importer, exporter or vendor of the food shall be declared.

4.5 COUNTRY OF ORIGIN

4.5.1 The country of origin of the food shall be declared if its omission would mislead or deceive the consumer.

4.5.2 When a food undergoes processing in a second country which changes its nature, the country in which the processing is performed shall be considered to be the country of origin for the purposes of labelling.

4.6 LOT IDENTIFICATION

Each container shall be embossed or otherwise permanently marked in code or in clear to identify the producing factory and the lot.

4.7 DATE MARKING AND STORAGE INSTRUCTIONS

4.7.1 If not otherwise determined in an individual Codex standard, the following date marking shall apply:

  (i) The "date of minimum durability" shall be declared.

  (ii) This shall consist at least of:

  - the day and the month for products with a minimum durability of not more than three months;
  - the month and the year for products with a minimum durability of more than three months. If the month is December, it is sufficient to indicate the year.

  (iii) The date shall be declared by the words:

---

[5] The declaration of net contents represents the quantity at the time of packaging and is subject to enforcement by reference to an average system of quantity control.

[6] The declaration of drained weight is subject to enforcement by reference to an average system of quantity control.

- "Best before ..." where the day is indicated;
- "Best before end ..." in other cases.

(iv) The words referred to in paragraph (iii) shall be accompanied by:

- either the date itself; or
- a reference to where the date is given.

(v) The day, month and year shall be declared in uncoded numerical sequence except that the month may be indicated by letters in those countries where such use will not confuse the consumer.

(vi) Notwithstanding 4.7.1 (i) an indication of the date of minimum durability shall not be required for:

- fresh fruits and vegetables, including potatoes which have not been peeled, cut or similarly treated;
- wines, liqueur wines, sparkling wines, aromatized wines, fruit wines and sparkling fruit wines;
- beverages containing 10% or more by volume of alcohol;
- bakers' or pastry-cooks' wares which, given the nature of their content, are normally consumed within 24 hours of their manufacture;
- vinegar;
- food grade salt;
- solid sugars;
- confectionery products consisting of flavoured and/or coloured sugars;
- chewing gum.

4.7.2 In addition to the date of minimum durability, any special conditions for the storage of the food shall be declared on the label if the validity of the date depends thereon.

4.8 INSTRUCTIONS FOR USE

Instructions for use, including reconstitution, where applicable, shall be included on the label, as necessary, to ensure correct utilization of the food.

5. ADDITIONAL MANDATORY REQUIREMENTS

5.1 QUANTITATIVE LABELLING OF INGREDIENTS

5.1.1 Where the labelling of a food places special emphasis on the presence of one or more valuable and/or characterizing ingredients, or where the description of the food has the same effect, the ingoing percentage of the ingredient (m/m) at the time of manufacture shall be declared.

5.1.2 Similarly, where the labelling of a food places special emphasis on the low content of one or more ingredients, the percentage of the ingredient (m/m) in the final product shall be declared.

5.1.3 A reference in the name of a food to a particular ingredient shall not of itself constitute the placing of special emphasis. A reference in the labelling of a food to an ingredient used in a small quantity and only as a flavouring shall not of itself constitute the placing of special emphasis.

5.2 IRRADIATED FOODS

5.2.1 The label of a food which has been treated with ionizing radiation shall carry a written statement indicating that treatment in close proximity to the name of the food. The use of the international food irradiation symbol, as shown below, is optional, but when it is used, it shall be in close proximity to the name of the food.

5.2.2 When an irradiated product is used as an ingredient in another food, this shall be so declared in the list of ingredients.

5.2.3 When a single ingredient product is prepared from a raw material which has been irradiated, the label of the product shall contain a statement indicating the treatment.

6. EXEMPTIONS FROM MANDATORY LABELLING REQUIREMENTS

With the exception of spices and herbs, small units, where the largest surface area is less than 10 cm^2, may be exempted from the requirements of paragraphs 4.2 and 4.6 to 4.8.

7. OPTIONAL LABELLING

7.1 Any information or pictorial device written, printed, or graphic matter may be displayed in labelling provided that it is not in conflict with the mandatory requirements of this standard and those relating to claims and deception given in Section 3 - General Principles.

7.2 If grade designations are used, they shall be readily understandable and not be misleading or deceptive in any way.

8. PRESENTATION OF MANDATORY INFORMATION

8.1 GENERAL

8.1.1 Labels in prepackaged foods shall be applied in such a manner that they will not become separated from the container.

8.1.2 Statements required to appear on the label by virtue of this standard or any other Codex standards shall be clear, prominent, indelible and readily legible by the consumer under normal conditions of purchase and use.

8.1.3 Where the container is covered by a wrapper, the wrapper shall carry the necessary information or the label on the container shall be readily legible through the outer wrapper or not obscured by it.

8.1.4 The name and net contents of the food shall appear in a prominent position and in the same field of vision.

8.2 LANGUAGE

8.2.1 If the language on the original label is not acceptable, to the consumer for whom it is intended, a supplementary label containing the mandatory information in the required language may be used instead of relabelling.

8.2.2 In the case of either relabelling or a supplementary label, the mandatory information provided shall be fully and accurately reflect that in the original label.

ANNEX IV

CODE OF HYGIENIC PRACTICE FOR THE TRANSPORT OF FOOD IN BULK AND SEMI-PACKED FOOD

CAC/RCP 47-2001[1]

INTRODUCTION

Food may become contaminated or reach their destination in an unsuitable condition for consumption unless control measures are taken during transport. Such condition may occur even where adequate hygiene measures have been taken earlier in the food chain. Adequate transportation systems should be in place which will ensure that foods remain safe and suitable for consumption upon delivery and assist countries to assure continued trade.

Good communication between shipper/manufacturer, transporter and receiver of foods is essential. They share responsibility for food safety on this part of the food chain. Food manufacturers or receivers are responsible for communicating to transporters specific food safety control procedures required during transportation.

This document is formatted in accordance with the Recommended International Code of Practice - General Principles of Food Hygiene (CAC/RCP 1-1969, Rev.3 (1997)), which must be consulted in the use of this Code. Those sections of this Code that require specific food safety requirements beyond those contained in the Recommended International Code of Practice - General Principles of Food Hygiene (CAC/RCP 1-1969, Rev.3 (1997)), due to specific transportation characteristics, are noted and the specific requirements are detailed.

This code is not applicable to, and do not take precedence over, other Codex commodity - specific codes already in existence for such commodities in bulk, for example the Recommended International Code of Practice for the Storage and Transport of Edible Oils and Fats in Bulk (CAC/RCP 36-1987, Rev.1 – 1999).

SECTION I   OBJECTIVES

The code of hygienic practice for the transport of bulk and semi-packed foods:

- identifies additional requirements of food hygiene applicable to the Recommended International Code of Practice – General Principles of Food Hygiene (CAC/RCP 1-1969, Rev 3 (1997)) applicable to the condition of the food transportation unit and the loading, transport, in-transit storage and unloading of bulk and semi-packed foods to ensure that food remains safe and suitable for human consumption.

- indicates how to implement these controls, and

- provides ways to verify that these controls have been applied.

SECTION II   SCOPE, USE AND DEFINITIONS

2.1   SCOPE AND USE

This code of practice covers the condition of the food transportation unit, loading, transport, in-transit storage and unloading of bulk, semi-packed foods and fresh produce. This code covers food transportation unit and product from the points of shipment to the points of receipt. Examples of foods included in this code include:

---

[1] The Code of Hygienic Practice for the Transport of Food in Bulk and Semi-Packaged Food was adopted by the Codex Alimentarius Commission, 2001. The Code has been sent to all Member Nations and Associate Members of FAO and WHO as an advisory text, and it is for individual governments to decide what use they wish to make of the Guidelines.

- Food transported from the packaging or processing facility to a retail/distribution establishment,

- Food transported from one process/distribution facility to another or from a process/distribution facility to another or from a process/distribution facility to a retail establishment,

- Food transported from collection points, elevators, storage facilities, etc., to processing plants/distribution sites, or retail markets.

This code does not cover growing and gathering or fishing operations that occur prior to loading product into the food transportation unit for shipment, nor does it cover in-plant conveyance of product that occurs after unloading or after off-loading and emptying. Examples of foods excluded from this code are the following:

- On farm movement of a product,

- Movement from the field to collection facility, packaging facility, or storage facility.

The code's provisions are to be applied in addition to all applicable provisions of the Recommended International Code of Practice - General Principles (CAC/RCP 1-1969, Rev. 3 (1997)) including Section 8 that specifically addresses transportation.

2.2 DEFINITIONS

Food transportationunit: Includes food transport vehicles or contact receptacles (such as containers, boxes, bins, bulk tanks) in vehicles, aircraft, railcars, trailers and ships and any other transport receptacles in which food is transported.

Bulk: Means unpacked food in direct contact with the contact surface of the food transportation unit and the atmosphere (for example, powdered, granulated or liquid form).

Semi-packed food: Semi-packed food is a food which might come in direct contact with the food transportation unit or the atmosphere (e.g. vegetables and food in crates and bags).

SECTION III   PRIMARY PRODUCTION

All sub-sections of the provisions of the Recommended International Code of Practice - General Principles of Food Hygiene (CAC/RCP 1- 1969, Rev.3 (1997)) and, as appropriate, other Codex Codes of Hygienic Practice, shall be applied.

SECTION IV   ESTABLISHMENT: DESIGN AND FACILITIES

All sub-sections of the provisions of the Recommended International Code of Practice - General Principles of Food Hygiene (CAC/RCP 1- 1969, Rev.3 (1997)) and, as appropriate, other Codex Codes of Hygienic Practice, shall be applied.

SECTION V   CONTROL OF OPERATION

5.1   CONTROL OF FOOD HAZARDS

The provisions of the Recommended International Code of Practice - General Principles of Food Hygiene (CAC/RCP 1- 1969, Rev.3 (1997)) and, as appropriate, other Codex Codes of Hygienic Practice, shall be applied.

5.1.1 IDENTIFICATION OF POTENTIAL HAZARDS

It may be useful to refer to the listed questions (see Table 1) to identify and manage hazards during transport of bulk and semi-packed foods. Reference is made also to the HACCP approach.

Table 1

Is the food "ready for direct consumption"?
Are the conditions of the food transportation unit likely to introduce or support the increase of a hazard?
Is it likely, that a hazard is introduced or increased during loading?
Is it likely, that a hazard may increase during transport or storage in the food transportation unit?
Is it likely, that a hazard is introduced or increased during unloading?

5.1.2 RECORDS OF PRIOR CARGOES AND PRIOR CLEANING

The transporter should maintain records, readily available at the food transportation unit or as prescribed by the official agency having jurisdiction, of the three most recent prior cargoes and cleaning and disinfection, where necessary, method employed of the food transportation unit including volumes transported and make this information, on request, available to the food shipper, official control authorities and/or receiver/food manufacturers, for evaluation of potential hazards.

A complete record of previous cargoes should be kept over a period of six months by the transporter.

5.1.3 SOURCES OF HAZARDS

The possibility of a hazard should be considered from the following sources, cited as examples:

5.1.3.1 Hazards related to the food transportation unit

Unsuitability of the construction material and coating, lack of sealing/locking device, residues of previous cargoes, residues from cleaning and sanitizing materials.

Where appropriate consideration should be given to food transportation units dedicated to single commodity use.

5.1.3.2 Hazards related to loading and unloading

Increase/decrease of temperature of the food. Undesirable introduction of microbes, dust, moisture, or other physical contamination.

5.1.3.3 Hazards related to transport

Leakage of heating/cooling fluid. Break down of temperature control.

5.2 KEY ASPECTS OF HYGIENE CONTROL SYSTEMS

The provisions of the Recommended International Code of Practice - General Principles of Food Hygiene (CAC/RCP 1- 1969, Rev.3 (1997)) and, as appropriate, other Codex Codes of Hygienic Practice, shall be applied.

5.3 INCOMING MATERIAL REQUIREMENTS

The provisions of the Recommended International Code of Practice - General Principles of Food Hygiene (CAC/RCP 1- 1969, Rev.3 (1997)) and, as appropriate, other Codex Codes of Hygienic Practice, shall be applied.

5.4 PACKAGING

The provisions of the Recommended International Code of Practice - General Principles of Food Hygiene (CAC/RCP 1- 1969, Rev.3 (1997)) and, as appropriate, other Codex Codes of Hygienic Practice, shall be applied.

5.5 WATER

The provisions of the Recommended International Code of Practice - General Principles of Food Hygiene (CAC/RCP 1- 1969, Rev.3 (1997)) and, as appropriate, other Codex Codes of Hygienic Practice, shall be applied.

5.6 MANAGEMENT AND SUPERVISION

The provisions of the Recommended International Code of Practice - General Principles of Food Hygiene (CAC/RCP 1- 1969, Rev.3 (1997)) and, as appropriate, other Codex Codes of Hygienic Practice, shall be applied.

5.7 DOCUMENTATION AND RECORDS

Suitable controls can be formulated by shippers or receivers to ensure food safety during transport in particular cases (see questions in Table 1). Such controls should be communicated in writing. Documentation is an important tool for validation and for verification that the principles have been adhered to. This documentation may include food transportation unit number, registration of previous loads, temperature/time recordings and cleaning certificates. Such documentation should be available to the official agencies having jurisdiction. It should be noted that some food transportation unit s are intended for single use only.

5.8 RECALL PROCEDURES

The provisions of the Recommended International Code of Practice - General Principles of Food Hygiene (CAC/RCP 1- 1969, Rev.3 (1997)) and, as appropriate, other Codex Codes of Hygienic Practice, shall be applied.

5.9 DEDICATED TRANSPORT

Where appropriate, particularly bulk transport, containers and conveyances should be designated and marked for food use only and be used only for that purpose.

Bulk food in liquid, granulated or powder form must be transported in receptacles and/or containers/tankers reserved for the transport of food unless the application of principles such as HACCP demonstrates that dedicated transport for these products is not necessary to achieve the same level of food safety.

SECTION VI ESTABLISHMENT: MAINTENANCE AND SANITATION

All sub-sections of the Recommended International Code of Practice - General Principles of Food Hygiene (CAC/RCP 1- 1969, Rev.3 (1997)) and, as appropriate, other Codex Codes of Hygienic Practice, shall be applied.

Food transportation unit s, accessories, and connections should be cleaned, disinfected (where appropriate) and maintained to avoid or at least reduce the risk of contamination. It should be noted that depending on the commodity relevant, different cleaning procedures are applicable, which should be recorded. Where necessary, there should be disinfection with subsequent rinsing unless manufacturers instruction indicates on a scientific basis that rinsing is not required.

SECTION VII ESTABLISHMENT: PERSONAL HYGIENE

All sub-sections of the provisions of the Recommended International Code of Practice - General Principles of Food Hygiene (CAC/RCP 1- 1969, Rev.3 (1997)) and, as appropriate, other Codex Codes of Hygienic Practice, shall be applied.

The General Principles of Food Hygiene should apply to all personnel in contact with the food.

SECTION VIII   TRANSPORTATION

All sub-sections of the provisions of the Recommended International Code of Practice - General Principles of Food Hygiene (CAC/RCP 1- 1969, Rev.3 (1997)) and, as appropriate, other Codex Codes of Hygienic Practice, shall be applied.

8.4   FOOD TRANSPORTATION UNITS

The design of the food transportation unit should be such as to avoid cross contamination due to simultaneous or consecutive transport. Important aspect are cleanability and appropriate coatings.

Construction and design of the food transportation unit should facilitate inspection, cleaning, disinfection and when appropriate enable temperature control.

Use of means for cooling or heating should by design and construction be such as to avoid contamination. Although hot water and steam are preferred means of heating, other substances may be used on the basis of safety and risk evaluation and inspection procedures. Upon request by the competent authority, evidence may be required to demonstrate that the heating media employed have been properly evaluated and safely used.

Inner surface materials suitable for direct food contact should be used. These should be non-toxic, inert, or at least compatible with the transported food, and which do not transfer substances to the food or adversely affect the food. Stainless steel or surface coated with food-grade epoxy resins are most suitable. The interior design should eliminate areas that are difficult to access and clean.

The appropriate design of the food transportation unit should assist in preventing access of insects, vermin, etc, contamination from the environment, and when necessary, providing insulation against loss or gain of heat, adequate cooling or heating capacity, and facilitation of locking or sealing.

There should be appropriate facilities conveniently available for cleaning and, where appropriate disinfecting of the food transportation unit.

Auxiliary equipment should be (where appropriate) subjected to the above stated requirements.

To maintain sanitary conditions, facilities should be provided for the storage of pipes, hoses and other equipment used in the transfer of foods.

SECTION IX   PRODUCTION INFORMATION AND CONSUMER AWARENESS

All sub-sections of the provisions of the Recommended International Code of Practice - General Principles of Food Hygiene (CAC/RCP 1- 1969, Rev.3 (1997)) and, as appropriate, other Codex Codes of Hygienic Practice, shall be applied.

SECTION X   TRAINING

All sub-sections of the provisions of the Recommended International Code of Practice - General Principles of Food Hygiene (CAC/RCP 1- 1969, Rev.3 (1997)) and, as appropriate, other Codex Codes of Hygienic Practice, shall be applied.

It is important that personnel responsible for the transport are well aware of the nature of the foods that are being handled/transported and the possible extra precautionary measures that may be required. Personnel should be trained on food transportation unit inspection procedures for food safety.

ANNEX V

GUIDELINES FOR FOOD IMPORT CONTROL SYSTEMS
CAC/GL 47-2003

SECTION 1 - SCOPE

This document provides a framework for the development and operation of an import control system to protect consumers and facilitate fair practices in food trade while ensuring unjustified technical barriers to trade are not introduced. The Guideline is consistent with the Codex *Principles for Food Import and Export Inspection and Certification*[1] and provides specific information about imported food control that is an adjunct to the *Guidelines for the Design, Operation, Assessment and Accreditation of Food Import and Export Inspection and Certification Systems*[2].

SECTION 2 - DEFINITIONS [3]

Appropriate Level of Protection (ALOP) is the level of protection deemed appropriate by the country establishing a sanitary measure to protect human life or health within its territory. (This concept may otherwise be referred to as the "acceptable level of risk".)

Audit* is a systematic and functionally independent examination to determine whether activities and related results comply with planned objectives.

Certification* is the procedure by which official certification bodies and officially recognized bodies provide written or equivalent assurance that foods or food control systems conform to requirements. Certification of food may be, as appropriate, based on a range of inspection activities which may include continuous on-line inspection, auditing of quality assurance systems, and examination of finished products.

Inspection* is the examination of food or systems for control of food, raw materials, processing and distribution, including in-process and finished product testing, in order to verify that they conform to requirements.

Legislation* includes acts, regulations, requirements or procedures, issued by public authorities, related to foods and covering the protection of public health, the protection of consumers and conditions of fair trading.

Official accreditation* is the procedure by which a government agency having jurisdiction formally recognizes the competence of an inspection and/or certification body to provide inspection and certification services.

Official inspection systems and official certification systems* are systems administered by a government agency having jurisdiction empowered to perform a regulatory or enforcement function or both.

Officially recognized inspection systems and officially recognized certification systems* are systems which have been formally approved or recognized by a government agency having jurisdiction.

Requirements* are the criteria set down by the competent authorities relating to trade in foodstuffs covering the protection of public health, the protection of consumers and conditions of fair trading.

Risk assessment* A scientifically based process consisting of the following steps (i) hazard identification, (ii) hazard characterisation, (iii) exposure assessment, and (iv) risk characterisation.

---

[1] *Principles for Food Import and Export Inspection and Certification* (CAC/GL 20-1995)

[2] *Guidelines for the Design, Operation, Assessment and Accreditation of Food Import and Export Inspection and Certification Systems* (CAC/GL 26-1997).

[3] Definitions drawn from the *Guidelines for the Design, Operation, Assessment and Accreditation of Food Import and Export Inspection and Certification Systems* (CAC/GL 26-1997) are marked with * . Definitions drawn from Codex Alimentarius Commission, Procedural Manual (12th edition) are marked with * * .

Risk analysis* A process consisting of three components: risk assessment, risk management and risk communication.

SECTION 3 - GENERAL CHARACTERISTICS OF FOOD IMPORT CONTROL SYSTEMS

Food import control systems should have the following main characteristics:

- requirements for imported food that are consistent with requirements for domestic foods;
- clearly defined responsibilities for the competent authority or authorities;
- clearly defined and transparent legislation and operating procedures;
- precedence to the protection of consumers;
- provision of the importing country for recognition of the food control system applied by an exporting country's competent authority;
- uniform nationwide implementation;
- implementation that ensures the levels of protection achieved are consistent with those for domestic food.

REQUIREMENTS FOR IMPORTED FOOD THAT ARE CONSISTENT WITH REQUIREMENTS FOR DOMESTIC FOODS

Requirements are commonly expressed as end-point standards with specific limits and complementary sampling regimes. These requirements may consist of standards, provisions for sampling, process controls, conditions of production, transport, storage, or a combination of these.

The extent and stringency of requirements applied in specific circumstances should be proportionate to risk, noting that risk may vary from one source to another because of factors such as specific and/or similar situations in the region of origin, technology employed, compliance history, etc. and/or examination of relevant attributes of a sample of products at import.

As far as possible, requirements should be applied equally to domestically produced and imported food. Where domestic requirements include process controls such as good manufacturing practices, compliance may be determined or equivalence confirmed by auditing the relevant inspection and certification systems and, as appropriate, the facilities and procedures in the exporting country[4].

CLEARLY DEFINED RESPONSIBILITIES OF COMPETENT AUTHORITY OR AUTHORITIES

The competent authority(ies) involved in any of the imported food inspection functions at the point or points of entry, during storage and distribution and/or at point of sale, should have clearly defined responsibilities and authority. Multiple inspection and duplicative testing for the same analyte(s) on the same consignment should be avoided to the extent possible.

Some countries, for example those that are part of a regional economic grouping, may rely on import controls implemented by another country. In such cases, the functions, responsibilities, and operating procedures undertaken by the country which conducts the imported food control should be clearly defined and accessible to authorities in the country or countries of final destination with the aim of delivering an efficient and transparent import control system.

Where the competent authorities of an importing country use third party providers as officially recognised inspection bodies and/or officially recognized certification bodies to implement controls, such arrangements should be conducted in the manner discussed in CAC/GL 26-1997, Section 8, Official Accreditation. The functions that can be conducted by such providers may include:

---

[4] *Guidelines for the Design, Operation, Assessment and Accreditation of Food Import and Export Inspection and Certification Systems* (CAC/GL 26-1997), Para. 54.

- sampling of target consignments;
- analysis of samples;
- compliance evaluation of relevant parts or all of a quality assurance system that may be operated by importers in order to comply with official requirements.

CLEARLY DEFINED AND TRANSPARENT LEGISLATION AND OPERATING PROCEDURES

The object of legislation is to provide the basis and the authority for operating a food import control system. The legal framework allows for the establishment of the competent authority(ies) and the processes and procedures required to verify the conformity of imported products against requirements.

Legislation should provide the competent authority with the ability to:

- appoint authorised officers;
- require prior notification of the importation of a consignment of a foodstuff;
- require documentation;
- inspect, including the authority to enter premises within the importing country, physically examine the food and its packaging; collect samples and initiate analytical testing; inspection of documentation provided by an exporting country authority, exporter or importer; and verification of product identity against documentary attestations;
- apply risk-based sampling plans, taking into consideration the compliance history of the particular food, the validity of accompanying certification, and other relevant information;
- charge fees for the inspection of consignments and sample analysis;
- recognize accredited or accredit laboratories;
- accept; reject; detain; destroy; order to destroy; order reconditioning, processing, or re-export; return to country of export; designate as non-food use;
- recall consignments following importation;
- retain control over consignments in transit during intra-national transport or during storage prior to import clearance; and,
- implement administrative and/or judicial measures when the specific requirements are not satisfied.

In addition, the legislation may make provisions for:

- licensing or registration of importers;
- recognition of verification systems used by importers;
- an appeal mechanism against official actions;
- assessing the control system of the exporting country; and
- certification and/or inspection arrangements with competent authorities of exporting countries.

PRECEDENCE TO THE PROTECTION OF CONSUMERS

In the design and operation of food import control systems, precedence should be given to protecting the health of consumers and ensuring fair practices in food trade over economic or other trade considerations.

PROVISION OF THE IMPORTING COUNTRY FOR RECOGNITION OF THE FOOD CONTROL SYSTEM APPLIED BY AN EXPORTING COUNTRY'S COMPETENT AUTHORITY

Food import control systems should include provisions for recognition as appropriate of the food control system applied by an exporting country' s competent authority. Importing countries can recognise the food safety controls of an exporting country in a number of ways that facilitate the entry of goods, including the use of memoranda of understanding, mutual recognition agreements and equivalence agreements and unilateral recognition. Such recognition should, as appropriate, include controls applied during the production, manufacture, importation, processing, storage, and transportation of the food products, and verification of the export food control system applied.

UNIFORM NATION-WIDE IMPLEMENTATION

Uniformity of operational procedures is particularly important. Programmes and training manuals should be developed and implemented to assure uniform application at all points of entry and by all inspection staff.

IMPLEMENTATION THAT ENSURES THE LEVELS OF PROTECTION ACHIEVED ARE CONSISTENT WITH THOSE FOR DOMESTIC FOOD

As an importing country has no direct jurisdiction over process controls applied to food manufactured in another country, there may be a variation in approach to the compliance monitoring of domestic and imported food. Such differences in approach are justifiable provided they are necessary to ensure that the level of protection achieved is consistent with that of domestically produced food.

SECTION 4 - IMPLEMENTATION OF THE CONTROL SYSTEM

Operational procedures should be developed and implemented to minimize undue delay at the point or points of entry without jeopardizing effectiveness of controls to meet requirements. Implementation should take into account the factors listed in this section and the possibility of recognizing guarantees at origin that includes implementation of controls in the exporting countries.

POINT OF CONTROL

Control of imported food by the importing country can be conducted at one or more points including the points of :

- origin, where agreed upon with the exporting country;
- entry to the country of destination;
- further processing;
- transport and distribution;
- storage; and,
- sale, (retail or wholesale).

The importing country can recognize controls implemented by the exporting country. The application of controls by the exporting country, during production, manufacture and subsequent transit should be encouraged, with the aim of identifying and correcting problems when and where they occur, and preferably before costly recalls of food already in distribution are required.

Pre-shipment clearance is a possible mechanism for ensuring compliance with requirements of, for example, valuable bulk packed products that if opened and sampled upon entry, would be seriously compromised, or for products that require rapid clearance to maintain safety and quality.

If the inspection system encompasses pre-shipment clearance then the authority to conduct the clearance should be determined and procedures defined. The importing country's competent authority may choose to conduct pre-shipment clearance from an exporting country's official certification system or from officially recognised third party certification bodies working to defined criteria. The pre-shipment clearance should be based on the results of the documentary check on the consignments.

INFORMATION ABOUT FOOD TO BE IMPORTED [5]

The efficacy of the control system in applying efficient targeted control measures depends upon information about consignments entering the jurisdiction. Details of consignments that may be obtained include:

- date and point of entry;
- mode of transport;
- comprehensive description of the commodity (including for example product description, amount, means of preservation, country of origin and/or of dispatch, identifying marks such as lot identifier or seal identification numbers etc);
- exporter's and importer's name and address;
- manufacturer and/or producer, including establishment registration number;
- destination; and,
- other information.

FREQUENCY OF INSPECTION AND TESTING OF IMPORTED FOOD

The nature and frequency of inspection, sampling and testing of imported foods should be based on the risk to human health and safety presented by the product, its origin and the history of conformance to requirements and other relevant information. Control should be designed to account for factors such as:

- the risk to human health posed by the product or its packaging;
- the likelihood of non-compliance with requirements;
- the target consumer group;
- the extent and nature of any further processing of the product;
- food inspection and certification system in the exporting country and existence of any equivalence, mutual recognition agreements or other trade agreements; and,
- history of conformity of producers, processors, manufacturers, exporters, importers and distributors.

Physical checks of imported product, preferably using statistically based sampling plans, should represent valid methods for the verification of compliance with requirements by the product as established by the importing country, or in the case of importing a product for the purposes of re-exportation, verification should be made on the requirements of the country of final destination and said requirements should be specified in the certificate of re-exportation. Inspection procedures should be developed to include defined sampling frequencies or inspection intensities, including for re-exported product.

---

[5] *Generic Official Certificate Formats and the Production and Issuance of Certificates* (CAC/GL 38-2001)

Sampling frequency of products supplied from a source for which there is no or known poor compliance history may be set at a higher rate than for products with a good compliance history provided this is shown through transparent and objective criteria. The sampling process enables a compliance history to be created. Similarly, food from suppliers or imported by parties with a known poor compliance history should be sampled at higher intensity. In these cases, every consignment may need to be physically inspected, until a defined number of consecutive consignments meets requirements. Alternatively the inspection procedures can be developed to automatically detain product from suppliers with a known poor compliance history and the importer may be required to prove the fitness of each consignment through use of a laboratory (including official laboratory) recognized, accredited and/or listed by the competent authority until a satisfactory compliance rate is achieved.

SAMPLING AND ANALYSIS

The inspection system should be based on Codex sampling plans for the particular commodity/contaminant combination where available. In the absence of Codex sampling plans, reference should be made to internationally accepted or scientifically based sampling plans.

Internationally validated standard methods of analysis or methods validated through international protocols should be used where available. Analysis should be conducted in official or officially accredited laboratories.

DECISIONS

Decision criteria (without prejudice to the application of customs procedures) should be developed that determine whether consignments are given:

- acceptance;
- entry if cleared upon inspection or verification of conformance;
- release of non-conforming product after re-conditioning and/or corrective measures have been taken;
- rejection notice, with redirecting product for uses other than human consumption;
- rejection notice, with re-exportation option or return to country of export option at exporter expense;
- rejection notice with destruction order.

Results of inspection and, if required, laboratory analysis, should be carefully interpreted in making decisions relating to acceptance or rejection of a consignment. The inspection system should include decision-making rules for situations where results are borderline, or sampling indicates that only some lots within the consignment comply with requirements. Procedures may include further testing and examination of previous compliance history.

The system should include formal means to communicate decisions regarding clearance and status of consignments.[6] There should be an appeal mechanism and/or opportunity for review of official decisions on consignments.[7] When food is rejected because it fails to meet national standards of the importing country but conforms to international standards, the option of withdrawing the rejected consignment should be considered.

---

[6] Paragraph 4 of the *Guidelines for the Exchange of Information Between Countries on Rejections of Imported Food* (CAC/GL 25-1997) should be consulted in this regard.

[7] Paragraph 6 of the *Guidelines for the Exchange of Information Between Countries on Rejections of Imported Food* (CAC/GL 25-1997) should be consulted in this regard.

DEALING WITH EMERGENCY SITUATIONS

The responsible authority should have procedures that can respond appropriately to emergency situations. This will include holding suspect product upon arrival and recall procedures for suspect product already cleared and, if relevant, rapid notification of the problem to international bodies and possible measures to take.

If the food control authorities in importing countries detect problems during import control of foodstuffs which they consider to be so serious as to indicate a food control emergency situation, they should inform the exporting country promptly by telecommunication.[8]

RECOGNITION OF EXPORT CONTROLS

Consistent with paragraph 12 of these guidelines, the importing country should establish mechanisms to accept control systems in an exporting country where these systems achieve the same level of protection required by the importing country. In this regard, the importing country should:

- develop procedures to conduct assessment of the exporting country systems consistent with the Annex of the *Guidelines for the Design, Operation, Assessment and Accreditation of Food Import and Export Inspection and Certification Systems* (CAC/GL 26-1997);
- take into account the scope of the arrangement, for example, whether it covers all foods or is restricted to certain commodities or certain manufacturers;
- develop clearance procedures that achieve its appropriate level of protection if arrangements developed with an exporting country are limited in scope;
- provide recognition of export controls through, for example, exemption from routine import inspection;
- conduct verification procedures for example, occasional random sampling and analysis of products upon arrival. (Section 5 and Annex of CAC/GL 26-1997 deal with the provision and verification of systems that provide certification for food in trade);
- recognize that arrangements need not rely on the presentation of certificates or documentation with individual consignments, when such an approach is acceptable to both parties.

The competent authority of the importing country may, develop certification agreements with exporting country official certification bodies or officially recognized certification bodies, with the aim of ensuring requirements are met. Such agreements may be of particular value where, for example, there is limited access to specific facilities such as laboratories and consignment tracking systems.[9]

INFORMATION EXCHANGE

Food import control systems involve information exchange between competent authorities of exporting and importing countries. The information may include:

- requirements of food control systems;
- "hard copy" certificates attesting to conformity with requirements of the particular consignment;
- electronic data or certificates where accepted by the parties involved;
- details about rejected food consignment, such as destruction, re-exportation, processing, re-conditioning or redirection of consignment for uses other than human consumption;
- list of establishments or facilities that conform to importing country requirements.

---

[8] *Guidelines for the Exchange of Information in Food Control Emergency Situations* (CAC/GL 19-1995)
[9] *Guidelines for the Development of Equivalence Agreements Regarding Food Import and Export Inspection and Certification Systems* (CAC/GL 34-1999)

Any changes to import protocols, including specifications, which may significantly affect trade, should be promptly communicated to trading partners, allowing a reasonable interval between the publication of regulations and their application.

OTHER CONSIDERATIONS

The competent authority may consider developing alternative arrangements in lieu of routine inspection. This may include agreements where the competent authority assesses the controls that importers implement over suppliers and the procedures that are in place to verify compliance of suppliers. Alternative arrangements may include some sampling of product as an audit, rather than routine inspection.

The competent authority may consider developing a system where registration of importers is mandatory. Advantages include the ability to provide the importers and exporters with information about their responsibilities and mechanisms to ensure imported food complies with requirements.

If a product registration system exists or is implemented, a clear rationale for such product registration (e.g. specific and documented food safety concerns) should exist. Such product REGISTRATIONS SHOULD TREAT IMPORTED AND DOMESTIC PRODUCT IN THE SAME OR EQUIVALENT MANNER.

DOCUMENTING THE SYSTEM

A food import control system should be fully documented, including a description of its scope and operation, responsibilities and actions for staff, in order that all parties involved know precisely what is expected of them.

Documentation of an food import control systems should include:

- an organizational chart of the official inspection system, including geographical location and the roles of each level in the hierarchy;
- job functions as appropriate;
- operating procedures including methods of sampling, inspection and testing;
- relevant legislation and requirements that should be met by imported food;
- important contacts;
- relevant information about food contamination and food inspection; and,
- relevant information on staff training.

TRAINED INSPECTORATE

It is fundamental to have adequate, reliable, well trained and organised inspection staff, with supporting infrastructure, to deliver the food import control system. Training, communication, and supervisory elements should be organised to provide consistent implementation of requirements by the inspectorate throughout the food import control system.

Where third parties are officially recognised by the competent authority of the importing country to perform specified inspection work, the qualifications of the inspection staff should be at least the same as inspection staff of the competent authority who may carry out similar tasks.

The competent authority of the importing country responsible for conducting assessment of food control systems of exporting countries should engage personnel with appropriate qualifications, experience and training required of personnel assessing domestic food controls.

SYSTEM VERIFICATION

Verification should be carried out on the basis of Section 9 of the *Guidelines for the Design, Operation, Assessment and Accreditation of Food Import and Export Inspection and Certification Systems* (CAC/GL 26-1997) and the food import control system should be independently assessed on a regular basis.

SECTION 5 - FURTHER INFORMATION

The Food and Agriculture Organization of the United Nations *Manual of Food Quality Control. Imported Food Inspection* (Food and Nutrition Paper 14/15, 1993) and World Health Organization/Western Pacific Regional Center for the Promotion of Environmental Planning and Applied Science (PEPAS): *Manual for the Inspection of Imported Food* (1992) contribute valuable information for those engaged in the design and re-design of food import control systems.

BIBLIOGRAPHY

J. Abraham and E. Millstone, 'Food additive controls. Some international comparisons', 14 *Food Policy* (1989) pp. 43-57.

D. Abdel Motaal, 'The Agreement on Technical Barriers to Trade, the Committee on Trade and Environment, and Eco-labelling', in G.P. Sampson and W. Bradnee (eds.), *Trade, Environment and the Millennium* (Tokyo, United Nations University Press 1999).

D. Abdel Motaal, 'The 'Multilateral Scientific Consensus' and the World Trade Organization', 38 *Journal of World Trade* (2004) pp. 855-876.

G. Abi-Saab, *The concept of international organization* (Paris, UNESCO 1981).

D. Ahn, 'Linkages between International Financial and Trade Institutions', 34 *Journal of World Trade* (2000) pp. 1-35.

D. Ahn, 'Comparative Analysis of the SPS and the TBT Agreements', 8 *International Trade Law & Regulation* (2002) pp. 85-96.

D. Akande, 'The International Court of Justice and the Security Council: is there room for judicial control of decisions of the political organs of the United Nations?', 46 *International and Comparative Law Quarterly* (1997) pp. 309-343.

C.H. Alexandrowicz, *The Law-Making Functions of the Specialised Agencies of the United Nations* (Sydney, Angus and Robertson 1973)

Ph. Allott, 'Making the new international law: law of the sea as law of the future', 40 *International Journal* (1985) pp. 442-460.

Ph. Allott, 'European governance and the re-branding of democracy', 27 *European Law Review* (2002) pp. 60-71.

J.E. Alvarez, 'Judging the Security Council', 90 *American Journal of International Law* (1996) pp. 1-39.

J.E. Alvarez, International Organizations as Law-makers (Oxford, Oxford University Press 2005).

J.E. Alvarez, 'International Organizations: Then and Now' 100 *American Journal of International Law* (2006) pp. 324-347.

A. D'Amato, 'On Consensus', 8 *Canadian Yearbook of International Law* (1970) pp. 104-122.

C.F. Amerasinghe, 'Interpretation of texts in open international organizations', 65 *British Yearbook of International Law* (1995) pp. 175-209.

R. Ancuceanu, 'Maximum Residue Limits of Veterinary Medical Products and Their Regulation in European Community Law', 9 *European Law Journal* (2003) pp. 215-240.

Ch. Apostolidis, et al. (eds.), *L'Humanite face a la Mondialisation* (Paris, L'Harmattan 1997).

A.E. Appleton, *Environmental Labelling Programmes: International Trade Law Implications* (London, Kluwer Law International 1997).

A.E. Appleton, 'The Labeling of GMO Products Pursuant to International Trade Rules', 8 *N.Y.U. Environmental Law Journal* (2000) pp. 566-578.

A.E. Appleton and V. Heiskanen, 'The Sardines Decision: Fish without Chips?', in A.D. Mitchell (ed.), *Challenges and Prospects for the WTO* (London, Cameron May 2005), pp. 165-192.

A. Arnull and D. Wincott, *Accountability and Legitimacy in the European Union* (Oxford, Oxford University Press 2002).

J. Atik, 'Science and International Regulatory Convergence', 17 *Northwestern Journal of International Law & Business* (1996-1997) pp. 736-758.

N. Avery, et al., *Cracking the Codex* (London, National Food Alliance, 1993).

B. Balassa, 'The Tokyo Round and the Developing countries', 14 *Journal of World Trade Law* (1980) pp. 93-118.

J.J. Barcelo, 'Product Standards to Protect the Local Environment - the GATT and the Uruguay Round Sanitary and Phytosanitary Agreement', 27 *Cornell International Law Journal* (1994) pp. 755-776.

R. Barents and L.J. Brinkhorst, *Grondlijnen van Europees Recht*, 9th edn. (Deventer, Kluwer 1999).

L. Bartels, 'Applicable Law in WTO Dispute Settlement Proceedings', 35 *Journal of World Trade* (2001) pp. 499-519.

M. Bedjaoui, *Droit International. Bilan et perspectives,* Tome 1 (Paris, Éditions A. Pedone, Unesco 1991).

D. Beetham and Ch. Lord, *Legitimacy and the European Union* (London, Longham 1998).

Y. Beigbeder, *L'Organisation Mondiale de la Santé* (Paris, Presses Universitaires de France 1997).

M. Beise, *Die Welthandelsorganisation (WTO). Funktion, Status, Organisation* (Baden-Baden, Nomos Verlagsgesellschaft 1996).

P.H.F. Bekker, *The Legal Position of Intergovernmental Organizations. A Functional Necessity Analysis of Their Legal Status and Immunities* (Dordrecht, Martinus Nijhoff Publishers 1994).

W. Benedek, 'Relations of the WTO with other international organizations and NGOs', in F. Weiss, et al. (eds.), *International Economic Law with a Human Face* (The Hague, Kluwer Law International 1998), pp. 479-495.

J. Bengoetxea, *The Legal Reasoning of the European Court of Justice* (Oxford, Oxford University Press 1993).

R. Bernhardt (ed.), *Encyclopedia of public international law* (Max Planck Institute for Comparative Public Law and International law, Amsterdam, North Holland 1995).

M. Bettati and P.-M. Dupuy, *Les O.N.G. et le Droit International,* Collection Droit International (Paris, Economica 1986).

J. Beyers and G. Dierickx, 'The Working Groups of the Council of the European Union: Supranational or Intergovernmental Negotiations?', 36 *Journal of Common Market Studies* (1998) pp. 289-317.

J.N. Bhagwati and R.E. Hudec (eds.), *Fair Trade and Harmonization* (Cambridge, MA, The MIT Press 1996).

F. Biermann, 'The Rising Tide of Green Unilateralism in World Trade Law. Options for Reconciling the Emerging North-South Conflict', 3 *Journal of World Trade* (2001) pp. 421-448.

E.J. Bigwood and A. Gérard, *Fundamental Principles and Objectives of a Comparative Food Law* (Basel, S. Karger 1967).

J. Bizet, 'Sécurité alimentaire: le Codex Alimentarius', 450 *Les Rapports du Sénat* (1999-2000).

N.M. Blokker and H.G. Schermers (eds.), *Proliferation of International Organizations: legal issues* (The Hague, Kluwer Law International 2001).

D. Bodansky, 'The Legitimacy of International Governance: A Coming Challenge for International Environmental Law?', 93 *American Journal of International Law* (1999) pp. 596-624.
L. Boisson de Chazournes and Ph. Sands, *International Law, the International Court of Justice and nuclear Weapons* (Cambridge, Cambridge University Press 1999).
L. Boisson de Chazournes, et al. (eds.), *International organizations and international dispute settlement: trends and prospects* (Ardsely, New York, Transnational Publishers 2002).
P. Van den Bossche, 'Appellate Review in WTO Dispute Settlement', in F. Weiss (ed.), *Improving WTO Dispute Settlement Procedures. Issues and Lessons from the Practice of Other International Courts and Tribunals* (London, Cameron May 2000), pp. 305-319.
P. Van den Bossche, *The Law and Policy of the World Trade Organization. Text, Cases and Materials* (Cambridge, Cambridge University Press 2005).
P. Van den Bossche, et al., *WTO Rules on Technical Barriers to Trade,* Maastricht Working Papers, no. 6 (Maastricht, University of Maastricht 2005).
G. Bossis, 'Gestion des risques alimentaires et droit international: la prise en compte de facteurs non-scientifiques', 107 *Revue Générale de Droit International Public* (2003) pp. 694-713.
G. Bossis, *La sécurité sanitaire des aliments en droit international et communautaire. Rapports croisés et perspectives d'harmonisation* (Bruxelles, Bruylant 2005).
G. Bourgeois, et al., *La Pratique de l'Expertise Judiciaire* (Paris, Litec 1999).
J. Bourrinet and S. Maljean-Dubois, *Le commerce international des organismes génétiquement modifiés* (CERIC, Université d'Aix-Marseille, La documentation Française 2002).
E. Boutrif, 'The new role of Codex Alimentarius in the context of WTO/SPS agreement', 14 *Food Control* (2003) pp. 81-88.
D.W. Bowett, et al., *Bowett's law of international institutions,* 5th edn. (London, Sweet & Maxwell 2001).
J. Braithwaite and P. Drahos, *Global Business Regulation* (Cambridge, Cambridge University Press 2000).
M.E. Bredahl and K.W. Forsythe, 'Harmonizing Phyto-sanitary and Sanitary Regulations', 12 *The World Economy* (1989) pp. 189-206.
S. Brewer, 'Scientific Expert Testimony and Intellectual Due Process', 107 *The Yale Law Journal* (1998) pp. 1535-1681.
S. Breyer and V. Heyvaert, 'Institutions for Regulating Risk', in R.L. Revesz, et al. (eds.), *Environmental Law and the Economy and Sustainable Development* (Cambridge, Cambridge University Press 2000).
M.C.E.J. Bronckers, 'More power to the WTO?', 4 *Journal of International Economic Law* (2001) pp. 41-65.
O. Brouwer, 'Free movement of foodstuffs and quality requirements: has the Commission got it wrong?', 25 *Common Market Law Review* (1988) pp. 237-261.
E. Brown Weiss and J. Jackson, *Reconciling Environment and Trade* (New York, Transnational Publishers 2001).
J. Brunnée, 'COPing with Consent: Law-Making Under Multilateral Environmental Agreements', 15 *Leiden Journal of International Law* (2002) pp. 1-52.
D.E. Buckingham, 'A Recipe for Change: Towards an Integrated Approach to Food under International Law', 6 *Pace International Law Review* (1994) pp. 285-321.

Th. Buergenthal, 'Proliferation of International Courts and Tribunals: Is It Good or Bad?', 14 *Leiden Journal of International Law* (2001) pp. 267-275.

G. de Burca, 'The Principle of Proportionality and its Application in EC Law', 13 *Yearbook of European Law* (1993) pp. 105-150.

G. de Burca and J.H.H. Weiler (eds.), *The European Court of Justice* (Oxford, Oxford University Press 2001).

W.W. Burke-White, 'International Legal Pluralism', 25 *Michigan Journal of International Law* (2003) pp. 963-979.

C. Button, *The Power to Protect. Trade, Health and Uncertainty in the WTO*, Studies in International Trade Law, Vol. 2 (Oxford, Hart Publishing 2004).

B. Buzan, 'Negotiating by consensus: developments in technique at the United Nations Conference on the Law of the Sea', 75 *American Journal of International Law* (1981) pp. 324-348.

D. Chalmers, 'Food for Thought': Reconciling European Risks and Traditional Ways of Life', 66 *Modern Law Review* (2003) pp. 532-562.

J.I. Charney, 'Is International Law Threatened by Multiple International Tribunals?', 63 *Receuil des Cours* (1998).

S. Charnovitz, 'Triangulating the World Trade Organization', 96 *American Journal of International Law* (2002) pp. 28-55.

S. Charnovitz, 'Improving the Agreement on Sanitary and Phytosanitary Standards', in G.P. Sampson and W. Bradnee Chambers (eds.), *Trade, Environment and the Millenium* (Tokyo, United Nations University Press 1999), pp. 171-193.

A. Chayes and A.H. Chayes, 'On compliance', 47 *International Organization* (2001) pp. 175-205.

J.P. Chiaradia-Bousquet, *Legislation governing food control and quality certification*, FAO Legislative Study 54 (Rome, FAO 1995).

E. Chiti, 'The emergence of a Community administration: the case of European Agencies', 37 *Common Market Law Review* (2000) pp. 309-343.

E. Chiti, 'Administrative proceedings involving european agencies', 68 *Law and Contemporary Problems* (2004) pp. 219-236.

T. Christoforou, 'Settlement of Science-based Trade Disputes in the WTO: A Critical Review of the Developing Case Law in the Face of Scientific Uncertainty', 8 *New York University Environmental Law Journal* (2000) pp. 622-648.

A. Chua, 'The Precedential Effect of WTO Panel and Appellate Body Reports', 11 *Leiden Journal of International Law* (1998) pp. 45-61.

W.C. Clarke and G. Majone, 'The Critical Appraisal of Scientific Inquiries with Policy Implications', 10 *Science, Technology, & Human Values* (1985) pp. 6-19.

Ch.P. Cockerill, 'Agricultural Pesticides: The Urgent Need for Harmonization of International Regulation', 9 *California Western International Law Journal* (1979) pp. 111-138.

J.M. Coicaud, 'Reflections on international organisations and international legitimacy: constraints, pathologies, and possibilities', *International Social Science Journal*, UNESCO (2001) pp. 523-536.

J.M. Coicaud and V. Heiskanen (eds.), *The Legitimacy of international organizations* (Tokyo, United Nations University Press 2001).

C.A. Colliard and L. Dubouis, *Institutions Internationales*, 10th edn. (Paris, Dalloz 1995).

A. Cosbey, *A Force Evolution? The Codex Alimentarius Commission, Scientific Uncertainty and the Precautionary Principle* (Canada, International Institute for Sustainable Development 2000).

Th. Cottier, 'Risk management experience in WTO dispute settlement', in D. Robertson and A. Kellow (eds.), *Globalization and the Environment. Risk Assessment and the WTO* (Cheltenham, UK, Edward Elgar 1999), pp. 41-62.
P. Craig and G. de Burca, *EU Law: Texts, Cases and Materials*, 3rd edn. (Oxford, Oxford University Press 2003).
Crossley, S.J., *Consultant's report: Review of the Working Procedures of the Joint FAO/ WHO Meeting on Pesticide Residues (JMPR)*, 3 February 2002.
K.W. Dam, *The GATT Law and International Economic Organization* (Chicago, The University of Chicago Press 1970)
G.M. Danilenko, *Law-Making in the International Community* (Dordrecht, Martinus Nijhoff Publishers 1993).
R.J. Dawson, 'The role of the Codex Alimentarius Commission in setting food standards and the SPS agreement implementation', 6 *Food Control* (1995) pp. 261-265.
V. Deckmyn and I. Thomson (eds.), *Openness and Transparency in the European Union* (Maastricht, European Institute of Public Administration 1998).
M. Diez de Velasco Vallejo, *Les organisations internationales*, Collection Droit International (Paris, Economica 2002).
D. Dinan, *Ever Closer Union. An Introduction to European Integration* (NewYork, Palgrave 1994).
J.P. Dobbert, 'Le Codex Alimentarius vers une nouvelle méthode de réglementation internationale', 15 *Annuaire Français de Droit International* (1969) pp. 677-717.
J.P. Dobbert, 'Decisions of International Organizations-Effectiveness in Member States. Some Aspects of the Law and Practice of FAO', in S.M. Schwebel (ed.), *The Effectiveness of International Decisions - Papers of a conference of The American Society of International Law and the Proceedings of the conference*, (Leyden, Sijthoff 1971).
W.Th. Douma, 'How Safe is Safe? The EU, the USA and the WTO Codex Alimentarius Debate on Food Safety Issues', in V. Kronenberger (ed.), *The EU and the International Legal Order* (The Hague, Asser Press 2001), pp. 181-197.
P.-M. Dupuy, 'The danger of fragmentation or unification of the international legal system and the International Court of Justice', 31 *New York University Journal of International Law and Politics* (1999) pp. 791-807.
M. Echols, *Food Safety and the WTO. The interplay of culture, science and technology* (The Hague, Kluwer Law International 2001).
C.D. Ehlermann and L. Ehring, 'Decision-Making in the World Trade Organization. Is the consensus practice of the World Trade Organization adequate for making, revising and implementing rules on international law?', 8 *Journal of International Economic Law* (2005) pp. 51-75.
A. Fischer-Lescano and G. Teubner, 'Regime-collisions: the vain search for unity in the fragmentation of global law', 25 *Michigan Journal of International Law* (2003) pp. 999-1046.
J.M. Finger and A. Olechowski, *The Uruguay Round. A Handbook on the Multilateral Trade Negotiations* (Washington D.C., The World Bank 1987).
M. Footer, 'The Role of Consensus in GATT/WTO Decision-making', 17 *Northwestern Journal of International Law and Business* (1996-1997) pp. 653-680.
M. Footer, *An Institutional and Normative Analysis of the World Trade Organization* (Leiden, Martinus Nijhoff Publishers 2006).
Th.M. Franck, *The Power of Legitimacy among Nations* (New York, Oxford University Press 1990).

J.P. Frawley, 'Codex Alimentarius – Food Safety – Pesticides', 42 *Food, Drug and Cosmetic Law Journal* (1987) pp. 168-173

P. Gray, 'Food law and the internal market, 15 *Food Policy* (1990) pp. 111-121.

J. Groetzinger, 'The New GATT Code and the International Harmonization of Products Standards', 8 *Cornell International Law Journal* (1975) pp. 168-188.

H.F.W.M van Haastert, *Het Internationaal Landbouw Instituut (I.I.A.) en de Organisatie voor Voedsel en Landbouw (F.A.O.)*, ('s-Hertogenbosch, N.V. Zuid-Nederlandsche Drukkerij 1947).

G. Hafner, 'Pros and Cons ensuing from Fragmentation of International Law', 25 *Michigan Journal of International Law* (2003) pp. 849-863.

R. Haigh, 'The activities of the Scientific Committee for Food of the Commission of the European Communities', in C.L. Galli, et al. (eds.), *Chemical Toxicology of Food* (Amsterdam, Elsevier 1978), pp. 81-97.

R. Hankin, 'The role of Scientific Advice in the Elaboration and Implementation of the Community's Foodstuffs Legislation', in Ch. Joerges, et al. (eds.), *Integrating Scientific Expertise into Regulatory Decision-Making,* (Baden-Baden, Nomos Verlagsgesellschaft 1997) pp. 141-167.

M. Haq, et al. (eds.), *The UN and the Bretton Woods Institutions. New Challenges for the Twenty-First Century* (Basingstoke, MacMillan 1995)

S. Hathaway, 'Management of food safety in international trade', 10 *Food Control* (1999) pp. 247-253.

R. van Havere, 'Codex Alimentarius', in R. Kruithof, *Levensmiddelenrecht* (Brussel, Ced-Samson 1979).

R. van der Heide, 'The Codex Alimentarius on Food Labelling', *European Food Law Review* (1991) pp. 291-300.

S. Henson, et al., *Review of Developing Country Needs and Involvement in International Standards-setting Bodies* (University of Reading 2001).

M.A. Hermitte, 'L'Expertise scientifique à finalité réflexions sur l'organisation et la responsabilité des experts', 8 Justices (1997) pp. 79-103.

A. Herwig, 'Legal and institutional aspects in the negotiation of the Codex Alimentarius Convention', 2 *Zeitschrift für das gesamte Lebensmittelrecht* (2001).

E. Hey, 'Sustainable development, normative development and the legitimacy of decision-making', 34 *Netherlands Yearbook of International Law* (2003) pp. 3-53.

H.C. Von Heydebrand u.d. Lasa, 'Free Movement of Foodstuffs, Consumer protection and Food Standards in the European Community: Has the Court of Justice Got It Wrong?', 16 *European Law Review* (1991) pp. 319-415.

V. Heyvaert, 'Reconceptualizing Risk Assessment', 8 *Review of European Community and international environmental law* (1999) pp. 135-143.

A.J. Hoekema, *Legitimiteit door legaliteit, over het recht van de overheid* (Nijmegen, Ars Aequi Libri 1991).

H. Horn and J.H.H. Weiler, 'European Communities – Trade Description of Sardines: Textualism and its Discontent', in H. Horn and P. Mavroidis (eds.), *The WTO Case Law of 2002,* (Cambridge, Cambridge University Press 2005), pp. 248-275.

L.R. Horton, 'Risk analysis and the law: international law, the World Trade Organization, Codex Alimentarius and national legislation', 18 *Food Additives and Contaminants* (2001) pp. 1057-1067.

R. Howse, 'Democracy, science and free trade: risk regulation on trial at the World Trade Organization', 98 *Michigan Law Review* (2000) pp. 2329-2356.

R. Howse, 'The Sardines Panel and AB Rulings - Some Preliminary Reactions', 29 *Legal Issues of Economic Integration* (2002) pp. 247-254.
J. Hu, 'The Role of International Law in the Development of WTO Law', 7 *Journal of International Economic Law* (2004) pp. 143-167.
R. Hudec, 'The New WTO Dispute Settlement Procedure', 8 *Minnesota Journal of Global Trade* (1999) pp. 1-53.
I. Hurd, 'Legitimacy and Authority in International Politics', 53 *International Organization* (1999) pp. 379-408.
D.R. Hurst, 'Hormones: European Communities - Measures Affecting Meat and Meat Products', 9 *European Journal of International Law* (1998) pp. 182-214.
Y. Iwasawa, 'WTO Dispute Settlement as Judicial Supervision', 5 *Journal of International Economic Law* (2002) pp. 287-305.
J.H. Jackson, *The World Trading System. Law and Policy of International Economic Relations*, 2nd edn. (Cambridge, Massachusetts, MIT Press 2000).
J.H. Jackson, *The Jurisprudence of GATT and the WTO. Insights on treaty law and economic relations* (Cambridge, Cambridge University Press 2000).
S. Jasanoff, *Risk Management and Political Culture* (United States of America, Russell Sage Foundation 1986).
S. Jasanoff, *The Fifth Branch* (Cambridge, Massachusetts, Harvard University Press 1990).
Ch. Joerges, *Integrating Scientific Expertise into Regulatory Decision-Making. Scientific Expertise in Social Regulation and the European Court of Justice: Legal Frameworks for Denationalized Governance Structures*, EUI Working Paper RSC 96/10, (San Domenico, European University Institute, 1996).
D. Jukes, 'The role of science in international food standards', 11 *Food Control* (2000) pp. 181-194.
K.H. Kaikobad, *The International Court of Justice and Judicial Review. A Study on the Court's Powers with Respect to Judgments of the ILO and UN Administrative Tribunals* (The Hague, Kluwer Law International 2000).
A. Kantrowitz, 'Controlling Technology Democratically', *American Scientist* (1975) pp. 505-509.
J. Klabbers, *An Introduction to International Institutional Law* (Cambridge, Cambridge University Press 2002).
S. Krapohl, 'Credible Commitment in Non-Independent Regulatory Agencies: A Comparative Analysis of the European Agencies for Pharmaceuticals and Foodstuffs', 10 *European Law Journal* (2004) pp. 518-538.
A.O. Krueger, *The WTO as an international organization* (Chicago, University of Chicago Press 1998).
M. Kumm, 'The Legitimacy of International Law: A Constitutionalist Framework of Analysis', 15 *European Journal of International Law* (2004) pp. 907-931.
T. Larsson, 'How open can a government be? The Swedish experience', in V. Deckmyn and I. Thomson, (eds.), *Openness and Transparency in the European Union* (Maastricht, European Institute of Public Administration 1998).
D.L. Leive, *International Regulatory Regimes* (Lexington, Lexington Books 1976).
K. Lenaerts and M. Desomer, 'New models of constitution-making in Europe: the quest for legitimacy', 39 *Common Market Law Review* (2002) pp. 1217-1253.
Ch. Lewis and A. Randell, 'Nutrition labelling of foods: comparisons between US regulations and Codex guidelines', 7 *Food Control* (1996) pp. 285-293.
P. Lichtenbaum, 'Procedural Issues in WTO Dispute Resolution', 19 *Michigan Journal of International Law* (1998) pp. 1195-1272.

A. Lindroos, 'Addressing Norm Conflicts in a Fragmented Legal System: The Doctrine of *Lex Specialis*', 74 *Nordic Journal of International Law* (2005) pp. 27-66.

Ch. Lister, *Regulation of Food Products by the European Community*, Current EC Legal Developments Series (London, Butterworth 1992).

Ch. Lister, 'The naming of foods: the European Community's rules for non-brand food product names', 18 *European Law Review* (1993) pp. 179-201.

M.A. Livermore, 'Authority and Legitimacy in Global Governance: Deliberation, Institutional Differentiation, and the Codex Alimentarius', 81 *New York University Law Review* (2006) pp. 766-801.

J. Lodge, 'Transparency and Democratic Legitimacy', 32 *Journal of Common Market Studies* (1994) pp. 343-368.

O. Long, *Law and its Limitations in the GATT Multilateral Trade System* (Dordrecht, Martinus Nijhoff 1987).

C. Lopez-Hurtado, 'Social Labelling and WTO Law', 5 *Journal of International Economic Law* (2002) pp. 719-746.

F.C. Lu, 'The Joint FAO/WHO Food Standards Programme and the Codex Alimentarius', 24 *WHO Chronicle* (1970) pp. 198-205.

A. Mahiou and F. Snyder (eds.), *La sécurité alimentaire. Food Security and Food Safety*, Hague Academy of International Law (Leiden, Martinus Nijhoff Publishers 2006).

G. Majone, 'Science and Trans-science in Standard Setting', 9 *Human Values* (1984) pp. 15-22.

G. Majone, *International Economic Integration, National Autonomy, Transnational Democracy: An Impossible Trinity?*, EUI Working Paper RSC 2002/48 (San Domenico, European University Institute 2002).

T. Makatsch, *Gesundheitsschutz im Recht der Welthandelsorganisation (WTO). Die WTO und das SPS-Ubereinkommen im lichte von Wissenschaftlichkeit, Verrechtlichung und Harmonisierung* (Berlin, Duncker & Humblot 2004).

P. Malanczuk, *Akehurst's modern introduction to international law*, 7th edn. (Routledge, London 1997).

G. Marceau, 'A Call for Coherence in International Law. Praises for the Prohibition Against "Clinical Isolation" in the WTO Dispute Settlement', 33 *Journal of World Trade* (1999) pp. 87-152.

G. Marceau and J. Trachtman, 'The Technical Barriers to Trade Agreement, the Sanitary and Phytosanitary Measures Agreement, and the General Agreement on Tariffs and Trade. A Map of the World Trade Organization law of Domestic Regulation of Goods', 36 *Journal of World Trade* (2002) pp. 811-881.

Ch. Martini, 'States' Control over New International Organization', 6 *Global Jurist Advances* (2006), pp. 1-25.

J. Mathis, 'WTO Panel Report, European Communities - Trade Descriptions of Sardines, WT/DS231/R, 29 May 2002', 29 *Legal Issues of Economic Integration* (2002) pp. 335-347.

M. Matsushita, et al., *The World Trade Organization* (Oxford, Oxford University Press 2003).

M.D. Matthee, 'Regulating Scientific Expertise on Risks deriving from Genetically Modified Organisms: Procedural rules on Risk Assessment Committees under European Community and International Law', in V. Kronenberger (ed.), *The EU and the International Legal Order* (The Hague, Asser Press 2001) pp. 199-220.

M. Matthee, 'Co-ordination as Means to Promote a Coherent System of Intergovernmental Institutions Dealing with Food Security', in A. Mahiou and F. Snyder (eds.), *La sécurité alimentaire. Food Security and Food Safety,* Hague Academy of International Law (Leiden, Martinus Nijhoff Publishers 2006) pp. 675-702.

W. Mattli, 'The politics and economics of international institutional standards setting: an introduction', 8 *Journal of European Public Policy* (2001) pp. 328-344.

G. Mayeda, 'Developing disharmony? The SPS and TBT Agreements and the impact of harmonization on developing countries', 7 *Journal of International Economic Law* (2004) pp. 737-764.

J. McDonald, 'Domestic regulation, international standards, and technical barriers to trade', 4 *World Trade Review* (2005) pp. 249-274.

R. McKay, 'The Codex Food Labelling Committee - maintaining International Standards relevant to changing consumer demands', *European Food Law Review* (1992) pp. 70-80.

J.A. McMahon, 'Learning from Experience? The SPS Agreement and European Community Law', 10 *International Trade Law and Regulation* (2004) pp. 11-17.

C. McMaoláin, 'Ethical food labelling: the role of European Union freetrade in facilitating international fairtrade', 39 *Common Market Law Review* (2002) pp. 295-314.

N. McNelis, 'The role of the Judge in the EU and the WTO. Lessons from the BSE and Hormones Cases', 4 *Journal of International Economic Law* (2001) pp. 189-208.

D. McRae, 'What is the Future of WTO Dispute Settlement?', 7 *Journal of International Economic Law* (2004) pp. 3-21.

J.G. Merrills, *International Dispute Settlement*, 3rd edn. (Cambridge, Cambridge University Press 1998).

R.W. Middleton, 'The GATT Standards Code', 14 *Journal of World Trade Law* (1980) pp. 201-219.

E. Millstone and P. van Zwanenberg, 'Politics of expert advice: lessons from the early history of the BSE saga', 28 *Science and Public Policy* (2001) pp. 99-112.

E. Millstone and P. van Zwanenberg, 'The evolution of Food Safety Policy-making Institutions in the UK, EU and Codex Alimentarius', 36 *Social Policy and Administration* (2002) pp. 593-609.

G. Moore and A. Tavares, 'Mesures récentes prises par l'Organisation des Nations Unies pour l'alimentation et l'agriculture en conséquence de l'accord de l'Organisation Mondiale du Commerce sur l'application des mesures sanitaires et phytosanitaires', 43 *Annuaire Français de Droit International* (1997) pp. 544-550

F. Morgenstern, 'Legality in international organizations', 48 *British Yearbook of International Law* (1976-1977) pp. 241-257.

K.J. Mortelmans, 'SANDOZ-arrest. De Warenwet tussen DE-regulering en EG-regulering', 33 *Ars Aequi* (1984) pp. 100-109.

K.J.M. Mortelmans, 'De interne markt en het facettenbeleid na het Keck-arrest: nationaal beleid, vrij verkeer of harmonisatie', 4 *Sociaal Economische Wetgeving* (1994) pp. 236-250.

R. Muñoz, 'La Communauté entre les mains des normes internationales: les conséquences de la décision Sardines au sein de l'OMC', 4 *Revue du Droit de l'Union Européenne* (2003) pp. 457-484.

J. Neumann and E. Türk, 'Necessity Revisited: Proportionality in World Trade Organization Law After Korea-Beef, EC-Asbestos and EC-Sardines', 37 *Journal of World Trade* (2003) pp. 199-233.

K. Nicolaïdis and J.L. Tong, 'Diversity or Cacophony? The Continuing Debate over New Sources of International Law', 25 *Michigan Journal of International Law* (2003) pp. 1349-1375.

Ch. Noiville, 'Principe de précaution et Organisation mondiale du commerce. La cas du commerce alimentaire', 127 *Journal du Droit international* (2000) pp. 263-297.

J. Nusbaumer, 'The GATT Standards Code in Operation', 18 *Journal of World Trade Law* (1984) pp. 542-552.

M. Oesch, 'Standards of review in WTO dispute resolution', 6 *Journal of International Economic Law* (2003) pp. 635-659.

R.H. van Ooik en W.T. Eijsbouts, 'De wonderbaarlijke vermenigvuldiging van Europese agentschappen. Verklaring, analyse, perspectief', 16 *Sociaal Economische Wetgeving* (2006) pp. 102-111.

E. Osieke, 'The Exercise of the Judicial Function with respect to the International Labour Organization', 47 *Britisch Yearbook of International Law* (1974-1975) pp. 315-340.

D.A. Osiro, 'GATT/WTO Necessity Analysis: Evolutionary Interpretation and its Impact on the Autonomy of Domestic Regulation', 29 *Legal Issues of Economic Integration* (2002) pp. 123-141.

A.A. Ostrovsky, 'The New Codex Alimentarius Commission standards for food created with modern biotechnology: implication for the EC GMO Framework's compliance with the SPS Agreement', 25 *Michigan Journal of International Law* (2004) pp. 813-843.

D. Palmeter and P.C. Mavroidis, 'The WTO Legal System: Sources of Law', 92 *American Journal of International Law* (1998) pp. 398-413.

D. Palmeter and P.C. Mavroidis, *Dispute Settlement in the World Trade Organization. Practice and Procedure* (The Hague, Kluwer Law International 1999).

S. Pardo Quintillan, 'Free Trade, Public Health Protection and Consumer Information in the European and WTO Context', 33 *Journal of World Trade* (1999) pp. 149-193.

J. Pauwelyn, 'The WTO Agreement on Sanitary and Phytosanitary (SPS) Measures as applied in the first three SPS disputes. EC-Hormones, Australia-Salmon and Japan-Varietals', 2 *Journal of International Economic Law* (1999) pp. 641-664.

J. Pauwelyn, 'The role of public international law in the WTO. How far can we go?', 95 *American Journal of International Law* (2001) pp. 535-578.

J. Pauwelyn, 'Bridging fragmentation and unity: international law as a universe of interconnected islands', 25 *Michigan Journal of International Law* (2003) pp. 903-916.

J. Pauwelyn, 'A typology of multilateral treaty obligations: are WTO obligations bilateral or collective in nature?', 14 *European Journal of International Law* (2003) pp. 907-951.

E.U. Petersmann, *The GATT/WTO Dispute Settlement System: International Law, International Organisations and Dispute Settlement* (Dordrecht, Kluwer Law International 1998).

E.U. Petersmann, 'From 'negative' to 'positive' integration in the WTO: time for 'mainstreaming human rights' into WTO law?', 37 *Common Market Law Review* (2000) pp. 1363-1382.

E.U. Petersmann, 'From Negative to Positive Integration in the WTO: The TRIPs Agreement and the WTO Constitution, in Th. Cottier and P. Mavroidis (eds.), *Intellectual property: trade, competition and sustainable development,* Studies in international economics, Vol. 3 World Trade Forum (Michigan, Michigan University Press 2003), pp. 21-52.

M.C.W. Pinto, 'Democratization of international relations and its implications for development and application of international law', 5 *Asian Yearbook of International Law* (1997) pp. 111-124.
S. Poli, 'The European Community and the Adoption of International Food Standards within the Codex Alimentarius Commission', 10 *European Law Journal* (2004) pp. 613-630.
D. Prévost and M. Matthee, 'The SPS Agreement as a Bottleneck in Agricultural Trade between the European Union and Developing Countries: How to Solve the Conflict', 29 *Legal Issues of Economic Integration* (2002) pp. 43-59.
D. Prévost and P. Van den Bossche, 'The Agreement on Sanitary and Phytosanitary measures', in P.F.J. Macrory, et al. (eds.), *The World Trade Organization: legal, economic and political analysis*, Vol. I (New York, Springer 2005), pp. 231-370.
R. Quick and A. Bluthner, 'Has the Appellate Body erred? An appraisal and criticism of the ruling in the WTO Hormones Case', 2 *Journal of International Economic Law* (1999) pp. 603-639.
N. Rees and D. Watson (eds.), *International Standards for Food Safety* (Gaithersburg, Maryland, Aspen Publishers 2000).
A. Reich, 'The WTO as a Law-Harmonizing Institution', 25 *University of Pennsylvania Journal of International Economic Law* (2004) pp. 321-382.
B. Reinalda and B. Verbeek (eds.), *Decision Making Within International Organizations* (London, Routlegde 2004).
O. Ribbelink, *Opvolging van internationale organisaties. Van Volkenbond - Verenigde Naties tot ALALC - ALADI.* (The Hague, T.M.C. Asser Institute 1988).
D. Roberts, 'Preliminary Assessment of the Effects of the WTO Agreement on Sanitary and Phytosanitary Trade Regulations', 1 *Journal of International Economic Law* (1998) pp. 377-405.
D. Roberts, et al., 'Sanitary and phytosanitary barriers to agricultural trade: progress, prospects, and implications for developing countries', in M. Ingco and A. Winters (eds.), *Agriculture and the New Trade Agenda* (Cambridge, Cambridge University Press 2004), pp. 329-358.
C.P.R. Romano, 'The proliferation of international judicial bodies: the pieces of the puzzle', 31 *New York University Journal of International Law and Politics* (1999) pp. 709-751.
R. Romi, 'Codex Alimentarius: de l'ambivalence à l'ambiguïté', *Revue Juridique de l'Environnement* (2001) pp. 201-213.
L. Rosman, 'Public participation in international pesticide regulation: when the Codex Commission decides, who will listen?', 12 *Virginia Environmental Law Journal* (1993) pp. 329-365.
S.J. Rothberg, 'From Beer to BST: Circumventing the GATT Standards Code's Prohibition on Unnecessary Obstacles to Trade', 75 *Minnesota Law Review* (1990) pp. 505-536.
H. Rothstein, et al., 'Regulatory Science, Europeanization, and the Control of Agrochemicals', 24 *Science, Technology and Human Values* (1999) pp. 214-264.
L. Salter, *Mandated Science, Science and Scientists in the Making of Standards* (Dordrecht, Kluwer Academic Publishers 1988).
G. Sander, 'Gesundheitsschutz in der WTO - eine neue Bedeutung des Codex Alimentarius im Lebensmittelrecht?', 3 *Zeitschrift für Europarechtliche Studien* (2000) pp. 335-375.
H. Schepel, *The Constitution of Private Governance. Product Standards in the Regulation of Integrating Markets* (Oxford, Hart Publishing 2005).

H.G. Schermers and N.M. Blokker, *International Institutional Law*, 4th edn.(The Hague, Martinus Nijhoff Publishers 2003).

R. von Schomberg, *Argumentatie in de context van een wetenschappelijke controverse. Een analyse van de discussie over de introductie van genetisch gemodificeerde organismen in het milieu*, (Delft, Uitgeverij Eburon 1997).

L.J. Schuddeboom, 'Food inspection and EEC policy', 2 *Food Policy* (1977) pp. 17-26.

J. Scott, 'International Trade and Environmental Governance: Relating Rules (and Standards) in the EU and the WTO', 15 *European Journal of International Law* (2004) pp. 307-354.

S.A. Shapiro, 'International Trade Agreements, Regulatory Protection, and Public Accountability', 54 *Administrative Law Review* (2002) pp. 435-458.

M.N. Shaw, *International Law*, 5th edn. (Cambridge, Cambridge University Press 2003)

S. Shubber, 'The Codex Alimentarius Commission under International Law', 21 *International and Comparative Law Quarterly* (1972) pp. 631-655.

S. Shubber, *The International Code of Marketing of Breast-Milk Substitutes. An International Measure to Protect and Promote Breast-feeding* (The Hague, Kluwer Law International 1998).

B. Simma, 'Self-Contained Regimes', 16 *Netherlands Yearbook of International Law* (1985) pp. 112-136.

B. Simma, 'Fragmentation in a positive light', 25 *Michigan Journal of International Law* (2003) pp. 845-847.

G. Skogstad, 'The WTO and Food Safety Regulatory Policy Innovation in the European Union', 39 *Journal of Common Market Studies* (2001) pp. 485-505.

P.J. Slot, 'Harmonisation', 21 *European Law Review* (1996) pp. 378-397.

F. Snyder, 'Governing Globalisation', in M. Likosky (ed.), *Transnational legal processes* (London, Butterworths LexisNexis 2002), pp. 65-97.

F. Snyder, 'The Gatekeepers: The European Courts and WTO Law', 40 *Common Market Law Review* (2003) pp. 313-367.

G.E. Spencer, et al., 'Effects of Codex and GATT', 9 *Food Control* (1998) pp. 177-182.

C. Stahn, 'Governance beyond the state: Issues of Legitimacy in International Territorial Administration', 2 *International Organizations Law Review* (2005) pp. 9-56.

G. Stanton, 'A review of the operation of the Agreement on Sanitary and Phytosanitary Measures', in M. Ingco and A. Winters (eds.), *Agriculture and the New Trade Agenda* (Cambridge, Cambridge University Press 2004) pp. 101-110.

J. Steffek, *The Power of Rational Discourse and the Legitimacy of International Governance*, EUI Working Paper RSC 2000/46 (San Domenico, European University Institute 2000).

E. Stein, 'International integration and democracy: no love at first sight', 95 *American Journal of International Law* (2001) pp. 489-534.

S.M. Stephenson, *Standards and Conformity Assessment as Nontariff Barriers to Trade*, Policy Research Working Paper 1826 (Washington D.C., World Bank 1997).

T.P. Stewart and D.S. Johnson, 'The SPS Agreement of the World Trade Organization and International Organizations: The Roles of the Codex Alimentarius Commission, the International Plant Protection Convention, and the International Office of Epizootics', 26 *Syracuse Journal of International* (1999) pp. 26-53.

R. Streinz, 'Diverging risk assessment and labelling', 2 *European Food Law Review* (1994) pp. 155-179.

S. Suppan, *Governance in the Codex Alimentarius Commission* (Consumers International 2005).

L.E. Susskind, *Environmental Diplomacy. Negotiating More Effective Global Agreements* (New York, Oxford University Press 1994)

R.E. Sweeney, 'Technical Analysis of the Technical Barriers to Trade Agreement', 12 *Law & Policy in International Business* (1980) pp. 179-217.

A. Swinbank, 'The EEC's policies and its food', 17 *Food Policy* (1992) pp. 53-64.

A.O. Sykes, *Product Standards for Internationally Integrated Goods Market* (Washington D.C., The Brookings Institution 1995).

A.O. Sykes, 'The (limited) role of regulatory harmonization in international goods and services markets', 2 *Journal of International Economic Law* (1999) pp. 49-70.

C. Tietje, 'Global Governance and Inter-Agency Co-operation in International Economic Law', 36 *Journal of World Trade* (2002) pp. 501-515.

M. Trebilcock and R. Howse, *The Regulation of International Trade*, 2nd edn. (London, Routlegde 1999).

C.C. van der Tweel, *Een nieuw GATT-Verdrag en de Codex Alimentarius. Een bestuurskundig onderzoek naar internationale beleidsvorming en nationale beleidsruimte*, (Leiden: Rijksuniversiteit Leiden 1993).

F. Veggeland and S.O. Borgen, *Changing the Codex: The Role of International Institutions*, Working Paper 2002-16 (Norwegian Agricultural Economics Institute 2002).

F. Veggeland and S.O. Borgen, 'Negotiating International Food Standards: The World Trade Organization's Impact on the Codex Alimentarius Commission', 18 *Governance: An International Journal of Policy, Administration, and Institutions* (2005) pp. 675-708.

W.H Vermeulen, 'De Codex Alimentarius Commissie en de Europese Commissie ter bestrijding van mond- en klauwzeer, twee volwassen dochterinstellingen van de Wereldvoedselorganisatie', 10 *Sociaal Economische Wetgeving* (1983) pp. 594-610.

B. Vesterdorf, 'Transparency-not just a vogue word', 22 *Fordham International Law Journal* (1999) pp. 902-929.

D.G. Victor, *Effective Multilateral Regulation of Industrial Activity: Institutions for Policing and Adjusting Binding and Nonbinding Legal Commitments*, Ph.D. Thesis (Massachusetts, Harvard University, Institute of Technology 1997).

D.G. Victor, 'Risk Management and the World Trading System: Regulating International Trade Distortions Caused by National Sanitary and Phytosanitary Policies', *Incorporating Science, Economics, and Sociology in Developing Sanitary and Phytosanitary Standards in International Trade: Proceedings of a Conference* (National Academies Press, 2000), pp. 118-169.

D. Victor, 'WTO Efforts to Manage Differences in National Sanitary and Phytosanitary Policies', in D. Vogel and R. Kagan, *Dynamics of Regulatory Change: How Globalization Affects National Regulatory Policies* (University of California Press 2002).

D. Vignes, 'The Harmonisation of National Legislation and the EEC', 15 *European Law Review* (1990) pp. 358-374.

K. Vincent, 'Mad Cows and Eurocrats-Community Responses to the BSE Crisis', 10 *European Law Journal* (2004) pp. 499-517.

D. Vogel, 'Environmental Regulation and Economic Integration', 3 *Journal of International Economic Law* (2000) pp. 265-279.

W. van de Voorde, 'De E.E.G. als lid van een internationale instelling: het geval F.A.O.', 45 *Studia Diplomatica* (1992) pp. 49-73.

E. Vos, *Institutional Frameworks of Community Health and Safety Regulation: Committees, Agencies and Private Bodies* (Oxford, Hart Publishing 1999).

E. Vos, 'Reforming the European Commission: What role to play for EU Agencies', 37 *Common Market Law Review* (2000) pp. 1113-1134.

E. Vos, 'European Food Safety Regulation in the Aftermath of the BSE Crisis', 23 *Journal of Consumer Policy* (2000) pp. 227-255.

E. Vos, 'Antibiotics, the Precautionary Principle and the Court of First Instance', 11 *Maastricht Journal of European and Comparative Law* (2004) pp. 187-200.

E. Vos, 'Risicobeheersing door de EU: het nieuwe beleid op het gebied van voedselveiligheid', in G. van Calster and E.Vos (eds.), *Risico and voorzorg in de rechtsmaatschappij* (Antwerpen, Intersentia 2005).

B. Vroom-Cramer, *Productinformatie over levensmiddelen. Etiketteringsvraagstukken naar Europees en Nederlands recht* (Amsterdam, Koninklijke Vermande 1998).

V.R. Walker, 'Keeping the WTO from Becoming the "World Trans-science Organization": Scientific Uncertainty, Science Policy, and Factfinding in the Growth Hormones Dispute', 31 *Cornell International Law Journal* (1998) pp. 251-320.

H.M. Wehr, 'Update on Issues before the Codex Alimentarius', 52 *Food and Drug Law Journal* (1997) pp. 531-536.

J. Weiler and J. Modrall, 'Institutional Reform: Consensus or Majority?', 10 *European Law Review* (1985) pp. 316-327.

J.H.H. Weiler, *The Constitution of Europe. "Do the new clothes have an emperor?" and other essays on European integration* (Cambridge, Cambridge University Press 1999).

J.H.H. Weiler, *The EU, the WTO and the NAFTA* (New York, Oxford University Press 2000).

F. Weiss, *Improving WTO Dispute Settlement Procedures: Issues & Lessons from the Practice of Other International Courts & Tribunals* (London, Cameron May 2000).

D. Welch, 'From 'Euro Beer' to 'Newcastle Brown', A Review of European Community Action to Dismantle Divergent 'Food' Laws', 22 *Journal of Common Market Studies* (1983) pp. 47-70.

D. Welch, 'Alcoholic beverage legislation. Harmonizing EEC laws', 10 *Food Policy* (1985) pp. 41-54.

N.D. White, *The law of international organizations* (Manchester, Manchester University Press 1996).

J. Wiers, *Trade and Environment in the EC and the WTO. A Legal Analysis* (Groningen, Europa Law Publishing 2002).

D.A. Wirth, 'The Role of Science in the Uruguay Round and NAFTA Trade Disciplines', 27 *Cornell International Law Journal* (1994) pp. 817-859.

J. Wouters, et al., 'Democracy and international law', 24 *Netherlands Yearbook of International Law* (2003) pp. 139-197.

K. Zemanek, 'Majority Rule and Consensus Technique in Law-Making Diplomacy', in R. St. J. MacDonald and D.M. Johnston (eds.), *The Structure and Process of International Law* (Dordrecht, Martinus Nijhoff Publishers 1983), pp. 857-887.

K. Zemanek, 'The Legal Foundations of the International System: General Course on Public International Law', 62 *Receuil des Cours: Collected Courses of the Hague Academy of International Law* (1997).

P. van Zwanenberg and E. Millstone, 'Beyond Skeptical Relativism: Evaluating the Social Constructions of Expert Risk Assessments', 25 *Science, Technology and Human Values* (2000) pp. 259-282.

TABLE OF CASES

Judgments and Opinions of the European Court of Justice
Haegeman, Case 181/73 [1974] ECR 449
Procureur du Roi v. *Benoit and Gustave Dassonville*, Case 8/74 [1974] ECR 631
Rewe-Zentral AGV Federal Monopoly Administration for Spirits (Cassis de Dijon), Case 120/78 [1979] ECR 649
Officier van Justitie v. *Koninklijke Kaasfabriek Eyssen BV* (preliminary ruling requested by the Gerechtshof, Amsterdam), Case 53/80 [1981] ECR 409
Kupferberg, Case 104/81 [1982] ECR 3641
Criminal proceedings v. *Léon Motte*, Case 247/84 [1985] ECR 3887
Commission v. *Italy (health checks on imports of curds)*, Case 35/84 [1986] ECR 545
Ministère public v. *Claude Muller and Others*, Case 304/84 [1986] ECR 1511
Commission v. *Germany (Reinheitsgebot)*, Case 178/84 [1987] ECR 1227
Demirel, Case 12/86 [1987] ECR 3719
Proceedings for compulsory reconstruction v. *Smanor SA (Smanor)*, Case 298/87 [1988] ECR 4512
Ministère public v. *Gérard Deserbais*, Case 286/86 [1988] ECR 4907
Criminal Proceedings against Michel Debus, Joined Cases C-13/91 and C-113/91 [1992] ECR I-3617
Commission v. *Italian Republic (nitrates)*, Case C-35/89 [1992] ECR I-4545
Commission v. *Italy (fish containing nematode larvae)*, Case C-228/91 [1993] ECR I-02701
Affish BV v. *Rijksdienst voor de keuring van Vee en Vlees*, Case-183/95 [1997] ECR I-4315
Portugal v. *Council*, Case C-149/96 [1999] ECR I-8395
Verein gegen Unwesen in Handel und Gewerbe Köln eV v. *Adolf Darbo AG*, Case C-465/98 [2000] ECR I-02297
Fábrica de Queijo Eru Portuguesa Ld^a v. *Tribunal Técnico Aduaneiro de Segunda Instância*, Case C-42/99 [2000] ECR I-07691
Criminal Proceedings v. *Jean-Pierre Guimont*, Case C-448/98 [2000] ECR I-10663
Portugal v. *Commission*, Case C-356/99 [2001] ECR I-05645
Criminal proceedings against Walter Hahn, Case C-121/00 ECR [2002] I-09193
Biret International SA v. *Council*, Case T-174/00 [2002] ECR II-00017
Pfizer Animal Health SA v. *Council*, Case T-13/99 [2002] ECR II 03305
Monsanto Agricoltura Italia SpA and Others v. *Presidenza del Consiglio dei Ministri and Others*, Case C-236/01 [2003] ECR I-08105
Commission v. *Denmark*, Case C-192/01 [2003] ECR I-09693
CEVA Santé animale SA v. *Commission*, Joined Cases T-344/00 and T-345/00 [2003] ECR II 00229
Ministère public v. *John Greenham and Léonard Abel*, Case C-95/01 [2004] ECR I-01333
Sachsenmilch AG v. *Oberfinanzdirektion Nürnberg*, Case C-196/05 [2006] ECR I-5161

Judgments and Opinions of the International Court of Justice
Legality of the Use by a State of Nuclear Weapons in Armed Conflict, 8 July 1996, ICJ, Advisory Opinion, Preliminary Objections, I.C.J. Reports 1996

WTO Panel Reports

Report of the WTO Panel, *EC-Measures Concerning Meat and Meat Products (Hormones)*, Complaint by the United States, WT/DS26/R/USA (1997), Report of the WTO Panel, *EC-Measures Concerning Meat and Meat Products (Hormones)*, Complaint by Canada, WT/DS48/R/CAN (1997)

Report of the WTO Panel, *Australia-Measures Affecting the Importation of Salmon*, WT/DS18/R (1998)

Report of the WTO Panel, *India-Quantitative Restrictions on Imports of Agricultural, Textile and Industrial Products*, WT/DS90/R (1999)

Report of the WTO Panel, *United States – Section 211 Omnibus Appropriations Act of 1998*, WT/DS176/R (2001)

Report of the WTO Panel, *European Communities – Trade Description of Sardines*, WT/DS231/R (2002)

Report of the WTO Panel, *Japan-Measures Affecting the Importation of Apples*, WT/DS245/R (2003)

WTO Appellate Body Reports

Report of the Appellate Body, *United States-Standards for Reformulated and Conventional Gasoline*, WT/DS2/AB/R (1996)

Report of the Appellate Body, *Japan-Taxes on Alcoholic Beverages*, WT/DS8/AB/R, WT/DS10/AB/R, WT/DS11/AB/R (1996)

Report of the Appellate Body, *European Communities – Regime for the Importation, Sale and Distribution of Bananas ('EC – Bananas III')*, WT/DS27/AB/R (1997)

Report of the Appellate Body, *India – Patent Protection for Pharmaceuticals and Agricultural Chemical Products*, WT/DS50/AB/R (1998)

Report of the Appellate Body, *EC-Measures Concerning Meat and Meat Products (Hormones)*, WT/DS26/AB/R, WT/DS48/AB/R (1998)

Report of the Appellate Body, *Australia-Measures Affecting the Importation of Salmon*, WT/DS18/AB/R (1998)

Report of the Appellate Body, *Korea – Definitive Safeguard Measure on Imports of Certain Dairy Products ('Korea – Dairy')*, WT/DS98/AB/R (2000)

Report of the Appellate Body, *European Communities – Trade Description of Sardines*, WT/DS231/AB/R (2002)

INDEX

A

Acceptable Daily Intakes (ADIs) 43, 64, 104, 113, 170-172, 193, 196, 223
ALARA principle 65

C

Codex Alimentarius
   'Code of Hygienic Practice for Refrigerated Packaged Food with Extended Shelf Life' 63
   'Codex General Standard for Cheese' 64, 67, 87
   'Codex Guidelines for the Establishment of a Regulatory Programme for Control of Veterinary Drug Residues in Foods' 64
   'Codex International Individual Standard for Gouda' 54, Annex I
   'Codex Standard for Butter' 88-89, 262
   'Codex Standard for Canned Sardines and Sardine-type Products' 88-89, 125,143-144, 152, 163-165, 179, 186-187, 190, 238, 272
   'Codex Standard for Chocolate' 66-67, Annex II
   'Codex Standard for Edible Fats and Oils not Covered by Individual Standards' 54
   'Codex Standard on Brie' 53, 66
   'Codex Standard on Emmentaler' 53, 66
   'Codex Standard on Mineral Waters' 57,61, 71, 124-125, 229, 232, 234
   Commodity standards 36, 52-55, 59-61, 63, 66-67, 71, 80, 84-85, 87-90, 210-212, 214, 262-263
   'European standard on fresh fungus 'chanterelle' 57
   'European standard for Mayonnaise' 57
   'General Standard for Contaminants and Toxins in Foods' 65
   'General Standard for Food Additives' 55, 63, 65, 196
   'General Standard for Irradiated Foods' 54-55
   'General Standard for the Labelling of Food Additives when Sold as Such' 54, 67, 88
   'General Standard for the Labelling of, and Claims for, Pre-packaged Foods for Special Dietary Uses' 55, 61, 88
   Horizontal approach 55-56, 59-60,93, 99 215, 256
   'Recommended International Code of Practice General Principles of Food Hygiene' 59, 107
   Regional standards 39, 53, 56-58, 71, 75, 93, 224
   Vertical approach 53-56, 59-60
Codex Alimentarius Commission
   Acceptance procedure 1, 12, 58, 60, 73-75, 83, 85-87, 89-94, 96, 102-103, 130, 136, 175-182, 195, 199, 201, 228, 232, 235, 237, 261, 270-273, 277-278, 281
   Budget 30, 33, 40, 49
   Executive Committee of the 26, 29, 31-33, 40, 74, 76-78, 145, 208-209, 211-217, 230, 233, 240-241,245, 252-253, 257-260, 263-265, 274, 282
   Mandate of the 9, 12-14, 21-24, 34, 49-51, 56, 85, 95, 129, 156, 200-201, 203-205, 213, 240, 269, 273-274
   Medium-Term Plan 20, 212-213
   Meetings 26, 28-29, 48
   Membership 25-27, 130-131, 157, 178, 225-226, 228, 242, 265, 275, 281
   Observers 27-28, 40, 77-78, 81, 129, 131, 133,224-225, 231, 248-253, 256, 259, 266-267
   Secretariat of the 30, 33-34, 49, 73, 76, 77, 85, 129, 176, 192-193, 197, 209, 226, 241, 254, 269, 273, 275, 283, 284
Codex Alimentarius Europaeus 1, 15-16, 56, 95
Codex Committees and Task Forces
   Adjourned *sine die* 36
   Host countries 25, 37, 40-41, 50, 77, 78, 201, 245-246, 266, 276
   Codex Co-ordinating Committees 26, 32, 36, 38-39, 40, 41, 57, 214, 253-254, 258
   Inter-Committee relationship 209-210, 214-217, 264, 281
   Codex Committee for Asia 32, 37, 39, 57, 214

INDEX

Codex Committee for Europe 32, 37, 39, 56-57
Codex Committee on Contaminants in Foods (CCCF) 36, 37, 41, 47, 208
Codex Committee on Fats and Oils (CCFO) 37, 38, 262
Codex Committee on Fish and Fishery Products 35, 37, 81, 107, 210
Codex Committee on Food Additives (CCFA) 36, 37, 41, 47, 55, 208, 246
Codex Committee on Food Additives and Contaminants (CCFAC) 40, 41, 82, 208, 214, 220, 226, 235, 240, 246
Codex Committee on Food Export and Import Inspection and Certification Systems (CCFICS)37, 236
Codex Committee on Food Hygiene (CCFH) 37, 47, 207, 209, 210, 213, 215, 246
Codex Committee on Food Labelling (CCFL) 36, 37 67, 105, 106, 215,252
Codex Committee on Fresh Fruits and Vegetables 37, 225-226, 281
Codex Committee on Methods of Analysis and Sampling (CCMAS) 36, 37, 206, 210
Codex Committee on Milk and Milk Products (CCMMP) 37, 38, 42, 215, 227
Codex Committee on General Principles (CCGP) 31, 36, 37, 41, 72, 210, 225, 241, 254, 256, 258, 283
Codex Committee on Pesticide Residues (CCPR) 36, 37, 41, 47, 206, 208, 210, 219, 223, 246, 251
Codex Committee on Processed Fruit and Vegetables (CCPFV) 37, 38, 57
Codex Committee on Residues of Veterinary Drugs in Food (CCRVDF) 37, 47, 208, 209, 220, 246
Committee of government experts on the Code of Principles concerning Milk and Milk Products 15, 16, 41, 53, 98, 227
Consensus 9, 26, 70, 77, 78, 80-82, 93, 184, 186-187, 228, 230, 231-235, 236, 237, 238, 239, 255, 265, 276, 280, 281
Consensus-building 81-82, 93, 230-231, 236, 237, 241, 245, 255, 265, 278, 280

D
Dispute Settlement Understanding 151, 184-185, 200, 273
    Article 3.2   189, 190-191
    Article 13   189, 192-194

E
European Commission and its relationship with the Codex Alimentarius Commission 96, 127, 129-132
European Court of Justice (ECJ) and Codex standards 108-120
European Food Safety Authority (EFSA) 127-128

F
Food and Agriculture Organization (FAO) 1, 12, 13, 14, 15, 18, 20, 21, 22, 26, 29, 30, 33-34, 35, 41, 42, 44, 45, 46,47, 48, 53, 58, 95, 101, 129, 202, 204, 205, 223, 224, 265, 279, 284
Fragmentation of international law 3, 4-5, 277-279

G
General Food Law (Regulation 178/2002) 3, 95, 101, 127
    Article 5(3)   120-123, 125, 126, 133, 271, 278

H
Harmonisation 3, 10-12, 51-52, 62, 86-87, 90, 93-94, 96-102, 135, 136-140, 144, 161, 163, 177-178, 181, 182, 260, 277
Hazard Analysis and Critical Control Point System (HACCP) 59, 63, 72

I
International Dairy Federation (IDF) 1, 14, 76, 226-228, 250, 265, 281
International Institute for Agriculture (IIA) 14
International Governmental Organisations (IGOs) 27
International Non-Governmental Organisations (INGOs) 27, 202, 241-242, 248-253, 258, 266, 275-276

J
Joint FAO/WHO Consultative Group for the Trust Fund 48-49, 50, 269
Joint FAO/WHO Expert Committee on Food Additives (JECFA)42, 43, 45-48, 72, 76, 98, 104, 106, 110, 111, 113, 116, 118, 123, 126, 128, 169-172, 194, 199, 208, 222, 226, 239
Joint FAO/WHO Expert Meeting on Pesticide Residues (JMPR) 8, 42-43, 45-48, 59, 66, 76, 98, 104, 126, 128, 202, 208, 216, 219-223, 226, 264
Joint FAO/WHO Expert Meetings on Microbiological Risk Assessment (JEMRA) 42-43, 45-47, 128, 220

INDEX

L
Latin American Food Code 1, 15, 35
Legality 8-9, 117, 119-121, 123
Legitimacy 6-7, 99, 101, 201-202
   Institutional legitimacy   8-9, 203, 207, 274-275
   Procedural legitimacy   9-10, 130, 185-188, 229-230, 275-276
   Substantive legitimacy   10-12, 260, 277

M
Maximum Residue Levels (MRLs) 58-60, 64, 66, 68, 69, 71, 72, 83, 90, 91, 98, 99, 102, 113, 117, 124, 128, 162-163, 172, 174-175, 180, 193, 208, 216, 219, 261

O
Organisation for Economic Co-operation and Development (OECD) 1, 35
'other legitimate factors' 72-73, 94

P
Participation 183-184, 241
   Codex developing member countries 242-248
   Rights 248-250
Principles for the Development of International Standards, Guides and Recommendations 183, 184, 186, 187
Procedural Manual of the Codex Alimentarius Commission
   Criteria for Establishing Work Priorities 76, 211, 264
   Format for Codex Commodity Standards 52, 53, 54, 60, 61, 63
   General Principles of the Codex Alimentarius 24, 51, 52, 58, 62-63, 84, 86, 91, 94, 187, 260
   Guidelines on Co-operation with Other International Intergovernmental Organisations 23, 75, 227-228, 265, 275, 281
   Procedure for the Elaboration of Codex Standards and related Texts 57, 78, 207, 237
   Rules of Procedure 8, 18, 20-27, 31, 32, 33, 35, 40, 56, 57, 74, 75, 77, 78, 130, 131, 206, 233, 255, 257, 265, 276, 281, 283
   Statements of Principles concerning the Role of Science in the Codex Decision-Making Process and the Extent to which other Factors are taken into Account 72, 94, 229, 235, 281
   Statutes 21-24, 25, 31, 32, 33, 34, 40, 58, 203-205, 225, 257, 283
   Working Principles for Risk Analysis for Application in the Framework of the Codex Alimentarius 23, 42, 218, 220-221, 257, 264

R
Reciprocity 91-92
Recommended Codes of Practice 58, 59, 60, 63, 83, 93, 88, 176, 177, 262, 263, 269, 272
Risk assessment 23-24, 42, 64, 112, 147, 166-172, 194, 217-218, 220, 257-258, 275
Risk assessment policy 172, 199, 219-220, 264, 282
Risk management 23, 42, 64, 100, 206, 218, 220-221, 257, 264, 275, 280
Roll-call vote 26, 185, 234

S
Secret ballot vote 234
SPS Agreement
   Preamble 135, 137-139, 144, 155, 173
   Article 2.1 138
   Article 3.1 119, 122, 140-148, 150, 157, 160-161, 181, 186, 188, 198
   'based on' 141, 142-144, 161-163
   Article 3.2 142, 153, 155, 158, 173, 199
   Presumption of consistency 140, 153-155, 158-159, 169, 173
   Article 3.3 122, 144-148, 150, 155, 162, 165, 198, 205
   Article 3.4 173
   Article 5.1 166-169, 172
   Article 5.2 167-168, 174
   Article 5.6 166, 173-175, 199
   Article 5.8 149
   Article 7 148
   Article 11.2 192
   Article 12 178, 195, 196, 197
   Annex A(3) 156
SPS Committee and Codex measures 176, 178, 195-197, 200, 209, 273-274, 278
Standard-setting procedure of the Codex Alimentarius Commission
   Decentralisation 206, 263
   Steps of the 74, 75-80, 83
   structure of two readings 81-82, 93, 211, 241, 255, 270, 279-281

T
Task Force on Animal Feeding Practices 38
Task Force on Antimicrobial resistance 37, 38

Task Force on Foods derived from Biotechnology 37, 38, 42
Task Force on the Processing and Handling of Quick Frozen Foods 37, 38
TBT Agreement
  Preamble   135, 137-140, 144, 149
  Article 2.2   153, 156, 157, 159
  Article 2.4 122, 141-144, 149-152, 155, 157, 158, 161, 163-165, 181, 186, 188, 198, 205
  'based on' 141, 142-144, 161-163
  Presumption of consistency 153-155, 158-159
  'ineffective' and 'inappropriate' 149-150, 152, 164-165, 198
  Article 2.5   151, 153, 155, 158
  Article 5.1.2   139
  Article 5.4   141, 158
  Annex 1.2   179, 186, 187

TBT Code (former TBT Agreement resulting from the Tokyo Round) 99, 136-137, 156-157, 182
Transparency   7, 184, 234, 254-258, 266, 275, 280, 282, 283
Trust Fund   48-49, 101, 245-248, 252, 266, 269, 276

U
United Nations Economic Commission for Europe (UNECE) 1, 35, 213, 226, 228, 258, 265, 281

W
World Health Organization (WHO) 1, 6, 12, 13, 15, 16, 18, 20, 22, 24, 29, 30, 33, 34, 42, 43, 44, 45, 47, 48, 49, 91, 101-102, 203, 204, 205, 222, 223, 224, 252, 265, 269, 275, 279, 284

SUMMARY

THE CODEX ALIMENTARIUS COMMISSION AND ITS STANDARDS

During the last decades, the Codex Alimentarius Commission and its standards have received increased attention. This renewed interest is the consequence of the references to international standards in the WTO Agreements (the *SPS Agreement* and the *TBT Agreement*), which entered into force in 1995. Indeed, many sources affirm that references to the Codex standards entail an increased status of these standards. This book is research into the legal aspects of the Codex Alimentarius Commission and its standards, and aims to define more precisely the actual status of the standards, both within the framework of the WTO Agreements as well as within the context of the EC legal order. In addition, this research aims to define the consequences of the new status of the Codex standards for the legitimacy of the institutional framework of the Codex Alimentarius Commission, its standard-setting procedure, and the Codex standards themselves. In order to evaluate the new status of Codex standards and the consequences thereof, this research starts by analysing the development of the Codex Alimentarius Commission and its standards over the years.

In 1962, the Codex Alimentarius Commission was established by the FAO and the WHO as subsidiary organ of both organisations. The reason behind this was the emerging need for internationally harmonised food requirements as a result of the expanding international trade in food products. Since its establishment, the Codex Alimentarius Commission has been charged with the creation of the Codex Alimentarius: a collection of uniformly-defined food standards. Its capacity to give effect to its mandate depends upon its institutional framework. The authority of the Codex Alimentarius Commission is restricted by its position as a subsidiary body of the FAO and the WHO. This means that, for budgetary issues and the management of the personnel of its Secretariat, the Codex Alimentarius Commission depends on its parent organisations, in particular on the FAO. However, despite its subsidiary position, the Codex Alimentarius Commission has some important powers, such as the power to establish subsidiary bodies and the co-ordination of international food standard activities, which has allowed it to construct an important institutional framework. Additionally, with regard to its normative powers, the Codex Alimentarius Commission operates independently from its parents organisations, as adopted standards do not need approval from the latter organisations. This institutional framework has enabled the Codex Alimentarius Commission to give content, *ratione materiae*, to its mandate.

The most important task attributed to the Codex Alimentarius Commission is the creation of the Codex Alimentarius: a collection of uniformly-defined food standards which aims to assist to the harmonisation of food requirements. Over the years, the Codex Alimentarius has become a complex system of food requirements which aim to protect human health and to ensure fair practice in the food trade, and consists of approximately 216 standards, 47 recommended codes of practice, 40 guidelines, maximum residue levels (MRLs) of pesticides residues for over 360 commodities and MRLs for 47 different veterinary drugs. These measures are often detailed in nature and cover a wide range of issues. They are formulated in a way that restricts the discretion of Codex members to an important extent. Both the flexible standard-setting procedure consisting of two readings and the voluntary acceptance procedure have contributed to facilitating agreement on these standards. Within the Codex acceptance procedure, only few standards became binding on only few Codex members. However, even though the Codex acceptance procedure never functioned as intended and was abolished in 2005, the consequences of its existence, as reflected in the detailed character and the high level of harmonisation contained in the Codex standards, are still present.

The first system in which references to Codex standards can frequently be found is the EC legal system. During the first years of the EC, up to 1987, reference to Codex standards and their scientific basis can frequently be found in both EC food legislation and in the jurisprudence of the European Court of Justice. The frequent use of Codex standards by EC institutions can most probably be explained by the fact that, during this period, the Codex Alimentarius Commission was able to adopt more standards than the EC institutions. It is therefore not surprising that, as from 1987, when some new institutional and procedural settings were introduced that allowed the EC institutions to adopt secondary legislation more easily, one can perceive a diminishing influence of Codex standards on the EC food legislation. However, the entry into force of the WTO agreements and the *EC-Hormones* decisions, which clearly demonstrated the consequences of the entry into force of the WTO agreements, meant a change in this development. In 2002, with the adoption of Regulation 178/2002, also referred to as the 'General Food Law', an obligation to take Codex standards into account, by both member states as well as the Community institutions, when adopting and preparing food law, was established. This means that Codex measures have acquired a status as 'reference point' in the justification of measures, and recently adopted EC food legislation illustrate that the EC institutions have indeed taken the Codex measures or their scientific basis into account in the legislative process.

Reference to international standards in the *SPS Agreement* and the *TBT Agreement* and the recognition of Codex standards as such have raised questions regarding the status of the standards. The references are included in the harmonisation provisions of the *SPS Agreement* and the *TBT Agreement*, and consist of an encouragement to conform national measures with international standards and an obligation to use international standards as the basis of national measures. The application of these provisions both by panels and by the Appellate Body illustrates that the status of Codex standards within the context of the WTO agreements amounts to a *de facto* binding force for some of their elements. This *de facto* binding force is not direct, as it only functions through the *SPS Agreement* and the *TBT Agreement*. Furthermore, it only concerns some – not all – of the elements of the Codex standards. Given the fact that the scope of the objectives of the WTO agreements is restricted *vis-à-vis* that of the Codex Alimentarius (the former aims to further trade liberalisation, whereas the Codex also aims to protect human health and to ensure fair practice in the food trade), it becomes clear that, within the context of the WTO agreements, Codex standards do not function as minimum platform standards which aim to set a minimum protection level. This also means that the consequence of the abolition of the Codex acceptance procedure in 2005 is that the Codex Alimentarius Commission has done away with the only 'compliance' mechanism to promote the Codex standards as minimum platform standards.

The references to international standards and the recognition of Codex measures as such have also had institutional consequences. As the WTO is not a standardising body itself and relies on 'outside' international standard-setting bodies – such as the Codex Alimentarius Commission – for the adoption of harmonisation measures, the relationship between the WTO and the Codex Alimentarius Commission can be characterised by a separation of powers: the Codex Alimentarius Commission being responsible for the 'legislative' acts on the one hand, and the WTO being responsible for the promotion and enforcement of the application of harmonisation measures on the other. Two aspects relating to the institutional rules of the

WTO may potentially obstruct the functioning of the Codex Alimentarius Commission. First, the rules regulating the WTO dispute settlement mechanism (*DSU*) do not ensure that panels and the Appellate Body function adequately as adjudicating bodies when it comes to disputes regarding the proper interpretation and application of Codex measures. Second, the function of the SPS Committee, as 'regular forum for consultation' also overlaps with the mandate of the Codex Alimentarius Commission, which, at this moment, does not lead to negative consequences, as the *SPS Agreement* contains provisions that provide for inter-institutional co-operation, and because there is no hierarchal structure between the SPS Committee and the Codex Alimentarius Commission.

With the increased status of the Codex standards, the authority of the Codex Alimentarius Commission has also increased, which has led scholars and organisations to question its legitimacy. When discussing the concerns of legitimacy in Chapter V of this book, a division into three types of concerns has been made: concerns of institutional legitimacy, of procedural legitimacy and of substantive legitimacy. Concerns of institutional legitimacy include, amongst others: 1. a delegation of the powers of the Codex Alimentarius Commission to strong subsidiary bodies which, if not carefully supervised and co-ordinated, can lead to an incoherent functioning of the Codex; 2. a delegation of its tasks to 'outside' international standard-setting institutions which are not bound by the internal rules of the Codex Alimentarius Commission; and 3. the mandated scientific consultation of expert bodies does not always come with a clearly-defined task. Concerns regarding the standard-setting procedure include: 1. the way of final decision-making by the Codex Alimentarius Commission is not always perceived as legitimate and preference has been expressed to emphasise the importance of taking decisions by consensus; 2. attention has been drawn to a dominant influence of the developed Codex members and the industry-interest groups on the standard-setting procedure *vis-à-vis* the developing Codex members and the public-interests NGOs; and 3. the transparency of the procedure and access to relevant documents is not always adequately ensured. Concerns with regard to substantive legitimacy concentrate on the ways and techniques incorporated in Codex measures to advance the harmonisation of food requirements. The Codex standards themselves consists of many strict and detailed provisions which do not leave Codex Members with much discretion to respond to domestic concerns. This raises the question as to whether Codex standards rightly address certain concerns or whether these concerns are more appropriately addressed at national level.

Recent developments, including the evaluation of the Codex Alimentarius Commission conducted by its parent organisations, have addressed concerns relating to the legitimacy of the Codex Alimentarius Commission. As a result, considerable improvements have been made. Examples include the establishment of a Trust Fund to enhance the participation of developing Codex Members, the codification in the Rules of Procedure of the importance to take decisions by consensus, the codification of the responsibilities of both the risk assessors (expert bodies) and the risk managers (subsidiary Codex bodies), the appointment of the Executive Committee as the critical reviewer of new work proposed by subsidiary Codex bodies. Although they are important steps towards the legitimisation of the Codex Alimentarius Commission, most of these modifications unfortunately only partially address the relevant concerns. These recent efforts to legitimise the Codex Alimentarius Commission are subject to a more fundamental criticism; they are based upon an 'issue-oriented' approach. These efforts mainly result in modifications which, despite addressing concerns in the short run, do not suffice to guarantee the legitimisation of the Codex Alimentarius Commission in the long run. A more structural approach is required. This structural approach would address the adequate functioning of the fundamental mechanisms that are responsible for the operation of the Codex Alimentarius Commission: the Procedural Manual as a codification of the internal rules and the structure of two readings of the Codex standard-setting procedure.

SUMMARY IN DUTCH / SAMENVATTING

DE CODEX ALIMENTARIUS COMMISSIE EN HAAR NORMEN

Gedurende de laatste decennia, is de belangstelling rond de Codex Alimentarius Commissie en haar normen toegenomen. Deze toenemende belangstelling is het gevolg van verwijzingen naar de Codex normen in de WTO akkoorden (het SPS Akkoord en het TBT Akkoord), die in 1995 in werking traden. Verwijzing naar Codex normen brengt volgens de algemene opvatting een toegenomen status van deze normen met zich mee. Dit boek heeft als doel om deze status nader te definiëren, zowel in het kader van de WTO Akkoorden, als ook in het kader van de Europese rechtsorde. Tevens wordt bekeken welke juridische gevolgen de nieuwe status van de normen met zich meebrengt en wat dit betekent voor de legitimiteit van het institutionele kader van de Codex Alimentarius Commissie, van haar besluitvormingsprocedure en van de inhoud van de Codex normen. Ten einde de nieuwe status van de Codex normen en de gevolgen daarvan te beoordelen, wordt allereerst aandacht geschonken aan de ontwikkeling van de Codex Alimentarius Commissie en haar normen gedurende de jaren van haar bestaan.

De Codex Alimentarius Commissie is in 1962 als ondergeschikt orgaan door de FAO en de WHO gezamenlijk opgericht. Reden hiervoor was de groeiende behoefte voor geharmoniseerde voedingsvoorschriften in verband met de toenemende internationale handel in voedselproducten. Sinds haar oprichting is de Codex Alimentarius Commissie belast met het opstellen van de Codex Alimentarius: een verzameling van eenduidig gedefiniëerde voedselnormen. De mogelijkheid om uitvoering te geven aan haar taken is uiteraard sterk afhankelijk van het institutionele kader. De autoriteit van de Codex Alimentarius Commissie wordt echter enigszins beperkt door het feit dat zij een ondergeschikt orgaan van de FAO en WHO is. Dit brengt met zich mee dat zij voor het budget en voor het management van de personeelszaken van haar Secretariaat afhankelijk is van haar 'moeder' organisaties, in het bijzonder van de FAO. Ondanks haar subsidiaire positie heeft de Codex Alimentarius Commissie echter een aantal belangrijke bevoegdheden, waarvan ze tijdens haar bestaan goed gebruik heeft gemaakt. Zo heeft ze de bevoegdheid om haar eigen besluitvormingsprocedure vorm te geven en te wijzigen en ondergeschikte Codex comités op te stellen die haar kunnen bijstaan bij het uitvoeren van haar taken. Van deze bevoegdheid heeft de Codex Alimentarius Commissie ruimschoots gebruik gemaakt door een stevig netwerk van subsidiaire comités op te zetten, die haar in staat stelt om vorm te geven aan de Codex Alimentarius. Ook op het gebied van haar regelgevende bevoegdheid opereert zij onafhankelijk en zij heeft voor het tot stand brengen en het aannemen van normen en andere maatregelen geen toestemming van haar 'moeder' organisaties nodig.

De belangrijkste taak die gedelegeerd is aan de Codex Alimentarius Commissie is de totstandkoming van de Codex Alimentarius: een verzameling van eenduidig gedefiniëerde voedselnormen, die de ondersteuning van de harmonisatie van voedselvoorschriften beoogt. Gedurende de jaren, is de Codex Alimentarius tot een complex systeem van voedselvoorschriften gegroeid, die zowel de bescherming van de volksgezondheid, als de bescherming van eerlijke handelspraktijken beoogt. Het bestaat momenteel uit ongeveer 216 standaarden, 47 gedragscodes, 40 richtlijnen, maximum residue niveau's voor pesticide residuen in meer dan 360 voedselprodukten en MRLs voor 47 verschillende veterinaire medicijnproducten. Deze voorschriften zijn vaak gedetailleerd en omvatten een aanzienlijk aantal aspecten die onderdeel van het harmonisatieproces zijn. De formulering van de voorschriften duidt op een aanzienlijke beperking van de soevereiniteit van de Codex lidstaten op dit gebied. Zowel de flexibele besluitvormingsprocedure, die uit twee lezingen bestaat, als de vrijwillige acceptatieprocedure van de Codex Alimentarius Commissie hebben bijgedragen om overeenstemming over de inhoud van de standaarden te vergemakkelijken. Slechts enkele standaarden zijn in het kader van de acceptatieprocedure bindend geworden en dan nog maar voor een klein aantal Codex lidstaten. Ondanks het feit dat de acceptatieprocedure nooit echt goed heeft gelopen en in 2005 afgeschaft is, blijven de gevolgen van haar bestaan voortduren. Deze zijn te herkennen in het gedetailleerde karakter van de voorschriften van de Codex standaarden en het hoge niveau van harmonisatie dat door de standaarden beoogt wordt.

Een eerste systeem waarbinnen veelvuldig verwijzigingen naar Codex standaarden terug te vinden zijn, is de Communautaire rechtsorde. Gedurende haar eerste jaren tot aan 1987 zijn deze verwijzingen regelmatig terug te vinden in de Communautaire voedselregelgeving en in de jurisprudentie van het Europese Hof van Justitie. Het veelvuldig gebruik van de Codex standaarden door de Communautaire instellingen kan hoogst waarschijnlijk verklaart worden door het feit dat gedurende deze periode de Codex Alimentarius Commission ten opzichte van de Communautaire instellingen beter in staat was voedingsstandaarden aan te nemen. Het is daarom dan ook niet verwonderlijk dat vanaf 1987, het moment dat nieuwe institutionele en procedurele regelgeving in de EG werd geïntroduceerd, die het totstandbrengen van regelgeving op Communautair niveau bevorderde, men een afnemende invloed van de Codex standaarden kan waarnemen. Echter, de inwerkingtreding van de WTO akkoorden en de uitspraak van het WTO Panel in de zaak *EC-Hormones*, die duidelijk de gevolgen van de inwerkingtreding naar voren bracht, betekende een ommekeer in deze ontwikkeling. Verordening 178/2002, totstandgekomen in 2002 en ook bekend als de 'General Food Law', bevat een verplichting voor zowel Communautaire instellingen als nationale bevoegde instanties om bij het ontwikkelen en aannemen van voedselregelgeving Codex standaarden in acht te nemen. Dit betekent dat Codex standaarden een status van 'richtsnoer' in de rechtvaardiging van regelgeving hebben verworven. De recentelijk tot stand gekomen EG voedselregelgeving toont aan dat Codex maatregelen of hun wetenschappelijke basis, inderdaad in acht worden genomen door de EG instellingen gedurende de wetgevingsprocedure.

Verwijzing naar internationale standaarden in het SPS Akkoord en het TBT Akkoord en de erkenning van Codex standaarden als dusdanig, heeft geleid tot de vraag wat dat voor de status van deze standaarden betekent. De verwijzingen zijn onderdeel van de harmonisatievoorschriften van het SPS Akkoord en het TBT Akkoord. Deze voorschriften bestaan uit een aanmoediging om nationale maatregelen conform te maken met de internationale standaarden en een verplichting om nationale maatregelen op internationale standaarden te baseren. Uit de toepassing van deze bepalingen door de panels en het beroepsorgaan van de WTO blijkt dat de status van Codex standaarden in het kader van de WTO akkoorden neerkomt op een *de facto* bindende kracht van sommige elementen van de standaarden. Deze *de facto* bindende kracht is niet direct, aangezien deze slechts werkt door het SPS Akkoord en het TBT Akkoord. Daarnaast betreft de *de facto* bindende kracht slechts enkele elementen van de standaard en niet de standaard in zijn geheel. Gezien het feit dat de reikwijdte van de doeleinden van de WTO Akkoorden ten opzichte van die van de Codex Alimentarius Commissie beperkt is (WTO beoogt handelsliberalisatie, terwijl de Codex naast handelsliberalisatie ook de volksgezondheid en eerlijke handelspraktijken beoogt te beschermen), is duidelijk dat in het kader van de WTO Akkoorden de Codex standaarden niet zullen worden toegepast als minimum platform standaarden die een minimum beschermingsniveau beogen. Dit betekent tevens dat met de afschaffing van de Codex acceptatieprocedure in 2005, de Codex Alimentarius Commissie afstand heeft gedaan van het enige 'nalevingsmechanisme' welke het gebruik van Codex standaarden als minimum platform standaarden bevorderde.

De verwijzing naar internationale standaarden heeft ook institutionele gevolgen. De WTO is niet zelf een standaardiseringsorgaan, maar vertrouwt voor deze taak op andere instellingen, zoals de Codex Alimentarius Commissie. Dit betekent dat de relatie tussen de WTO en de Codex Alimentarius Commissie wordt gekenmerkt door een scheiding van bevoegdheden. De Codex Alimentarius Commissie is verantwoordelijk voor de 'wetgevende' handelingen, terwijl de WTO verantwoordelijk is voor het bevorderen van en toezien op de toepassing van de harmonisatie maatregelen. In dit verband zijn er twee aspecten te noemen die het functioneren van de Codex Alimentarius Commissie nadelig kunnen beïnvloeden. Allereerst is dat het feit dat de bepalingen die het WTO geschillenbeslechtingsmechanisme regelen, op dit moment niet een adequate interpretatie en toepassing van de Codex maatregelen door de panels en het beroepsorgaan verzekeren. Ten tweede komt het mandaat van het SPS Comité, als forum voor consultatie, gedeeltelijk overeen met de taken van de Codex Alimentarius Commissie. Op dit moment leidt dit nog niet tot negatieve gevolgen, omdat het SPS Akkoord voorziet in een samenwerking tussen de twee instellingen en omdat de relatie tussen de twee instellingen geen hierachische struktuur bevat.

Doordat de Codex standaarden in status zijn toegenomen is ook de autoriteit van de Codex Alimentarius Commissie in belang gestegen. Dit heeft ertoe geleid dat onderzoekers en organisaties zich vragen stellen omtrent haar legitimiteit. Hoofdstuk V brengt verschillende aspecten van legitimiteit onder de aandacht en berust op een indeling van drie verschillende dimensies: institutionele legitimiteit, procedurele legitimiteit en materiële legitimiteit. Elementen van institutionele legitimiteit betreffen bijvoorbeeld, 1. de delegatie van bevoegdheden van de Codex Alimentarius Commissie aan sterke ondergeschikte Codex Comités, hetgeen in het geval deze niet zorgvuldig wordt gecontroleerd en gecoördineerd, kan leiden tot een onsamenhangend functioneren van de Codex als geheel, 2. de delegatie van bevoegdheden aan externe organen, die niet zijn gebonden aan de interne regels van de Codex, en 3. het verzoek tot wetenschappelijk advies van wetenschappelijke organen die niet altijd gepaard gaat met een duidelijk gedefinieerde taakomschrijving. Elementen die betrekking hebben op de procedurele legitimiteit omvatten aspecten zoals 1. het feit dat de besluitvormingsprocedure van de Codex Alimentarius Commissie niet altijd als legitiem wordt beschouwd en een duidelijke voorkeur uitgaat naar besluitvorming gebaseerd op consensus, 2. de aandacht is gevestigd op het feit dat geïndustrialiseerde Codex lidstaten en organisaties die de industrie vertegenwoordigen een grotere invloed hebben op de besluitvormingsprocedure ten opzichte van ontwikkelingslidstaten en consumentenorganisaties, 3. openbaarheid en transparantie van de procedure en toegang tot relevante documenten is niet altijd verzekerd. Vervolgens gaat het bij de materiële legitimiteit, ofwel de legitimiteit van de standaarden, voornamelijk om het gedetailleerde karakter van de standaarden, welke in beginsel beperkende ruimte laat aan Codex lidstaten om nationale belangen in acht te nemen. Dit leidt tot de vraag of Codex standaarden terecht bepaalde belangen beogen te harmoniseren of dat de bescherming van deze belangen beter kan worden overgelaten aan de volledige beleidsruimte van de Codex lidstaten.

Recentelijke ontwikkelingen, zoals de evaluatie van de Codex Alimentarius Commissie door de FAO en de WHO, hebben geleid tot belangrijke aanpassingen. Voorbeelden zijn de oprichting van het Trust Fund ter bevordering van de participatie van ontwikkelings lidstaten, het opnemen van een bepaling in de Rules of Procedure welke het belang om besluiten te nemen met consensus expliciet erkent, het opnemen van een bepaling welke de verantwoordelijkheden van de risicobeoordelaars enerzijds en van de 'risicomanagers' anderzijds duidelijk neerlegt, en de aanwijzing van het Executief Comité als 'critical reviewer' van initiatieven van nieuw werk, welke worden voorgesteld door de Codex Comités. Hoewel deze aanpassingen een belangrijke stap vormen ter legitimering van de Codex Alimentarius Commissie, komen zij echter slechts gedeeltelijk de bovengenoemde aandachtspunten tegemoet en zijn aan de volgende kritiek onderhevig. De aanpassingen zijn in de meeste gevallen gericht op het wegnemen van één bepaald aandachtspunt. Ze kunnen wel degelijk bijdragen tot het herstellen van de legitimiteit van de Codex Alimentarius Commissie op korte termijn, maar om de legitimiteit op lange termijn te garanderen schieten ze waarschijnlijk tekort. Daarvoor is een beter gestructureerde aanpak vereist. Deze benadering zou zich richten op het optimaliseren van een aantal fundamentele instrumenten die verantwoordelijk zijn voor het functioneren van de Codex Alimentarius Commissie: de 'Procedural Manual' als codificatie van interne regelgeving en de 'twee lezingen' structuur van de besluitvormingsprocedure. Als codificatie van interne regels, is de 'Procedural Manual' een essentieel instrument om de interne en externe transparantie van de Codex Alimentarius Commissie en haar besluitvormingsprocedure te garanderen. De 'Procedural Manual' is echter geen voorbeeld van duidelijk gestructureerde regelgeving. Door zijn totstandkoming en de ingevoerde wijzigingen gedurende de afgelopen 40 jaren, zijn verschillende gedeelten onsamenhangend, verouderd en onvolledig. Recentelijke wijzigingen hebben deze situatie niet verbeterd. Het behoeft geen nadere uitleg dat een grondige revisie van de 'Procedural Manual' als topprioriteit mag worden beschouwd op de weg naar de legitimiteit van de Codex Alimentarius Commissie. Op dezelfde manier, zou ook een duidelijkere definitie van de twee lezingen structuur van de besluitvormingsprocedure en de rol van de Codex Alimentarius Commissie daarin, bijdragen tot betere mogelijkheden tot het controleren van het werk van de ondergeschikte Codex Comités en de betrokken externe organen.

CURRICULUM VITAE

Name	Maria Elisabeth Dominica Matthee
Date of Birth	5 September 1965
Place of Birth	De Bilt, The Netherlands

1983 - 1986	VWO, Schoonoord, Zeist, The Netherlands
1987 - 1993	Dutch Law degree (meester in de rechten), equivalent to a LL.M., Erasmus University, Rotterdam
1995 - 1996	Legal Consultant, Economic Commission of Latin America and the Carribean (ECLAC), Santiago de Chile
1997 - 1998	*Diplomes d'Etudes Approfondies (DEA)*, specialised in Environmental Law, Paris I: Pantheon-Sorbonne (mention *assez bien*), Paris
1999 - 2004	Research Fellow, Asser Dissertation Programme, T.M.C. Asser Institute, The Hague
2003	Scholar, Center of Studies of The Hague Academy of International Law on the topic of 'International Law and Food Security', The Hague
2004 - 2005	Legislative Consultant, Legal Department of the Dutch Ministry of Agriculture, Food Safety and Nature, The Hague